Atomic Physics

The Manchester Physics Series

General Editors

F. MANDL : R. J. ELLISON : D. J. SANDIFORD

Physics Department, Faculty of Science, University of Manchester

Published

Properties of Matter: B. H. Flowers and E. Mendoza

Optics: F. G. Smith and J. H. Thomson

Statistical Physics: F. Mandl

Solid State Physics: H. E. Hall

Electromagnetism: I. S. Grant and W. R. Phillips

Atomic Physics: J. C. Willmott

In preparation

Electronics: J. M. Calvert and M. A. H. McCausland

ATOMIC PHYSICS

J. C. Willmott

Department of Physics,
University of Manchester.

John Wiley & Sons Ltd.

LONDON NEW YORK SYDNEY TORONTO

Library of Congress Cataloging in Publication Data:

Willmott, John Charles.
Atomic physics.

(Manchester physics series)
Bibliography: p.
1. Nuclear physics. 2. Quantum theory. I. Title.

QC171.2.W54 539.7 74-9580

ISBN 0 471 94930 2
ISBN 0 471 94931 0 (pbk.)

Printed in Great Britain by J. W. Arrowsmith Ltd.
Winterstoke Road, Bristol

Editors' Preface to the Manchester Physics Series

In devising physics syllabuses for undergraduate courses, the staff of Manchester University Physics Department have experienced great difficulty in finding suitable textbooks to recommend to students; many teachers at other universities apparently share this experience. Most books contain much more material than a student has time to assimilate and are so arranged that it is only rarely possible to select sections or chapters to define a self-contained, balanced syllabus. From this situation grew the idea of the Manchester Physics Series.

The books of the Manchester Physics Series correspond to our lecture courses with about fifty per cent additional material. To achieve this we have been very selective in the choice of topics to be included. The emphasis is on the basic physics together with some instructive, stimulating and useful applications. Since the treatment of particular topics varies greatly between different universities, we have tried to organize the material so that it is possible to select courses of different length and difficulty and to emphasize different applications. For this purpose we have encouraged authors to use flow diagrams showing the logical connection of different chapters and to put some topics into starred sections or subsections. These cover more advanced and alternative material, and are not required for the understanding of later parts of each volume.

Since the books of the Manchester Physics Series were planned as an integrated course, the series gives a balanced account of those parts of physics which it treats. The level of sophistication varies: '*Properties of Matter*' is for the first year, '*Solid State Physics*' for the third. The other volumes are intermediate, allowing considerable flexibility in use. '*Electromagnetism*', '*Optics*', '*Electronics*' and '*Atomic Physics*' start from first year level and progress to material suitable for second or even third year courses. '*Statistical Physics*' is suitable for second or third year. The books have been written in such a way that each volume is self-contained and can be used independently of the others.

Although the series has been written for undergraduates at an English university, it is equally suitable for American university courses beyond the Freshman year. Each author's preface gives detailed information about the prerequisite material for his volume.

In producing a series such as this, a policy decision must be made about units. After the widest possible consultations we decided, jointly with the authors and the publishers, to adopt SI units interpreted liberally, largely following the recommendations of the International Union of Pure and Applied Physics. Electric and magnetic quantities are expressed in SI units. (Other systems are explained in the volume on electricity and magnetism.) We did not outlaw physical units such as the electron-volt. Nor were we pedantic about factors on 10 (is 0.012 kg preferable to 12 g?), about abbreviations (while s or sec may not be equally acceptable to a computer, they should be to a scientist), and about similarly trivial matters.

Preliminary editions of these books have been tried out at Manchester University and circulated widely to teachers at other universities, so that much feedback has been provided. We are extremely grateful to the many students and colleagues, at Manchester and elsewhere, who through criticisms, suggestions and stimulating discussions helped to improve the presentation and approach of the final version of these books. Our particular thanks go to the authors, for all the work they have done, for the many new ideas they have contributed, and for discussing patiently, and frequently accepting, our many suggestions and requests. We would also like to thank the publishers, John Wiley and Sons, who have been most helpful in every way, including the financing of the preliminary editions.

Physics Department
Faculty of Science
Manchester University

F. MANDL
R. J. ELLISON
D. J. SANDIFORD

Author's Preface

Some years ago at Manchester we decided to try to define our physics syllabuses in terms of textbooks, or portions thereof. This exercise brought to light the absence of books at an appropriate level, and of an appropriate length, on certain topics. One of these topics was atomic physics at a level suitable for second year undergraduates in a British university. This textbook is an attempt to fill that gap.

Broadly speaking there are two possible approaches to the teaching of atomic physics and quantum mechanics. The first is to present the material in a roughly historical manner, while the second is to postulate the laws of quantum mechanics, in the same way as one postulates Newton's laws of motion, and to develop the subject from there. It is possible to use the latter technique in teaching classical mechanics because most people are familiar with the ideas of classical mechanics from everyday experience, even if they have not formalized them; the equality of action and reaction does not strike most people as extraordinary. There is nothing in everyday experience to suggest the laws of quantum mechanics. The advantage of proceeding from the postulates is that it makes clear the formal structure of quantum mechanics, but it is the author's view that the postulates of quantum mechanics are so extraordinary to the newcomer and contain so many unfamiliar concepts that it is better to lead the student, the first time, more gently explaining why classical physics will not do and introducing the new ideas as far as possible one at a time.

This book, therefore, leans towards the historical approach. The danger in this approach is a tendency to make it all seem inevitable and leave an impression that one can almost derive the Schrödinger equation. An attempt has been made to overcome this mainly by repeated assertion that derivation of quantum mechanics is not possible. Also, a certain amount of hindsight is employed to bring out points not obvious at the time; for example, the fact that without Planck's constant one could not construct a length with which to set the scale of atomic phenomena, and the inevitable failure of classical physics as a result, was surely not generally appreciated at the time.

The text is intended to form the basis of a course of some thirty lectures, each lasting fifty minutes. Needless to say, it would not be possible to cover the whole of the material in that time and it is assumed some selection will be made and some topics will be left to the student to read for himself. The first two chapters are concerned with the evidence for the atomic nature of matter, leading up to the Rutherford model of the atom. Much of this will be familiar to the student and may be omitted. Chapter 3 is concerned with the universal occurrence of Planck's constant in electromagnetic phenomena. Chapter 4 starts with electron diffraction, considers the nature of the wavefunction required to describe the phenomenon and goes directly to the Uncertainty Principle, it being regarded as important that the student realize that the association of wavelike phenomena with particle motion introduced an uncertainty independent of the exact form taken by the equations of motion. Chapter 5 is concerned with the Schrödinger equation and chapter 6 with angular momentum. After this the only new ideas introduced are the Pauli Exclusion Principle in Chapter 8, and parity in Chapter 10.

Each chapter is followed by a selection of problems. These vary from simple numerical problems intended to familiarize the student with the orders of magnitude involved in atomic physics to problems that go well beyond the scope of the book and are intended to stretch the better student.

An attempt has been made to avoid disposing of difficult material by phrases like '. . . it turns out that' but rather to try and give a plausible physical explanation. This is not always possible, but one of the attractions of physics is that there are simple situations which require the most advanced and penetrating analysis. It is hoped that this book will give its readers a taste for the subject and stimulate them to go on to more advanced texts.

J. C. WILLMOTT

April, 1974.
Manchester, England.

Contents

★ Starred sections may be omitted as they are not required later in the book.

9 ONE AND TWO ELECTRON SPECTRA

10 SPECTRAL LINES AND SELECTION RULES

11 CHARACTERISTIC X-RAY SPECTRA

CHAPTER 1

The atomic nature of matter and electricity

1.1 INTRODUCTION

During the nineteenth century a large amount of evidence had been accumulated that gave support to the idea that matter is composed of atoms.

We list below some of the chemical laws which receive a simple explanation on an atomic hypothesis.

(1) The law of constant proportions

When chemical elements are brought together under appropriate conditions to form a specific compound, the proportion in weight of the combining elements is always the same.

(2) The law of multiple proportions

When two elements combine together in different ways, to form different compounds, then the weights of one element that combine with a definite weight of the other always bear a simple ratio to each other.

On the atomic hypothesis these laws are explained as follows:

A quantity of a given chemical element consists of a large number of atoms of that element, each atom having the same weight, the weight being peculiar to that element (Fig. 1.1). When two elements combine to form a compound, the atoms of the elements combine in a simple ratio to form a *molecule* of the compound.

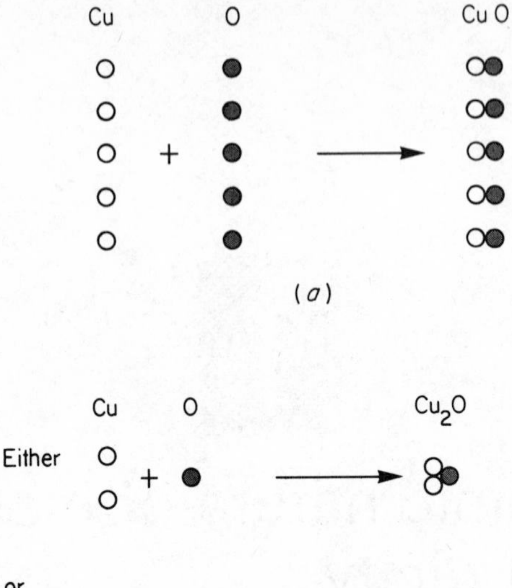

Fig. 1.1. (*a*) The law of constant proportions on the atomic hypothesis.
(*b*) The law of multiple proportions on the atomic hypothesis.

Thus, if we form the compound cupric oxide from copper and oxygen, we find that 63.5 g of copper always combine with 16 g of oxygen. Further, we can also form the compound cuprous oxide in which 63.5 g of copper combine with 8 g of oxygen.

The atomic hypothesis explains this by saying the atomic weights of copper and oxygen are in the ratio 63.5:16, and that cupric oxide is CuO and cuprous oxide Cu_2O.

With this simple hypothesis it was found possible to give a quantitative explanation of observed combining weights for the whole of simple inorganic chemistry.

It was found that in the gaseous state not only the weights but the volumes entering into chemical reactions were governed by simple laws.

(3) Gay–Lussac's Law

In every gas formed or decomposed, the volumes of the component and compound gases bear a simple ratio to one another.

In fact, this law only applies when the gases closely follow the perfect gas laws.

Consideration of Gay–Lussac's law with the previous laws leads to the conclusion that the *volume* of a gas is related to the number of particles in it, and we get

(4) Avogadro's Law

Equal volumes of different gases under the same conditions of temperature and pressure contain equal numbers of molecules.

It will be seen that this law implies that the molecular weight in grams of a gas will always occupy a specific volume, at a given temperature and pressure, whatever the gas. At normal temperature and pressure (0 °C and 760 mm of mercury) this volume is *22.4 litres.*

The *number* of molecules in one gram molecular weight is known as Avogadro's number. This number is

$$N_0 = 6.023 \times 10^{23} \ .$$

Avogadro's number may be measured in a variety of ways, the most accurate of which depends on the measurement of atomic spacings by X-ray diffraction. All of them naturally depend on the assumption of the atomic hypothesis, and the agreement of the various methods lends powerful support to this hypothesis.

1.2 KINETIC THEORY

Further support for the atomic hypothesis was provided by the great success of the kinetic theory of gases, as developed by Maxwell and Boltzmann. Perhaps the most dramatic support for an atomic hypothesis comes from the observation of *Brownian motion.*

Kinetic theory explains the pressure exerted by a gas as being due to the bombardment of the walls of the container by the molecules of the gas. If a small object is suspended in a gas, it too will be bombarded by the molecules, and because the number of molecules is finite, an exact balance will not occur at any one instant, and the object will undergo a random motion as a result. This was first observed by the botanist Brown.

As an example, we may consider a suspended mirror, such as a galvanometer mirror. Because the two halves of the mirror will not undergo bombardment by exactly equal numbers of molecules, it will suffer torsional motion. Kinetic theory not merely says this will happen, but makes assertions about the magnitude of the effect. According to the law of equipartition of energy (see F. Mandl, *Statistical Physics* (Manchester Physics Series), Wiley, London, 1971) every term in the expression for the total energy which enters in quadratic form has a mean energy equal to $\frac{1}{2}kT$ where T is the absolute temperature and k (known as Boltzmann's constant) is R/N_0.

R is the gas constant, and N_0 Avogadro's number, i.e. the number of molecules in one gram molecule.

Applying this to the galvanometer mirror, the expression for the energy will contain a term $\frac{1}{2}c\theta^2$ where θ is the deflection of the mirror and c the torsional constant. Thus we have

$$\tfrac{1}{2}c\overline{\theta^2} = \tfrac{1}{2}kT \tag{1.1}$$

or

$$\overline{\theta^2} = \frac{kT}{c} \tag{1.2}$$

where $\overline{\theta^2}$ is the mean square deflection.

Fig. 1.2 shows the results one obtains. These fluctuations provide impressive support for the atomic hypothesis.

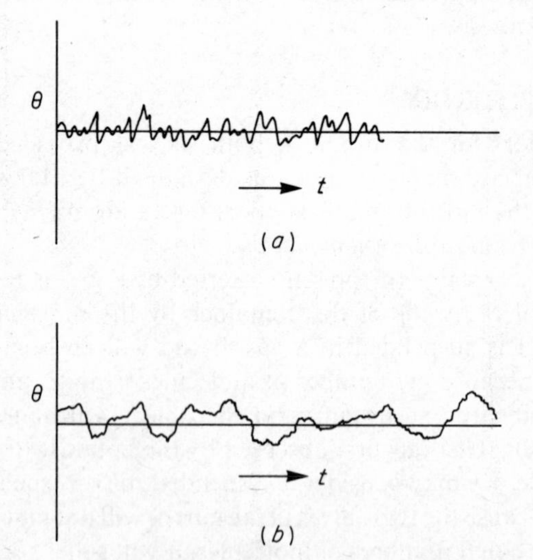

Fig. 1.2. The fluctuations of the angular position of a galvanometer mirror due to its Brownian motion (*a*) for a lightly damped galvanometer and (*b*) for a more heavily damped galvanometer. The mean square fluctuation is the same in both cases.

Measurements of fluctuations, in fact, provided the earliest values of Avogadro's number. The first experiments were done by Perrin in 1909 who studied the random motion of a suspension of gamboge particles.*

1.3 ATOMIC SIZES

Various experiments tell us that the value of Avogadro's number is $N_0 = 6 \times 10^{23}$. If we make the assumption that in a solid element (e.g. a metal) the atoms are packed together so that neighbouring atoms are touching, we get an estimate of their radii.

Let the radius of each atom be r cm. A length of 1 cm will therefore contain $1/2r$ atoms, and 1 cm^3 will contain $1/8r^3$ atoms. One gram atom of a substance contains $\sim 6 \times 10^{23}$ atoms. If it has a density ρ, one gram atom will occupy $A/\rho\ \mathrm{cm}^3$, where A is the atomic weight of the substance. Hence a volume of 1 cm^3 contains $\sim(6 \times 10^{23}\rho)/A$ atoms, from which

$$r = \tfrac{1}{2}\left(\frac{A}{6 \times 10^{23}\rho}\right)^{1/3}$$

We give the result for this calculation for some solid elements in the following table

Element	Atomic weight	ρ g/cm^3	r cm
Lithium	7	0.7	1.3×10^{-8}
Aluminium	27	2.7	1.3×10^{-8}
Copper	63	8.9	1.14×10^{-8}
Sulphur (rhombic)	32	2.07	1.48×10^{-8}
Lead	207	11.34	1.55×10^{-8}

Note: These radii are often expressed in ångstrom units where 1 Å = 10^{-8} cm.

Thus we see from this calculation that all atoms have a size that is a few times 10^{-8} cm. Other methods, e.g. from kinetic theory, give similar answers.

Thus a lot of evidence comes together to support the view that elements consist of atoms and these atoms have a radius of about 10^{-8} cm.

* In the course of this book details will be given of various direct methods of determining important physical quantities. It should be pointed out that the determination of the 'most accurate' values for these quantities is a complex business involving a careful statistical analysis of many different pieces of experimental information. It is often much easier to obtain really accurate results for quantities which are combinations of the quantities really required. The reader is most strongly urged to read the first twenty pages of the article 'Our present knowledge of the fundamental constants' by Jesse Du Mond and E. R. Cohen, *Reviews of Modern Physics*, Vol. 37, p. 537, 1965, and the similar article by B. N. Taylor, W. H. Parker and D. N. Langenberg, *Reviews of Modern Physics*, Vol. 41, p. 375, 1969.

In saying the above we have been careful not to define what we mean by the radius of an atom. This is the first of many cases where we shall have to investigate carefully what we mean by generally accepted terms, for which we usually have an intuitive understanding.

If given a micrometer and a high quality ball bearing we can perform an accurate measurement of the radius of the ball bearing. But two different people might obtain different answers, the difference being greater than the statistical spread in each set of results. This possibility is illustrated in Fig. 1.3. Further investigation might show that this was due to the fact that one

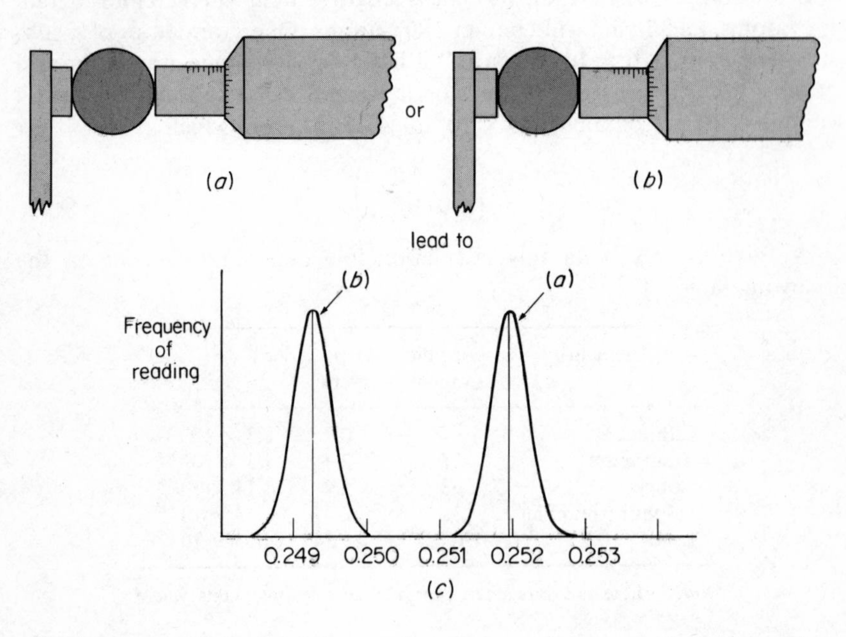

Fig. 1.3. The problem of defining a radius. Two people measure the radius of the same ball bearing using the same micrometer gauge. The person in (*a*) consistently uses a lighter pressure than (*b*) resulting in two sets of readings each with a standard deviation smaller than the separation of their means.

person consistently screwed up the micrometer more tightly than the other. Normally we would prefer the one that used less force, on the grounds that he distorted the ball bearings and micrometer less. But the point we want to make is that intuitively we are defining 'radius' through a sudden change in the interaction between the material of the ball bearing and the micrometer at a certain setting of the micrometer. If we had tried instead to measure the radius of a rubber ball, we should have to be much more careful, but the same general idea would prevail. Other objects may not have such a well defined radius. Thus the optical diameter of the sun is quite well defined, but is different from that determined by radio astronomy; and the outer envelope of the sun extends beyond the earth's orbit. So, in this case we would have to be careful what we were talking about. In the same way, as yet we do not know how the interaction between atoms varies as a function of distance (the variation is, in fact, quite abrupt), so we must admit that at the moment we are speaking very loosely when we talk about the radii of atoms.

1.4 THE ATOMICITY OF ELECTRIC CHARGE—ELECTROLYSIS

Experiments in the nineteenth century showed that the amount of a substance liberated at an electrode of an electrolytic cell by a given quantity of electric charge was proportional to the equivalent weight of the substance, i.e. its atomic weight divided by its valency. The constant of proportion is called the faraday. Thus we obtain

$$M = \frac{QA}{vF}$$

where M = the mass of substance liberated, Q = the quantity of charge passed, A = the atomic weight, v = the valency, F = the faraday.

The value of the faraday is

$$F = 96{,}500 \text{ coulombs/mole equivalent.}$$

The fact that the mass of material liberated is strictly proportional to the total quantity of charge suggests that the charge is carried by the atoms themselves. The simplest hypothesis is that each atom carries a charge qv, in which case the charge required to liberate one mole will be N_0qv where N_0 is Avogadro's number. Thus we have

$$M = N_0 qv \frac{A}{vF}$$

where $M = A$ for one mole. Hence

$$\begin{aligned} q &= F/N_0 \\ &= \frac{9.65 \times 10^4}{6 \times 10^{23}} \\ &= 1.60 \times 10^{-19}\,\mathrm{C}\ . \end{aligned}$$

Thus by combining the atomic hypothesis with the results of electrolysis we arrive at the idea that each atom is associated with a certain charge, equal to qv where q is 1.60×10^{-19} coulombs, and v is the valency. This phenomenon clearly receives a simple explanation if we assume that electric charge also has an atomic nature, and that in the electrolyte each ion carries a number of 'atoms' of electric charge equal to its valency.

1.5 ELECTRICAL DISCHARGES IN GASES

At normal temperatures and pressures, gases are essentially non-conducting, until the electric field strength is so high that a spark occurs.

If, however, a pair of electrodes are sealed into a glass vessel, and the pressure reduced to below about 10 mm of mercury, then, on the application of a few kilovolts between the electrodes, a steady *glow* discharge will be seen whose general structure is shown in Fig. 1.4. These discharges show

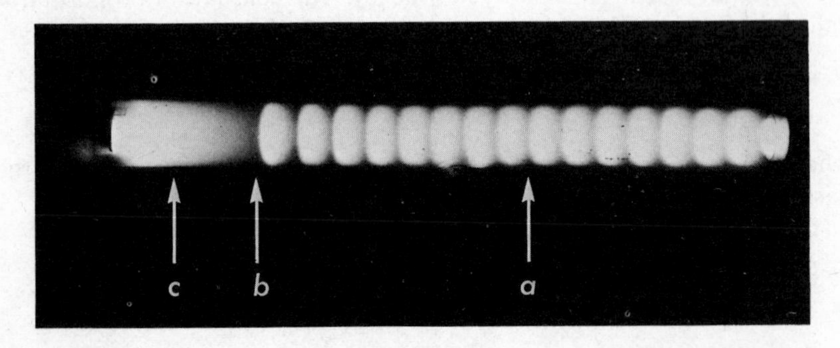

Fig. 1.4. A photograph of a glow discharge. The striations, *a*, are known as the positive column; *b* is the second dark space and *c* is known as the negative column. Between the negative column and the cathode are two further regions, not visible in this photograph, known as the Crookes dark space and the cathode glow.

remarkable colours and patterns, particularly if a steady D.C. source is used rather than in induction coil, as is commonly the case.

If the pressure is lowered still further, the dark region which starts close to the cathode extends until, at about a pressure of 10^{-3} mm, it fills the whole tube. Nevertheless the ammeter reading indicates that a current is still flowing. If the anode has a hole drilled in it, a green glow is observed on the wall of the glass tube beyond the hole in the anode. The agents responsible for this glow travel in straight lines from the hole in the anode, as can be shown by the fact that a sharp shadow is cast by an interposed object, as is shown in Fig. 1.5. If a very light paddle wheel is put in the way, it is set into rotation, showing that the agents also carry momentum. These agents were called 'cathode rays'.

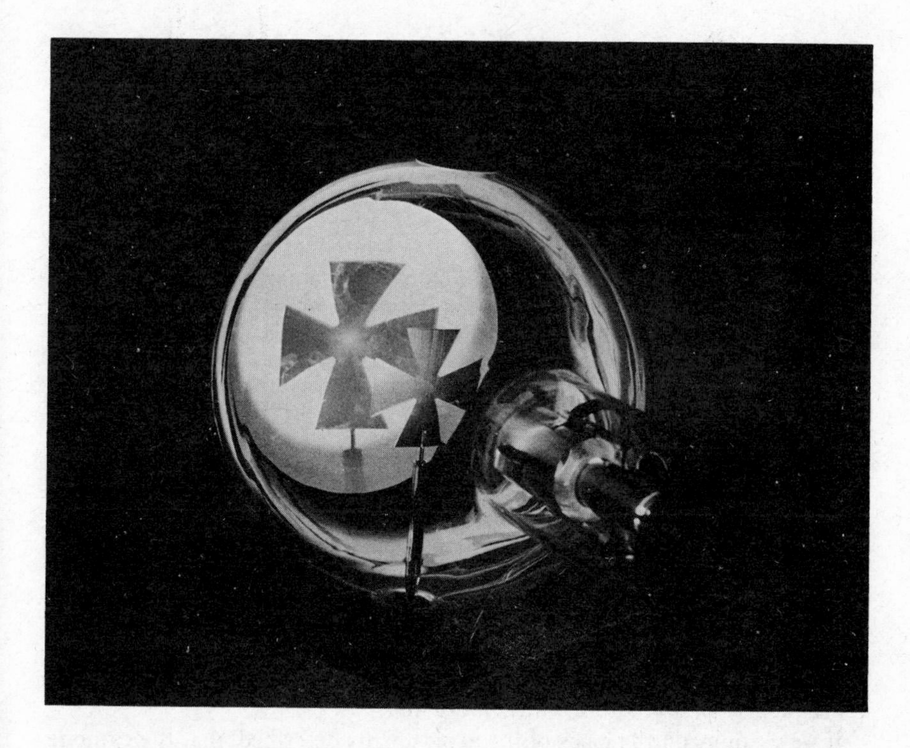

Fig. 1.5. A demonstration that cathode rays (electrons) move in straight lines in a field-free space. Electrons produced by the electron gun at the right of the photograph produce a bright light when they hit the fluorescent screen on the far side of the evacuated tube. A Maltese cross placed between the gun and the screen throws a sharp shadow on the screen.

1.6 THE PROPERTIES OF 'CATHODE RAYS'

If the anode in the discharge tube is made of two diaphragms, each with a fine hole in it, then the cathode rays emerge along the line joining the two holes in a fine pencil, and we may perform experiments on them to determine their properties. Such an arrangement was used by J. J. Thomson, whose apparatus is shown schematically in the accompanying diagram (Fig. 1.6).

The following experiments may be performed.

(*a*) A potential V may be applied across the plates. We observe that:

(i) The spot is deflected towards the positive plate, by an amount proportional to the applied voltage.

(ii) The spot does not spread out, i.e. all the cathode rays suffer the same deflection.

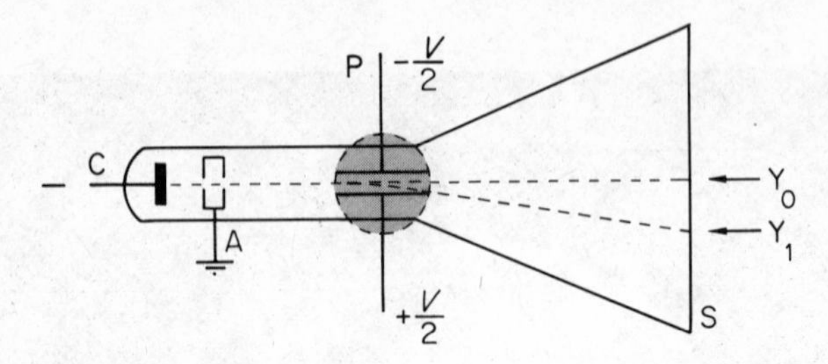

Fig. 1.6. Thomson's apparatus for measuring e/m for cathode rays. Cathode rays from a cathode C are accelerated through a hole in the anode A. They pass through a pair of plates P and finally impinge on a screen S which is at the same potential as A. When there is no potential difference between the plates the rays are undeviated and strike the screen at Y_0. If a potential difference V is applied between the plates, as shown, the rays are deflected and strike the screen at Y_1. The deflection can be cancelled by applying a magnetic field over the shaded area of appropriate magnitude and whose direction is out of the page.

From (i) we deduce that the cathode rays carry a negative charge, which we denote by $-e$, and are therefore presumably some sort of 'particle'. The consequences of (ii) require a little more analysis.

If we assume, on the basis of the experiments described, that the cathode rays are negatively charged particles, we may deduce their deflection as follows:

Let a particular particle emerge from the hole in the anode with a velocity v. On traversing the space between the plates, it will experience a transverse force which will give it a transverse component of velocity v_t, as illustrated in Fig. 1.7.

The deflection will be proportional to the ratio v_t/v. Now

$$mv_t = \text{force} \times \text{time}$$

$$= -eE\frac{l}{v}$$

Therefore

$$v_t = -\frac{eEl}{mv}$$

and

$$\frac{v_t}{v} = -\frac{eEl}{mv^2}. \tag{1.3}$$

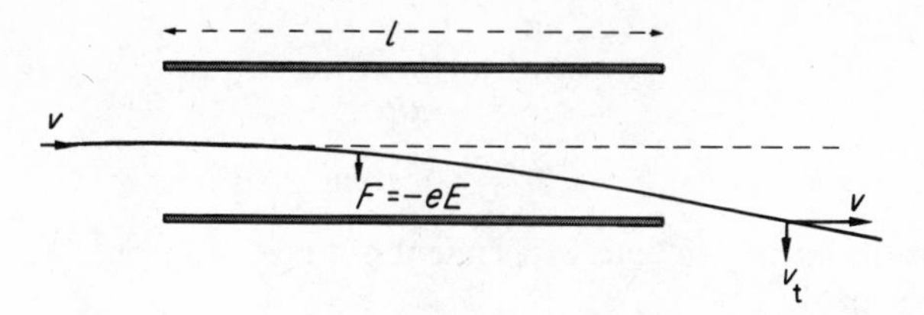

Fig. 1.7. Deflection of cathode rays by an electric field. A beam of cathode rays with velocity v enters the space between two metal plates which are at different potentials, and hence produce an electric field which deflects the rays. The initial line of the rays is parallel to the plates and midway between them. On emerging from the space between the plates the cathode rays have a transverse velocity component v_t. We have

$$mv_t = Ft, \quad F = -eE, \quad t = v/l \ .$$

Hence

$$v_t = -\frac{eEl}{mv}, \quad \frac{v_t}{v} = -\frac{eEl}{mv^2} \ .$$

Thus it appears that mv^2/e is constant. If the cathode rays originate from the cathode and travel to and beyond the anode without suffering any collisions, we would indeed expect, from conservation of energy, that

$$\tfrac{1}{2}mv^2 = -eV_0 \tag{1.4}$$

where V_0 is the potential difference between the cathode and anode.

(*b*) A magnetic field B may be applied over the length l. We observe that the cathode rays are deflected downwards if the field is at right angles to the plane of the diagram and going into it as would be expected if the cathode rays are negatively charged particles moving from left to right. Again, all the particles suffer the same deflection.

Inside a magnetic field charged particles move in an arc of a circle (see Fig. 1.8). Equating the centrifugal force to the magnetic force we have

$$-Bev = mv^2/r \ .$$

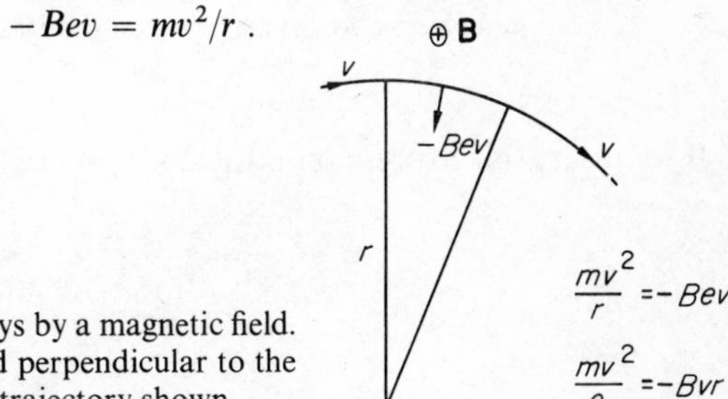

Fig. 1.8. Deflection of cathode rays by a magnetic field. There is a uniform magnetic field perpendicular to the paper over the whole of the trajectory shown.

Thus

$$mv^2/e = -Bvr \tag{1.5}$$

$$= \text{constant}$$

from the results of the previous experiment on the deflection of cathode rays by an electric field.

Thus we have proved, within the limits of experimental error, that the velocity v is the same for all the particles.

(c) We may arrange the electric and magnetic fields so the forces they exert exactly balance, as in Figure 1.9. In which case we have

$$-Bev = -Ee$$

whence

$$v = E/B\ . \tag{1.6}$$

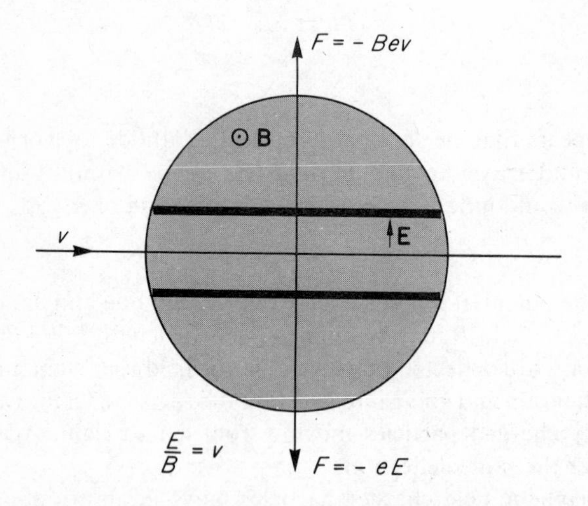

Fig. 1.9. The balance of electric and magnetic forces. The magnetic field is into the paper. When the magnetic and electric forces are equal and opposite the electron beam is undeflected.

If we insert this expression into equation (1.3) we have

$$\frac{v_t}{v} = -\frac{e}{m}\frac{ElB^2}{E^2}$$

$$= -\frac{e}{m}\frac{lB^2}{E}\ . \tag{1.7}$$

The deflection $y = Lv_t/v$ and $E = V/d$, where V is the potential applied across the plates and d their spacing (see Fig. 1.10).

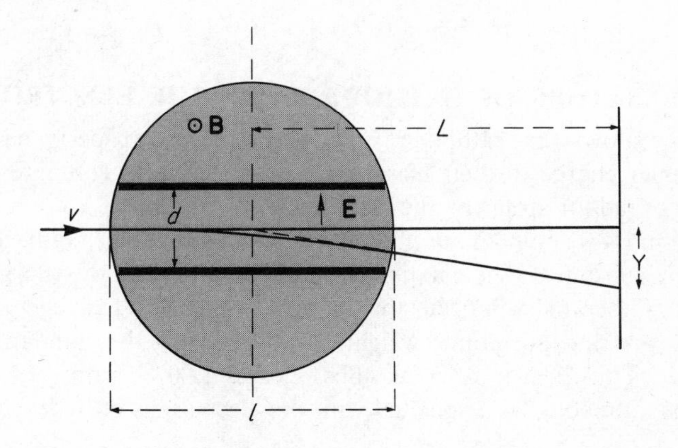

Fig. 1.10. The determination of e/m. The deflection on the screen, y, is given by

$$\begin{aligned} y &= Lv_t/v \\ &= -L\frac{eEl}{mv^2} \\ &= -L\left(\frac{e}{m}\right)\frac{lB^2d}{V}\,. \end{aligned}$$

where V is the potential difference between the plates and B the magnetic field required to reduce the deflection to zero.

Thus

$$y = -L\frac{e}{m}\frac{ldB^2}{V}$$

or

$$\frac{e}{m} = \frac{-yV}{LldB^2}\,\mathrm{C/kg} \tag{1.8}$$

where all quantities are measured in S.I. units.

Thomson found that the value he obtained was the same (within fairly wide limits of error) whatever the nature of the residual gas, or of the other conditions.* The mean value obtained by Thomson was 1.3×10^{10} C/kg.

A modern value is

$$= (1.7598 \pm 0.0004) \times 10^{10}\ \text{C/kg}\ .$$

1.7 THE NATURE OF CATHODE RAYS—THE ELECTRON

We have shown that cathode rays have one specific property, namely the ratio of their charge to their mass. How does this ratio compare with the value suggested for atoms by the results of electrolysis?

In section 1.4 we pointed out that one faraday releases one gram equivalent weight of a substance. The maximum value of the ratio of the charge passed to the mass liberated will occur for the lightest element, hydrogen, which is univalent, and has an atomic weight of 1.0078 using the standard atomic mass scale. Thus 9.65×10^4 coulombs releases 1.0078 grams of hydrogen. Hence the value of e/M associated with the electrolysis of hydrogen

$$= \frac{9.65 \times 10^4}{1.0078}$$

$$= 9.57 \times 10^4\ \text{C/g}\ .$$

This compares with e/m for cathode rays of 1.76×10^8 C/g. Hence the ratio

$$\frac{\left(\dfrac{e}{m}\right) \text{for cathode rays}}{\left(\dfrac{e}{M}\right) \text{for hydrogen}} = 1.84 \times 10^3\ .$$

Thus, either the cathode ray particles are much lighter than a hydrogen atom, or they carry about a thousand times the charge that a hydrogen ion carries in electrolysis. The latter proposition seems so unreasonable that J. J. Thomson made the assumption that, in fact, they all carried the same charge and that the cathode ray particles were much lighter than atoms. He called them *electrons* following a suggestion of Johnston–Stoney.

* This statement is not true if the potential V_0 across the discharge tube is very high. Due to the relativistic increase in mass with velocity the measured value of (e/m) is smaller than the limiting value obtained as $v \to 0$. The correction required is given in the following table.

Accelerating volts	Correction to be added to measured value
10 V	2×10^{-3}%
100 V	2×10^{-2}%
1,000 V	2×10^{-1}%
10,000 V	2%

★1.8 AN ACCURATE MEASUREMENT OF e/m FOR ELECTRONS

J. J. Thomson's method is of limited accuracy for a variety of reasons.

(i) It is difficult to account accurately for the effects of the fringing fields for the electric and magnetic fields.

(ii) Electric fields are usually computed from measured potential differences. But two metal surfaces have a contact potential between them, and this potential is a function of the nature of the surface, and in particular depends on layers of adsorbed gas. Thus the field may not be given by $E = V/d$ exactly, and similarly one cannot rely on a potential difference to give an exact value for the kinetic energy.

Hence methods must be devised which do not suffer from these uncertainties. One of the best of these is due to Dunnington (1937).

Dunnington's method for e/m

In Dunnington's method, shown diagrammatically in Fig. 1.11, electrons are forced to travel in an arc of a circle of specified radius by a series of slits

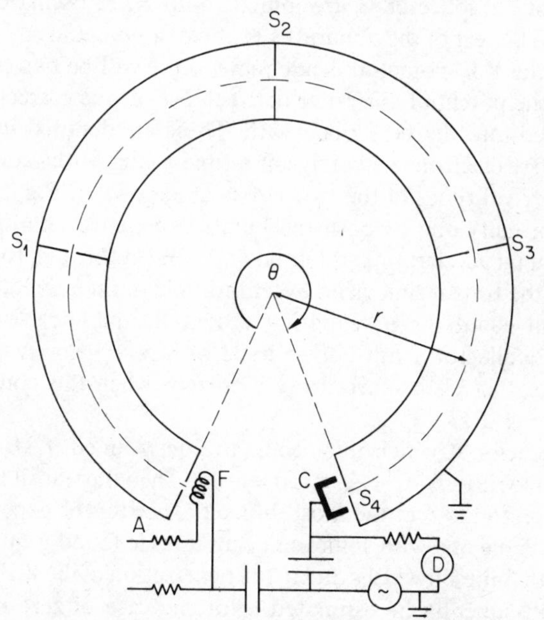

Fig. 1.11. Dunnington's apparatus for measuring e/m for electrons. A is earthed and contains slits S_1, S_2, S_3 and S_4 which define a circular path for the electrons of radius r. F is a filament acting as a source of electrons, C is the current collector and D measures the collected current. An A.C. generator is connected by condensers to both F and C. The apparatus is evacuated so that electrons do not scatter from air molecules.

in the apparatus which is placed in a uniform magnetic field. From the expression for the radius of curvature of a charged particle in a uniform field

$$r = \frac{mv}{Be}$$

and from the measured value of the field B one determines mv/e.

By measuring the time τ taken for the electrons to turn through an arc θ one obtains

$$r\theta = v\tau \ .$$

Eliminating v between these two expressions we have

$$\frac{e}{m} = \frac{\theta}{B\tau} \ . \tag{1.9}$$

The measurement of the transit time τ is achieved as follows. A heated filament F and a collector C are joined to an R. F. oscillator of variable frequency f. The rest of the apparatus is at earth potential.

Provided the R.F. potential is adequate, there will be two points in each cycle where the potential difference between F and A is correct for electrons to be accepted into the prescribed path. These are denoted by t_1 and t_2 in Fig. 1.12. These electrons will arrive at a time τ later at the collector.

Possible arrival times of the two bursts are shown in Fig. 1.12. It will be seen that normally one or both the bursts experience retarding fields less than the original accelerating potential and will get through to the collector. If, however, the transit time is an exact multiple of the period, the retarding potential will equal the accelerating potential, and the electrons should arrive at the collector, from both bursts, with zero velocity. Due to space charge effects, there is a sharp drop in current when this condition is met, as shown in Fig. 1.13.

The frequencies at which this occurs are determined, f_1, f_2, f_3 etc. These should be in the ratio $f_2/f_1 = 2, f_3/f_1 = 3$ etc. Then the transit time $\tau = 1/f_1$.

Dunnington found that the most difficult aspect of the experiment was to estimate the value of θ with sufficient accuracy. He found it quite impossible to do this with finite slit widths due to the penetration of the R.F. field through the slits. Consequently he estimated θ for the case of zero slit width and measured the value of the field B which gave a resonant dip as a function of slit width, and extrapolated the result to zero slit width.

Most of the other sources of error, such as contact potentials etc. merely affected the sharpness of the dip without altering its mean position. Dunnington did find that, in spite of goldplating the whole of the inside of his apparatus, it appeared to develop a surface charge which affected the results.

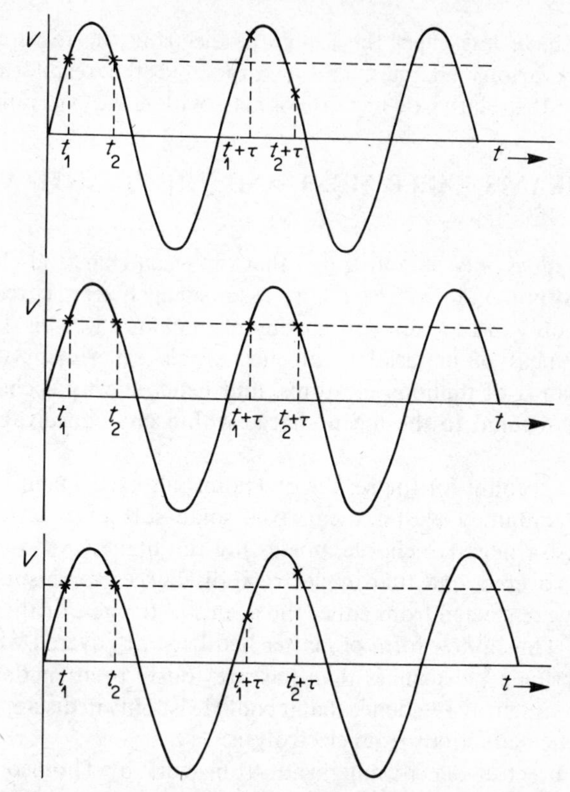

Fig. 1.12. Voltage-time relations in Dunnington's method. The three graphs show the voltage of the filament and collector with respect to earth (*a*) when $1/f$ is less than τ, (*b*) when $1/f$ equals τ, and (*c*) when $1/f$ is greater than τ, where τ is the time the electrons take to travel from the filament to the collector. The dotted line in each case represents the voltage level at which electrons will pass through the slit system. Electrons are therefore accepted into the apparatus at t_1 and t_2 in the first half cycle, and arrive at the collector at $t_1 + \tau$ and $t_2 + \tau$.

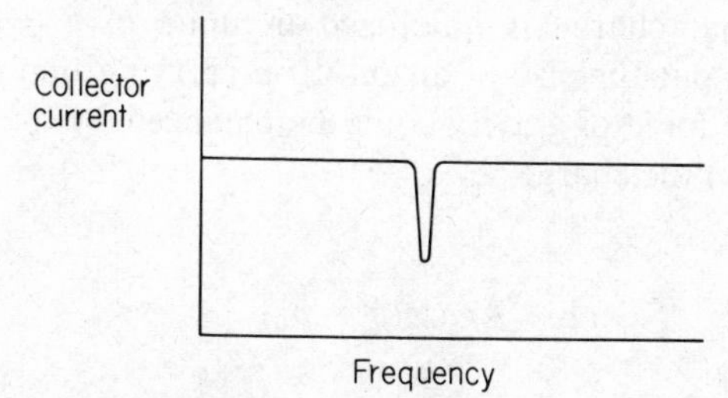

Fig. 1.13. Collector current as a function of frequency in Dunnington's experiment.

As this will have less effect the faster the electrons, he measured e/m for electrons of various energies, and extrapolated the result (making due allowance for the relativistic increase of mass with energy) to infinite energy.

1.9 MILLIKAN'S EXPERIMENT AND THE QUANTIZATION OF CHARGE

So far we have only demonstrated that the *mean* charge in electrolysis is quantized. Although, in section 1.4, we assumed each atom carried a charge qv, we need only have assumed that the mean charge was qv. The smallest measurable mass of material in an electrolysis experiment will contain many thousands of millions of atoms, and hence the total charge will be exactly proportional to the mean charge within any conceivable limits of error.

One could account for the results of Thomson's experiment by assuming that atoms contained within themselves some sort of electrical material which carried a negative charge, possessing the highest ratio of charge to mass yet discovered, and that in electrical discharges some small pieces of this matter were ejected from either the atoms of the gas, or the material of the cathode. Thus a new form of matter had been discovered which carried more charge for a given mass than has previously been encountered; but there was no conclusive evidence that it could exist only in discrete quantities, apart from the deductions from electrolysis.

The first direct evidence came again from work by Thomson. Thomson studied the positively charged particles that emerged from a long canal in the *cathode* of a discharge tube and was able to show that for any particular atomic species there existed discrete charge states, with charges in the ratios 1:2:3 etc. The unit of this charge sequence was equal to F/N_0 where F is the faraday and N_0 Avogadro's number.

Thomson's experiments were not of a very high accuracy and finally conclusive evidence was provided by Millikan.

Millikan's experiment

Let us assume that charge is quantized in units of $e = 1.6 \times 10^{-19}$ coulombs and work out the size of an oil drop carrying one quantum of charge for which the force of gravity could be balanced by the effect of an electric field acting on the charge.

The equation is

$$eE = \tfrac{4}{3}\pi a^3 g(\rho - \rho_0) \tag{1.10}$$

where a is the radius of the drop, ρ the density of the oil, ρ_0 the density of the atmosphere and g the acceleration due to gravity.

The electrical breakdown strength of air is about $3 \times 10^6\ \mathrm{V\,m^{-1}}$ between smooth spheres, so to avoid corona we take $E = 3 \times 10^5\ \mathrm{V\,m^{-1}}$. We take

$$\rho - \rho_0 = 10^3\ \mathrm{kg\,m^{-3}}$$

and put

$$\frac{4\pi g}{3} = 40\ \mathrm{m\,s^{-2}}\ .$$

With the value of $e = 1.6 \times 10^{-19}$ C from electrolysis we have from Eq. (1.10)

$$1.6 \times 10^{-19} \times 3 \times 10^5 = a^3 \times 40 \times 10^3\ .$$

Therefore

$$a^3 = 1.2 \times 10^{-18}\ \mathrm{m^3}$$

and

$$a \approx 10^{-6}\ \mathrm{m}\ .$$

Now let us calculate the terminal velocity for the free fall of such a drop due to the viscosity of air.

Stokes' law gives us

$$mg = 6\pi a\eta v \tag{1.11}$$

where v is the terminal velocity and η, the viscosity of air, equals $1.8 \times 10^{-5}\ \mathrm{N\,m^{-2}\,s}$. Hence

$$\begin{aligned} mg &= \tfrac{4}{3}\pi a^3 g \times 1{,}000 \\ &\approx 4 \times 10^{-14}\ \mathrm{N}\ . \\ 6\pi a\eta &\approx 19 \times 10^{-6} \times 1{\cdot}8 \times 10^{-5} \\ &\approx 3.4 \times 10^{-10}\ \mathrm{N\,m^{-1}\,s}\ . \end{aligned}$$

Hence $v \approx 10^{-4}\ \mathrm{m\,s^{-1}}$ or $0.01\ \mathrm{cm\,s^{-1}}$ which is very convenient for observation in a microscope.

A drop of this size is easily seen in a microscope when suitably illuminated. So if we can introduce charged drops of this size into the space between two plates and observe their motion with and without the presence of an electrical field we can deduce the charge they carry.

The apparatus used by Millikan is shown in Fig. 1.14.

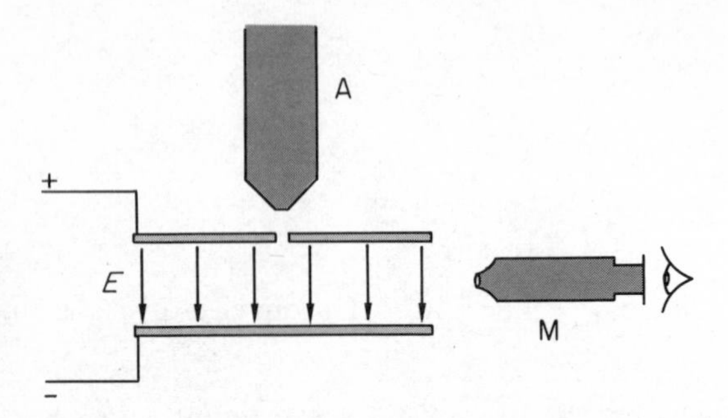

Fig. 1.14. Millikan's apparatus for the measurement of e. Electrified oil drops from the atomizer A pass through the hole to the region between the plates where there is an electric field E. Here they are observed through the microscope M.

Fine drops of oil were sprayed into the space between the plates by an atomizer. The process of atomizing is usually sufficient to produce electrification of the drops.

Those drops that are too large to use fall out of the field of vision quickly. A suitable drop is selected and its time of fall over a specified distance is measured, and hence its terminal velocity v_d is found.

Now a suitable electric field is imposed and a fresh value v_f is found. This may be upwards or downwards; we will assume it is upwards. In a large series of investigations of drops of various sizes and under various conditions Millikan found that at this small size of droplet it was necessary to modify Stokes' law and empirically deduced an expression for the viscous drag of $6\pi anv(1 + b/pa)$ where p is the air pressure and b an experimentally determined constant. No attempt was made to exactly balance the gravitational and electrical forces as this is too difficult (see Fig. 1.15).

Thus we have the equations:

(*a*) With no field

$$\tfrac{4}{3}\pi a^3(\rho - \rho_0)g = 6\pi a\eta v_d\left(1 + \frac{b}{pa}\right).$$

(b) With field

$$eE - \tfrac{4}{3}\pi a^3(\rho - \rho_0)g = 6\pi a\eta v_f\left(1 + \frac{b}{pa}\right).$$

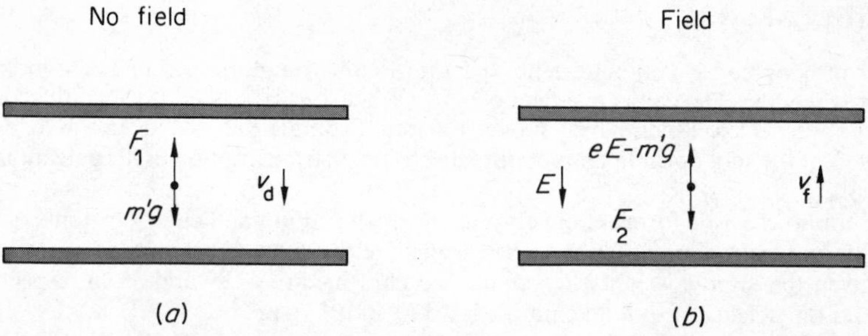

Fig. 1.15. Forces in Millikan's oil drop experiment: (*a*) With no field the gravitational force $m'g$ is balanced by the viscous force F_1, where m' is the effective mass allowing for buoyancy. Hence,

$$m'g = 6\pi a\eta v_d(1 + b/\rho a) .$$

(*b*) When the field is applied the viscous forces F_2 balance the sum of the gravitational and electrical forces, hence

$$eE - mg = 6\pi a\eta v_f(1 + b/\rho a) .$$

Adding these two equations we get

$$eE = 6\pi a\eta(1 + b/\rho a)(v_d + v_f) .$$

Adding these two together we have

$$eE = 6\pi a\eta\left(1 + \frac{b}{pa}\right)(v_d + v_f) . \tag{1.12}$$

From the first equation we find a value of a to put into this expression, and hence deduce e.

Millikan found that the charges he measured were always a multiple of the smallest charge held by a drop, thus proving the quantization of charge. The least accurate part of the experiment is the knowledge of the viscosity η. Millikan used a value about 1% too low. When the correct value for η is put into Millikan's results, we get for this quantum of charge the value:

$$e = (1.603 \pm 0.002) \times 10^{-19}\,\text{C}.$$

Combined with the measured value of e/m we get for the mass of the electron

$$m_e = (9.108 \pm 0.012) \times 10^{-28}\,\text{g}$$

$$= (9.108 \pm 0.012) \times 10^{-31}\,\text{kg} .$$

PROBLEMS 1

1.1 A moving coil galvanometer has a 50 turn coil wound on a frame 1 cm long and 0.5 cm wide. The coil is in a field of 0.5 T. When 1 μA is passed through the coil the spot on a scale 1 metre away moves 150 mm. Estimate the root mean square deviation of the spot at room temperature due to the thermal motion of the galvanometer coil.

1.2 Sodium chloride forms cubic crystals in which sodium and chlorine atoms occupy alternate sites along the three axes. Calculate the spacing between adjacent atoms given the atomic weights of sodium and chlorine are 22.99 and 35.46 respectively and the density of sodium chloride is 2.170×10^3 kg m^{-3}.

1.3 Calculate the radius of curvature of electrons of energy 1 eV, 100 eV, 10^6 eV and 10^8 eV in a field of 1 T. (See section 2.6 for the definition of eV.)

1.4 Electrons of 10 keV energy are passed between two plates 1 cm apart which have a potential difference of 1 kV between them. Calculate the magnetic field required for the electrons to be undeflected.

1.5 In a Millikan oil drop experiment one particular drop fell freely with a velocity of 2.26×10^{-4} ms^{-1} between horizontal plates 5 mm apart. On applying a potential difference of 1,600 V the drop rose steadily at 0.90×10^{-4} ms^{-1}. Find the radius of the drop and the charge on the drop. The viscosity of air is 1.80×10^{-5} Nsm^{-2}. The density of the oil was 900 kg m^{-3}.

1.6 Calculate the RMS speed of the oil drop in question 5 due to its thermal motion.

1.7 Given that e/m for an electron is 1.76×10^{11} C kg^{-1} calculate the velocity of an electron that has been accelerated through potential differences of 1 V, 100 V and 10^6 V.

CHAPTER 2

The constitution of the atom and the failure of classical physics

2.1 INTRODUCTION

In Chapter 1 we presented some of the evidence for the existence of atoms, and at the same time demonstrated the atomicity of electric charge. In so doing we also demonstrated the existence of a particle of much smaller mass than that possessed by any known atom.

The production of electrons in a gaseous discharge whatever the gas would suggest that they are a universal constituent of all atoms. Further evidence for this comes from the phenomenon of thermionic emission. If a metal is heated to a high temperature (around 2,000 °C) it copiously emits negatively charged particles, and experiments on these showed that they, too, were electrons. Richardson explained this phenomenon and gave an explicit formula for the number emitted per square centimetre, per unit time, by assuming the electrons partook of the thermal motion of the atoms.

Thus we have a picture in which matter consists of atoms which have a radius of about 10^{-8} cm, and these atoms are at least partly built up of electrons.

2.2 CONSEQUENCES OF NEUTRALITY

Because atoms are normally electrically neutral, it follows that the charge due to the electrons they contain must be balanced by an equal positive charge. But in experiments on positive rays it was shown that the charge to mass ratio was dependant on the gas in the discharge tube, and was equal to a small multiple of the electronic charge divided by the mass of the atoms forming the gas. In particular, no evidence at all was found for a positively charged equivalent of the electron. It would thus appear that the positive charge is in some way associated with the mass of the atom, and that positively charged atoms are ones in which one or more of their associated electrons are missing.

2.3 ATOMS AND X-RAYS. THE NUMBER OF ELECTRONS PER ATOM

We have indicated that all atoms contain electrons. The question arises, do the atoms of a particular atomic species contain a definite number of electrons? Evidence which points in this direction came from the scattering of X-rays.

X-rays were discovered in 1895 by Roentgen when conducting experiments on discharges in gases.

Evidence soon accumulated that X-rays were electromagnetic radiation of very short wave-length; for example:

(i) X-rays were produced when energetic electrons hit a solid object. Under these circumstances the electron will suffer a violent deceleration, and it is a consequence of electromagnetic theory that an accelerated or retarded electron will radiate electromagnetic radiation.

(ii) Haga and Wind showed in 1899 that X-rays could be diffracted by a very fine slit, showing them to be a wave phenomenon.

The size of the diffraction pattern indicated that the wavelength of the order of 10^{-8} cm.

(iii) In 1906 Barkla proved that the waves were *transverse* by showing that they can be polarized.

Now, if electromagnetic radiation passes by a charged particle, the particle will oscillate under the influence of the time varying electric field, and will therefore possess an acceleration. As we have already stated, an accelerated electric charge radiates, i.e. the electric charge scatters the incoming radiation.

The total intensity of radiation emitted by an accelerated charge is given by the formula:

$$R = \frac{2}{3}\frac{e^2a^2}{c}\frac{\mu_0}{4\pi}\ \mathrm{J\ s^{-1}}$$

$$= \frac{2}{3}\frac{e^2a^2}{c}10^{-7}\ \mathrm{J\ s^{-1}} \tag{2.1}$$

where e is the charge in coulombs, a the acceleration in metres per second squared and c the velocity of light in metres per second.

The important point is that the amount of energy emitted is proportional to the square of the acceleration.

In an electric field E the acceleration of a charge is given by eE/m. Thus the amount of energy scattered is proportional to e^4/m^2. Because electrons are so much lighter than any other atomic constituent, it follows that essentially all the scattering of X-rays by atoms will be due to the electrons in the atoms.

Under what circumstances then, might it be possible to use the scattering of X-rays to estimate the number of electrons in an atom?

Clearly, one condition is that the wavelength of the X-rays must be smaller than the spacing between the electrons. When this condition holds the oscillations of the different electrons will be random in phase, assuming that the electron spacing is not regular; hence the radiations emitted by the different electrons will also be random in phase, and we can just add the intensities together. The amount of scattering will thus be proportional to the number of electrons. A second condition is that the frequency of the X-rays must be much greater than any of the natural frequencies of oscillation the electrons might have in the atoms. This requirement is equivalent to being able to treat the electrons as if they were free. Remembering the charge on the electron is $-e$, the equation of motion of a harmonically bound electron is:

$$\frac{d^2x}{dt^2} + \omega_0^2 x = -\frac{eE}{m}\sin\omega t \tag{2.2}$$

where ω_0 is the natural frequency, ω the driving frequency.

The reader will verify that the solution of this equation is given by

$$x = \frac{eE}{m}\frac{\sin\omega t}{\omega^2 - \omega_0^2} \tag{2.3}$$

$$\approx \frac{eE}{m}\frac{\sin\omega t}{\omega^2} \tag{2.4}$$

for $\omega \gg \omega_0$, which is the solution for a free particle.

We may make an estimate of the highest of the natural frequencies as follows. If we bombard a substance with high energy electrons, we find the spectrum of X-rays emitted contains peaks which are characteristic of the elements in the substance. These wavelengths are presumably caused by oscillations of the electrons in the atom. The shortest wavelength line is called the K line, so we require that the wavelength of the X-rays we use should be shorter than the K line. We give the wavelengths of the K lines for some of the elements in Table 2.1, in units of 10^{-8} cm.

Table 2.1. Wavelengths of the *K*-line for light elements in Å

He	Li	C	O	Ne	Al	S
585	240	44	23	14	8	5.2

Thus we see that if we use X-rays of wavelength less than about 0.5×10^{-8} cm, we expect to get a fraction scattered proportional to the number of electrons present. This scattered radiation will reduce the intensity of the incident beam. Hence by measuring the attenuation of X-rays of suitable wavelength, we can hope to get an idea of the number of electrons in a particular atom.

We can calculate the attenuation without too much difficulty. We had (Eq. 2.1)

$$R = \frac{2}{3} \cdot \frac{e^2 a^2}{c} \times 10^{-7}\ \mathrm{J\,s^{-1}}\ . \tag{2.5}$$

The acceleration $a = eE/m$ where E is the electric field, therefore

$$R = \frac{2}{3}\frac{e^4 E^2}{m^2 c} \times 10^{-7}\ \mathrm{J\,s^{-1}}\ .$$

The mean radiated energy/sec $\bar{R}$ is therefore

$$\bar{R} = \frac{2}{3}\frac{e^4 \overline{E^2}}{m^2 c} \times 10^{-7}\ J\,\mathrm{s^{-1}}$$

where $\overline{E^2}$ is the mean square electric field.

The intensity of a plane electromagnetic wave is given by

$$I = \varepsilon_0 \overline{E^2} c\ J\,\mathrm{m^{-2}\,s^{-1}}\ . \tag{2.6}$$

Therefore the energy radiated per second by an electron under the influence of an electromagnetic wave of intensity I is

$$\bar{R} = \frac{2}{3}\frac{e^4}{m^2}\frac{I}{\varepsilon_0 c^2} \times 10^{-7}\ \mathrm{J\,s^{-1}}$$

$$= \frac{8\pi}{3}\left(\frac{e^2}{m}\right)^2 I \times 10^{-14}\ \mathrm{J\,s^{-1}} \tag{2.7}$$

when we insert $c^2 = 1/(\mu_0 \varepsilon_0)$ and substitute $\mu_0 = 4\pi \times 10^{-7}$ henries per metre.

The quantity

$$\frac{8\pi}{3}\left(\frac{e^2}{m}\right)^2 \times 10^{-14} = 6{\cdot}66 \times 10^{-29}\ \mathrm{m^2}$$

is known as the *Thomson cross-section of the electron.* If there are n electrons per $\mathrm{m^2}$ then the amount of energy radiated from a slab of material $1\ \mathrm{m^2}$ in cross-section and dx metres thick will be

$$n\,\mathrm{d}x \times \frac{8\pi}{3}\left(\frac{e^2}{m}\right)^2 I \times 10^{-14}\, J\,\mathrm{s}^{-1}$$

and this amount of energy per second will be removed from the beam.

Thus in a distance dx there will be a change in the intensity I equal to $-\mu I\,\mathrm{d}x$ where

$$\mu = n \times \frac{8\pi}{3}\left(\frac{e^2}{m}\right)^2 \times 10^{-14}\ .$$

Thus the equation for the intensity is

$$\frac{\mathrm{d}I}{\mathrm{d}x} = -\mu I, \text{ or } I = I_0\,\mathrm{e}^{-\mu x}. \tag{2.8}$$

Typical results for the attenuation per atom are shown in Fig. 2.1. We see that for hydrogen and helium we have a suitable result except for an unexpected fall in the attenuation for wavelength less than 0.2×10^{-8} cm.

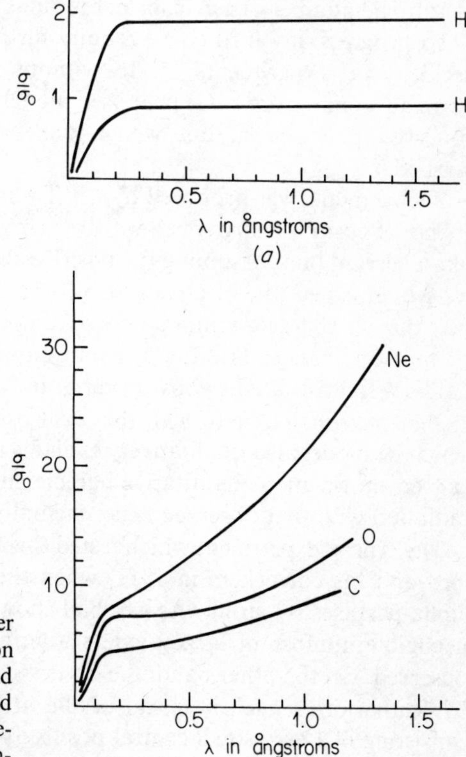

Fig. 2.1. The X-ray attenuation per atom, σ, in terms of the attenuation per electron σ_0 (*a*) for hydrogen and helium, (*b*) for carbon, oxygen and neon. If the electrons scatter independently σ/σ_0 should equal the number of electrons per atom.

For the heavier elements we see that at longer wavelengths there are clear signs of the influence of coherent effects between electrons, that is, the electrons in the atoms are oscillating nearly in step, and we have to add together the amplitudes of their radiations, rather than just the intensities. Nevertheless the results between 0.2 and 0.4×10^{-8} cm show clearly that the number of electrons per atoms is about half the atomic weight.

2.4 ATOMIC MODELS AND THE SCATTERING OF ALPHA PARTICLES

We have established the following facts about atoms:

(i) They have radii of the order of 10^{-8} cm.
(ii) They contain electrons which are negatively charged and are very much lighter than atoms.
(iii) They are neutral and therefore must contain a positive charge to balance the negative charge of the electrons.
(iv) This positive charge seems to be associated with the mass in some way.
(v) The number of electrons per atom is about half the atomic weight.

To progress any further we require further experiment. But one cannot decide what experiments to do without some guidance. This guidance normally comes from some proposed model, designed to account for already observed phenomena, but which makes predictions which are not yet verified.

Such a model was proposed by J. J. Thomson. He envisaged the atom as a sphere of positive charge of about 10^{-8} cm radius, with electrons buried in it, like a currant bun. Assuming the positive charge to be uniformly distributed, we would expect the electrons also to be uniformly distributed, for in this way the net charged within a sphere centred on the centre of the atom will be zero on average, and the distribution will be stable. We see that such a model will satisfy all the above criteria, and what is more should the positions of the electrons be disturbed, they will oscillate and so give off radiation. Thus the model also qualitatively explains the emission of light by atoms. It can be shown that quantitative agreement of the spectrum of the emitted radiation with that observed experimentally is most unlikely.

The crucial experiment which tested this model was performed by Rutherford and his coworkers in 1911, when they investigated the scattering of alpha particles by atoms. As we shall show, the Thomson model predicts a negligible number of large angle scattering events, contrary to what was observed. On the other hand, the observed data are very well accounted for by Rutherford's *nuclear model* of the atom, which pictures the atom as consisting of a very small central positively charged nucleus, surrounded by circulating electrons, such that the whole atom is neutral.

2.5 THE ALPHA PARTICLE

Alpha particles are positively charged helium ions, each particle carrying a charge numerically equal to that of two electrons, but of the opposite sign. They are emitted with a very high velocity by certain radioactive substances such as uranium and radium.

Let us summarize the evidence showing alpha particles to be doubly charged helium ions.

(i) The direction of their deflection in a magnetic field shows them to be positively charged.

(ii) Their behaviour in combined electric and magnetic fields gave a value of e/m which corresponded to doubly charged helium ions. The experiment is illustrated in Fig. 2.2.

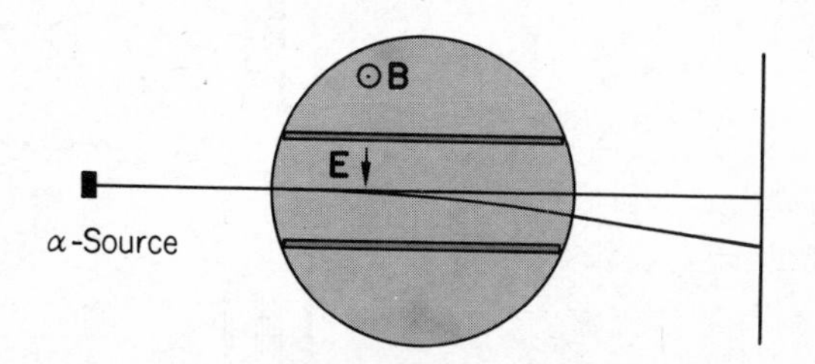

Fig. 2.2. An e/m measurement for alpha particles gives the value for doubly charged helium.

(iii) A measurement of their charge showed this to be approximately twice that of the electron. This was done by measuring the charge from a large number of alpha particles from a strong source, and then finding the disintegration rate by observing scintillations on a screen, as indicated in Fig. 2.3.

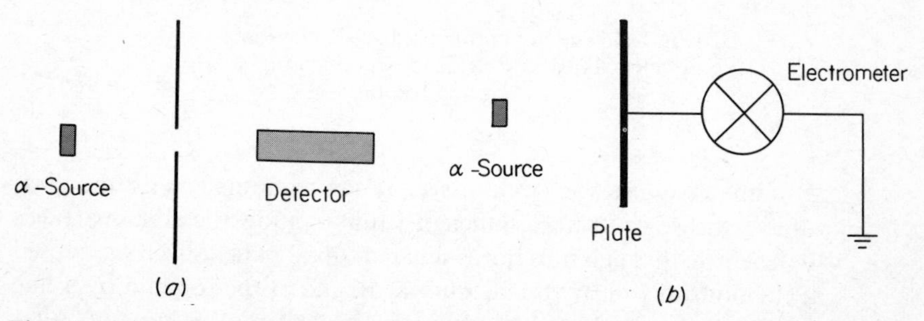

Fig. 2.3. (*a*) Count rate from a small solid angle determines the number of disintegrations per second. (*b*) An electrometer measures the charge emitted per second. Hence we determine the charge per alpha particle.

(iv) The final piece of evidence was obtained by Rutherford and Royd using the apparatus shown in Fig. 2.4.

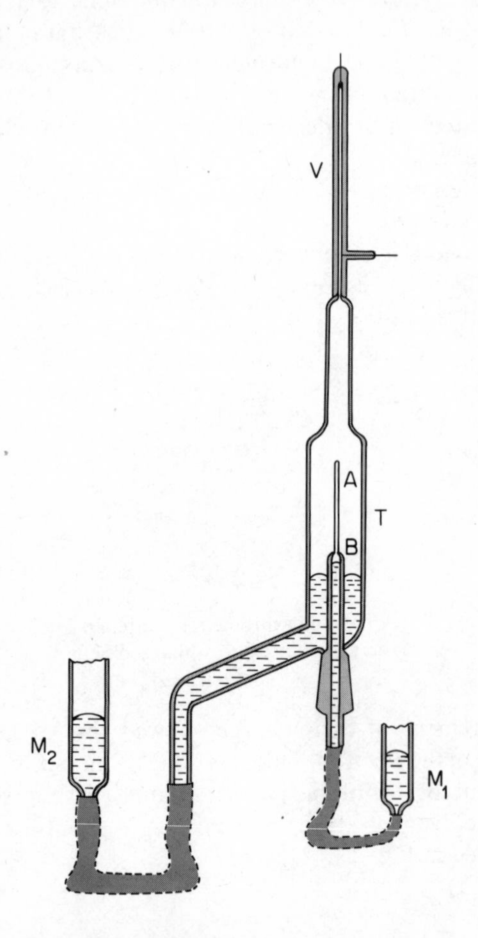

Fig. 2.4. The apparatus used by Rutherford and Royd to demonstrate the identity of alpha particles and helium.

A radioactive gas was compressed by the manometer into the glass tube A, whose walls were sufficiently thin to allow the alpha particles to penetrate through into the evacuated tube T. The tube T was closed at the bottom end by the mercury at B, and at the top end by a fine capillary tube V which had electrodes sealed in at either end. After several hours the level of the meniscus B was raised, using the manometer M_2, thus compressing all the gases in T into V. A discharge was then run in V, and the light from this discharge showed the characteristic spectral lines of helium.

A dummy experiment was performed in which the radioactive material in A was replaced by helium. In this case the helium lines did not appear in the discharge spectrum.

2.6 ALPHA PARTICLE SCATTERING ON THE THOMSON MODEL

If a collimated beam of α particles from a radioactive source is allowed to impinge on a very thin foil (for example, of gold about 10^{-4} cm thick), then it is found that nearly all of them penetrate right through the foil with a small loss of energy. In fact, the majority are found to travel within about 1° of their original direction (see Fig. 2.5).

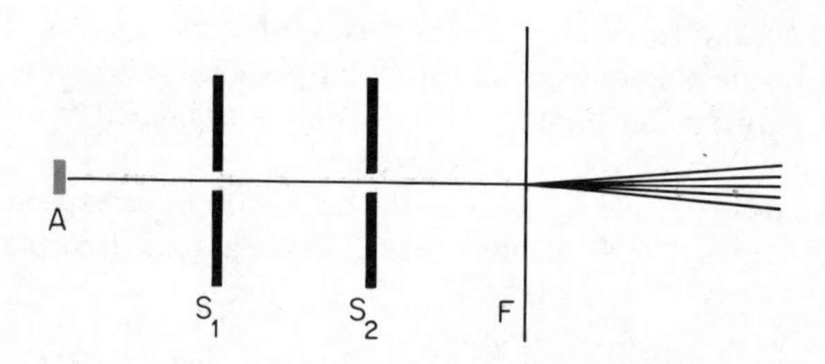

Fig. 2.5. Alpha particle scattering by a thin foil. A is the source, S_1 and S_2 are collimating slits, F the foil. Most particles stay within 1° of the original direction.

A few suffer much larger deflections, about 1 in 10^4 being deflected by 90° or more.

The fact that the alpha particles travel right through the foil means they travel *through* the atoms. On the way they will be deflected by the electric field due to the charges within the atoms.

Let us remind ourselves that on the Thomson model the atoms consist of a spherical positive charge about 10^{-8} cm. in radius, with the electrons embedded in this charge. These spheres are closely packed, hence if the foil is about 10^{-4} cm thick, the particle will pass through 10^{+4} atoms. Thus the final deflection will be the sum of the deflections in each atom. These will not always be in the same sense. Indeed, in general there will usually be about equal numbers of deflections in all senses. Thus the problem is one of multiple scattering, and we have to find a way of estimating the mean scattering angle, and the deviation from this mean. The situation is illustrated in Fig. 2.6.

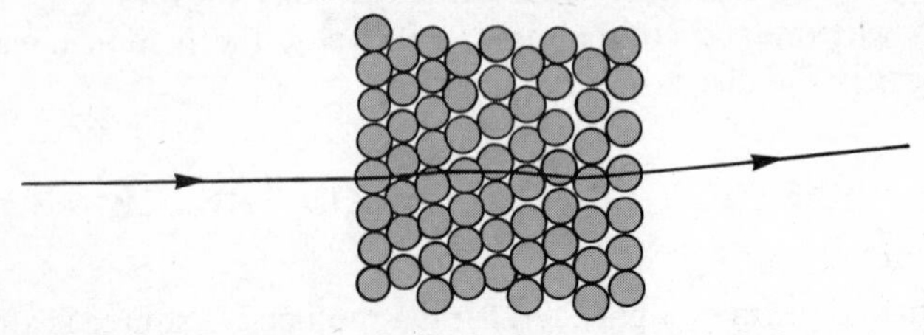

Fig. 2.6. Multiple scattering. A two-dimensional illustration of scattering on the Thomson model. The final deflection is the sum of many small deflections of both signs.

It should be emphasized that one only needs fairly crude estimates of the magnitudes involved as the calculation leads to a formula for the scattering which can be expressed in terms of measurable parameters. This is quite characteristic of phenomena due to multiple events, where each event is similar in nature. The final result is always dominated by the statistical probability of the way the events combine.

We make a *very crude* estimate of the deflection to be expected in a single encounter in a way we illustrate in Fig. 2.7. We assume that the atom exerts

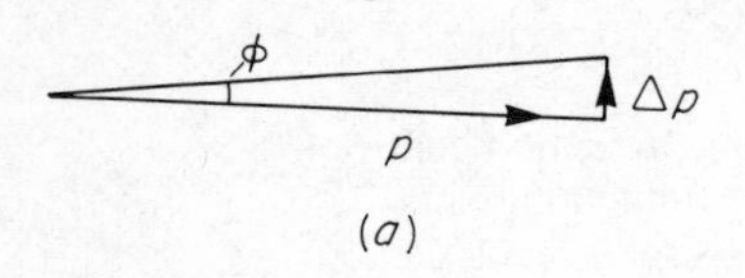

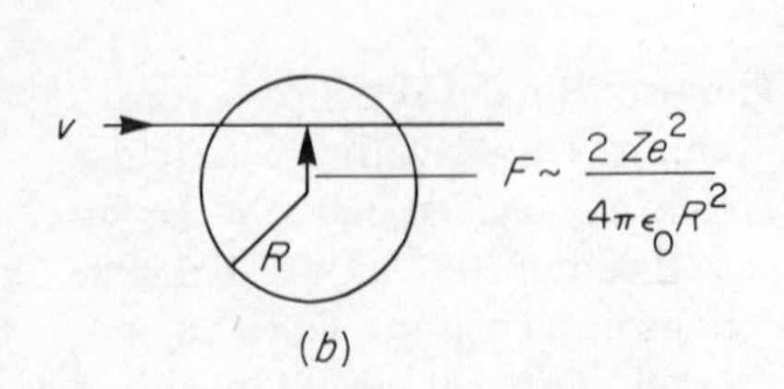

Fig. 2.7. Alpha particle scattering on the Thomson model. (*a*) The small deflection ϕ produced by a single atom equals $\Delta p/p$ where p is the incident momentum and Δp is the transverse momentum caused by the force F between the alpha particle and the atom acting for the time t it takes the alpha particle to cross the atom. Thus

$$\phi = Ft/p \ .$$

(*b*) The force F is of the order of $2Ze^2/4\pi\varepsilon_0 R^2$. The time taken for the alpha particle to cross the atom is of the order of R/v. Thus we have

$$\phi \approx \frac{2Ze^2}{4\pi\varepsilon_0 Rvp}$$

$$\approx \frac{2Ze^2}{4\pi\varepsilon_0 mv^2 R}$$

$$\approx \frac{Ze^2}{4\pi\varepsilon_0 ER}$$

where m and E are the mass and kinetic energy of the incident alpha particle.

a force F on the alpha particle which lasts for a time t. This changes the momentum of the alpha particle by an amount Δp which we assume to be at right angles to the original momentum p of the alpha particle. Hence the scattering angle ϕ will be given by

$$\phi \approx \frac{\Delta p}{p} = \frac{Ft}{p} . \tag{2.9}$$

The force F is of the order of $zZe^2/(4\pi\varepsilon_0 R^2)$ where Ze is the amount of positive charge in the atom whose radius is R; ze is the alpha particle's charge. The collision time t is approximately equal to R/v where v is the alpha particle velocity. Substituting for F and t in Eq. (2.9) we get

$$\begin{aligned} \phi &\approx \frac{zZe^2}{4\pi\varepsilon_0 Rvp} \\ &\approx \frac{zZe^2}{4\pi\varepsilon_0 mv^2 R} \\ &\approx \frac{Ze^2}{4\pi\varepsilon_0 ER} \end{aligned} \tag{2.10}$$

where $E = \frac{1}{2}mv^2$ is the kinetic energy of the alpha particle, m is its mass, and in the last line we substituted $z = 2$.

Now alpha particle energies, in common with other energies in atomic and nuclear physics are measured in *electron volts*. The units used are electron volts (eV), thousands of electron volts (keV) and millions of electron volts (MeV).

One electron volt is the energy acquired by an electron when it moves through a potential difference of 1 volt. As $e \simeq 1.6 \times 10^{-19}$ C we have

$$\begin{aligned} 1\ \text{eV} &\simeq 1.6 \times 10^{-19}\ \text{joules} \\ &\simeq 1.6 \times 10^{-12}\ \text{ergs} \end{aligned}$$

and

$$1\ \text{MeV} \simeq 1.6 \times 10^{-6}\ \text{ergs} .$$

Alpha particles emitted by radioactive nuclei have energies of a few MeV. Thus in Eq. (2.10) if we put $Z = 80$, $e = 1.6 \times 10^{-19}$ C, $R = 10^{-10}$ m and $E = 5$ MeV, we get

$$\begin{aligned} \phi &\approx \frac{80 \times (1.6 \times 10^{-19})^2}{4\pi \times 8.85 \times 10^{-12} \times 5 \times 1.6 \times 10^{-13} \times 10^{-10}} \\ &\approx 2 \times 10^{-4}\ \text{radians} . \end{aligned}$$

When we combine a large number of such deflections in a random manner in three dimensions, we arrive at the following expression for the probability $P(\Phi)\,\mathrm{d}\Phi$ of scattering through an angle between Φ and $\Phi + \mathrm{d}\Phi$

$$P(\Phi)\,\mathrm{d}\Phi = \frac{1}{2\pi\overline{\Phi^2}} \exp\left(-\frac{\Phi^2}{2\overline{\Phi^2}}\right) \cdot 2\pi\,\Phi\,\mathrm{d}\Phi \tag{2.11}$$

where $\overline{\Phi^2}$ is the mean square scattering angle, and is related to the deflection ϕ due to a *single* atom by the relation:

$$\overline{\Phi^2} = N\phi^2 \ . \tag{2.12}$$

Thus we can check (i) does the scattering follow Eq. (2.11), and (ii) is $\overline{\Phi^2} \approx N\phi^2$?

Rutherford and his collaborators Geiger and Marsden checked these results by scattering alpha particles from gold, and observing the scintillations produced when the alpha particles hit a zinc sulphide screen. They found that from 0° to 3° the number of particles did indeed follow Eq. (2.11) with a value of $(\overline{\Phi^2})^{1/2} \sim 1°$. This implies a deflection per atom of the order of 0.01° ($N \sim 10^4$ for a foil 10^{-4} cm thick), or 1.5×10^{-4} radians per collision; so in this region the agreement is excellent. But, as stated at the beginning of this section, about 1 in 10^4 alpha particles suffer deflections of 90° or greater.

While Eq. (2.11) is not exact for large angles, we may still use it to estimate the probability of large angle scattering occurring. If we put $(\overline{\Phi^2})^{1/2} = 1°$ we find from Eq. (2.11) that the probability of a scattering angle of greater than 5° occurring is 3.8×10^{-6}, and for angles greater than 10° it is 2×10^{-22}. Thus we are led to the conclusion that the small angle scattering could be accounted for by the Thomson model, but not the large angle scattering.

Slight modifications of the Thomson model will not do. If, for example, one reduces the radius of the positive charge, then Eq. (2.10) shows that the *single* scattering angle increases, but according to Eq. (2.12) the root mean square scattering angle is $\phi\sqrt{N}$ where N is the number of atoms that the alpha particle passes through. However a reduction of the radius will leave spaces between the positive spheres through which the alpha particle can travel. The probability that an alpha particle will traverse a particular atom is clearly proportional to the area the atom presents to the beam, as is indicated in Fig. 2.8. Hence the number of positive atomic cores traversed in the foil will be $N = N_t\pi R^2/\pi R_t^2$ where N_t is the number traversed for $R = R_t$ when the spheres touch. As, according to Eq. (2.10), ϕ is proportional to $1/R$, we get the result that $\phi\sqrt{N}$, and hence $\overline{\Phi^2}$, is constant independent of the value of R. Thus we arrive at the conclusion that *multiple scattering can never produce the observed large deflections.*

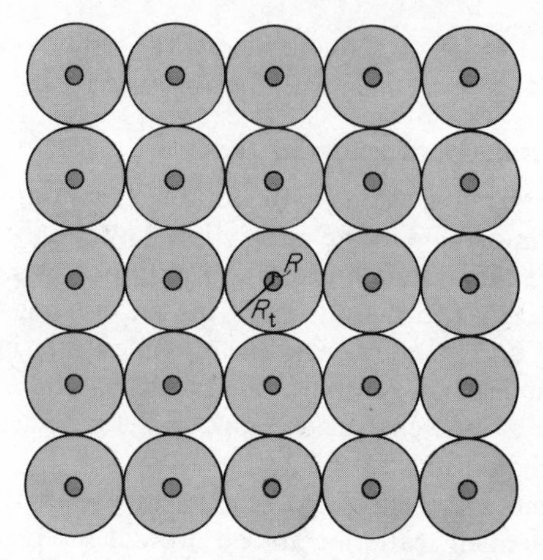

Fig. 2.8. A view of a plane of atoms with radius R_t such that they touch, and of smaller atoms of radius R. The areas occupied in the plane by two types are in the ratio R^2/R_t^2. The chances that an alpha particle moving perpendicularly to the plane will pass through an atom will be in the same ratio.

The condition that $\phi\sqrt{N}$ is constant clearly only holds down to radii such that $N \gg 1$. For smaller radii we have *single* scattering, and the deflection angle is just that due to a single event. Eq. (2.10) suggests this could occur for radii $\sim 10^{-12}$ cm, yet *all* methods of measuring atomic radii agree on a value of $\sim 10^{-8}$ cm. Clearly a different model is required.

2.7 THE RUTHERFORD MODEL OF THE ATOM

Rutherford proposed in 1911 that the atom consists of a minute central nucleus, carrying the positive charge and almost the whole of the mass, with electrons circling round it. In his treatment of alpha particle scattering the central core was to be treated as a point charge (Fig. 2.9).

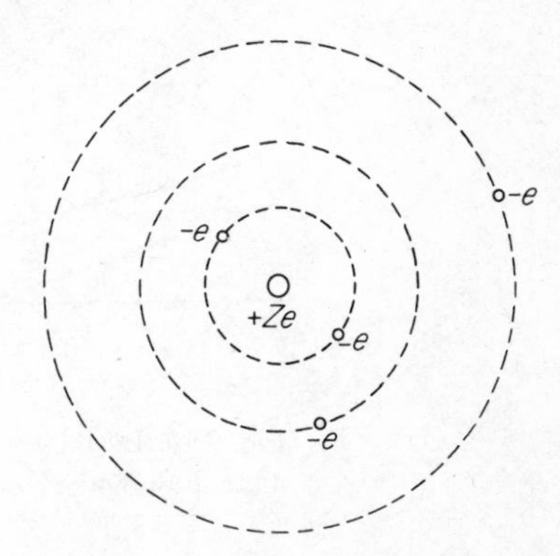

Fig. 2.9. The Rutherford model of the atom. A massive central nucleus carries a charge Ze, and Z electrons, each of charge $-e$, circle round it. No special significance should be attached to the fact that the orbits are shown to be circular and in the same plane.

From this model we shall derive a formula for alpha particle scattering which gives impressive agreement with the observed results.

Derivation of the Rutherford scattering formula

In deriving this formula, we make the following assumptions:

(i) The scattering is due to the interaction between the incoming alpha particle and the atomic nucleus and only occurs significantly if the path passes close to the nucleus. This means that scattering events occur rarely and the problem is one of single scattering.

(ii) The force acting between the alpha particle and the nucleus is solely due to their electric charges and follows the inverse square law down to a very small distance.

(iii) We can ignore the effect of the electrons. Clearly these will only introduce a small scattering spread around the predicted deflection angle.

(iv) We can treat the scattering as if the scattering nucleus were fixed in the laboratory.

The last assumption is made solely to simplify the calculation and can be dispensed with quite easily by working in a frame where the centre of mass of the alpha particle and the nucleus is at rest, instead of working in the laboratory frame of reference. In the centre of mass frame the problem becomes that of the scattering of a particle of mass $\mu = m_\alpha M_N/(m_\alpha + M_N)$ (where m_α and M_N are the alpha particle and nuclear masses respectively) by a scattering centre fixed at the origin. μ is called the reduced mass. The calculation which follows, therefore, is exact in the centre of mass frame if m is interpreted as the reduced mass rather than the alpha particle mass. Measurements are, of course, made in the laboratory system; it is a straightforward, if tedious, matter to convert these to the centre of mass frame.

Referring now to Fig. 2.10, we consider a nucleus fixed at the origin and an alpha particle approaching from the right with initial velocity $\mathbf{v}_i$. The alpha particle, if it were undeflected, would pass at a distance b from the nucleus.

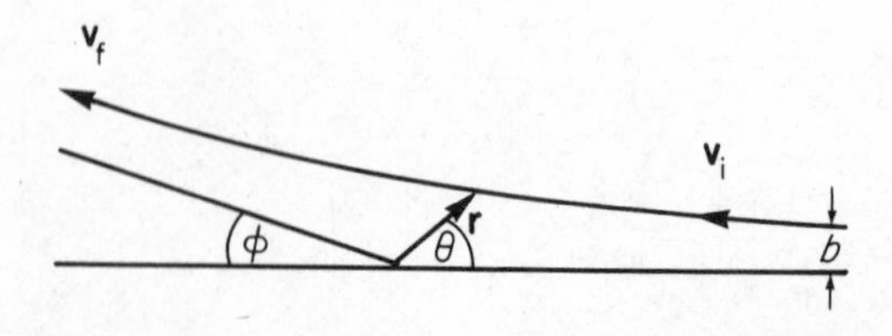

Fig. 2.10. Initial and final velocities, impact parameter and scattering angle in Rutherford scattering.

The distance b is known as the *impact parameter*. After scattering, the alpha particle has been deflected through an angle ϕ and has a final velocity $\mathbf{v}_f$. Because of assumption (iii) that the nucleus does not move the magnitudes of the initial and final velocities are the same, i.e.

$$|\mathbf{v}_f| = |\mathbf{v}_i| = v \ .$$

The Coulomb repulsion of the nucleus provides a force acting on the alpha particle along the vector $\mathbf{r}$ from the nucleus to the alpha particle

$$\mathbf{F} = \frac{zZe^2}{4\pi\varepsilon_0 r^2}\hat{\mathbf{r}}$$

where z and Z are the alpha particle and nuclear charges in units of e, r is their separation, and $\hat{\mathbf{r}}$ a unit vector in the direction of r.

Equating force to (mass) × (acceleration) we have

$$\frac{zZe^2}{4\pi\varepsilon_0 r^2}\hat{\mathbf{r}} = m\frac{\mathrm{d}\mathbf{v}}{\mathrm{d}t}$$

$$= m\frac{\mathrm{d}\mathbf{v}}{\mathrm{d}\theta}\frac{\mathrm{d}\theta}{\mathrm{d}t}$$

where θ is the polar angle of the position vector $\hat{\mathbf{r}}$. Hence

$$\frac{\mathrm{d}\mathbf{v}}{\mathrm{d}\theta} = \frac{zZe^2\hat{\mathbf{r}}}{4\pi\varepsilon_0 mr^2\, \mathrm{d}\theta/\mathrm{d}t} \ . \tag{2.13}$$

The denominator on the right-hand side is $4\pi\varepsilon_0$ times the angular momentum of the alpha particle, and this is conserved because the alpha particle is moving in a central field, as is shown in Fig. 2.11. Hence

$$mr^2\frac{\mathrm{d}\theta}{\mathrm{d}t} = L$$

$$= mvb$$

and

$$\frac{\mathrm{d}\mathbf{v}}{\mathrm{d}\theta} = \frac{zZe^2\hat{\mathbf{r}}}{4\pi\varepsilon_0 L} \ .$$

Therefore

$$\int \mathrm{d}\mathbf{v} = \frac{zZe^2}{4\pi\varepsilon_0 L}\int \hat{\mathbf{r}}\, \mathrm{d}\theta \ . \tag{2.14}$$

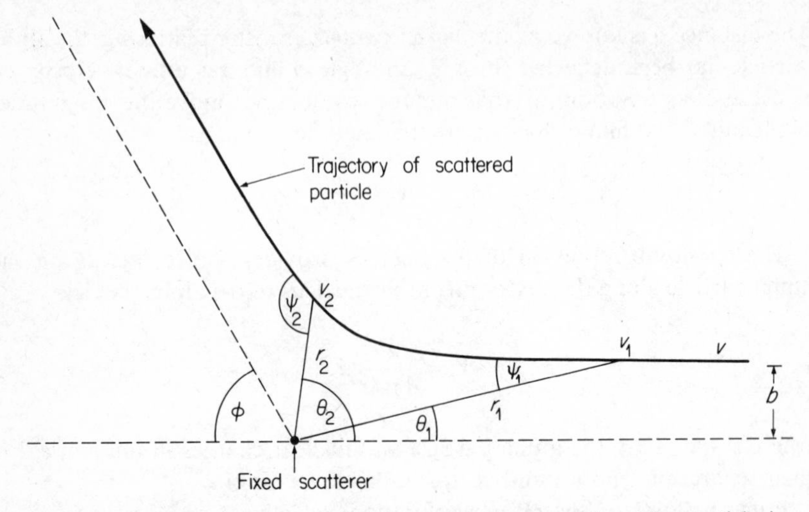

Fig. 2.11. Conservation of angular momentum in scattering. The particle is incident from the right with initial velocity v and impact parameter b, and is scattered through an angle ϕ. For *any* central law of force conservation of angular momentum requires that

$$mv_1r_1 \sin \psi_1 = mv_2r_2 \sin \psi_2 = mvb$$

where the positions 1 and 2 can occur anywhere on the trajectory. Now

$$v \sin \psi = r\frac{d\theta}{dt} .$$

Therefore

$$mvr \sin \psi = mr^2\frac{d\theta}{dt} = \text{constant} .$$

The left-hand integral is just $\mathbf{v}_f - \mathbf{v}_i$. From Fig. 2.12(*a*) we see that the magnitude of $|\mathbf{v}_f - \mathbf{v}_i|$ is given by

$$|\mathbf{v}_f - \mathbf{v}_i| = 2v \sin \phi/2 \tag{2.15}$$

and from Fig. 2.12(*b*) the unit vector in the direction of $\mathbf{v}_f - \mathbf{v}_i$ is

$$\mathbf{i} \sin \phi/2 + \mathbf{j} \cos \phi/2 \tag{2.15a}$$

where $\mathbf{i}$ and $\mathbf{j}$ are unit vectors along the x and y directions respectively.

The integral on the right hand side of Eq. (2.14) may be evaluated by making the substitution

$$\hat{\mathbf{r}} = \mathbf{i} \cos \theta + \mathbf{j} \sin \theta .$$

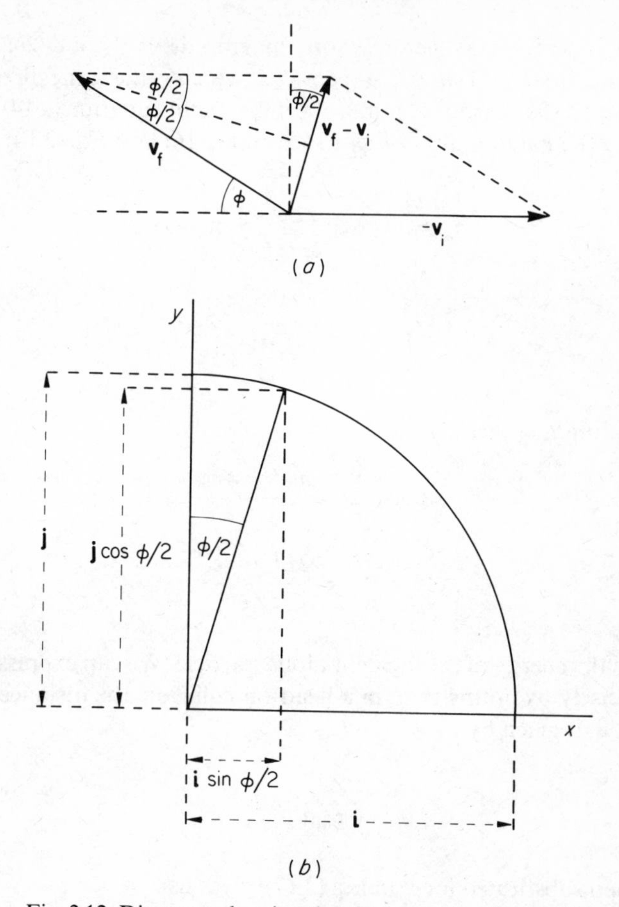

Fig. 2.12. Diagrams showing that (*a*) $|\mathbf{v}_f - \mathbf{v}_i| = 2v \sin \phi/2$ and (*b*) that the unit vector $\hat{\mathbf{r}}$ in the direction of $\mathbf{v}_f - \mathbf{v}_i$ is given by $\mathbf{i} \sin \phi/2 + \mathbf{j} \cos \phi/2$.

From Fig. 2.11 we see that the initial and final limits of integration are $\theta = 0$ and $\theta = \pi - \phi$. Hence we have

$$\begin{aligned}\int_0^{\pi-\phi} \hat{\mathbf{r}}\, d\theta &= \int_0^{\pi-\phi} (\mathbf{i}\cos\theta + \mathbf{j}\sin\theta)\, d\theta \\ &= [\mathbf{i}\sin\theta - \mathbf{j}\cos\theta]_0^{\pi-\phi} \\ &= \mathbf{i}\sin\phi + \mathbf{j}(\cos\phi + 1) \\ &= 2\mathbf{i}\sin\phi/2\cos\phi/2 + 2\mathbf{j}\cos^2\phi/2 \\ &= 2\cos\phi/2[\mathbf{i}\sin\phi/2 + \mathbf{j}\cos\phi/2] \ . \qquad (2.16)\end{aligned}$$

Eq. (2.16) represents a vector whose magnitude is $2\cos\phi/2$ and which points in the direction $\mathbf{i}\sin\phi/2 + \mathbf{j}\cos\phi/2$, which is the same direction as is given by Eq. (2.15a), as indeed it must be if the *vector* equation (2.14) is to hold.

Inserting the magnitudes of Eqs. (2.15) and (2.16) into Eq. (2.14) we obtain

$$2v\sin\phi/2 = \frac{zZe^2}{4\pi\varepsilon_0 L}2\cos\phi/2$$

whence

$$\cot\phi/2 = \frac{vL\cdot 4\pi\varepsilon_0}{zZe^2}\ .$$

Substituting $L = mvb$ we get

$$\cot\phi/2 = \frac{mv^2b\cdot 4\pi\varepsilon_0}{zZe^2}$$

$$= \frac{2Eb\cdot 4\pi\varepsilon_0}{zZe^2} \qquad (2.17)$$

where E is the energy of the incident alpha particle. We can express Eq. (2.17) more concisely by noting that, in a head-on collision, the distance of closest approach, a, is given by

$$\frac{zZe^2}{4\pi\varepsilon_0 a} = E \qquad (2.18)$$

which, when substituted for E in Eq. (2.17), gives us

$$\cot\phi/2 = 2b/a\ . \qquad (2.19)$$

Eq. (2.19) tells us the scattering angle in terms of the impact parameter b. To find the probability that an alpha particle is scattered through an angle ϕ all we have to do is to find the probability that the alpha particle has the corresponding impact parameter b, and to relate the two probabilities through Eq. (2.19). The way we do this is illustrated in Fig. 2.13.

Let us suppose the alpha particle hits the foil somewhere within a region of area A (we shall find this cancels out, but it makes visualization easier).

The alpha particle will be scattered through an angle between ϕ and $\phi + \mathrm{d}\phi$ provided it passes through an annulus centred on a nucleus with radii b and $b + \mathrm{d}b$ where from relation (2.19) we have

$$\tfrac{1}{2}\operatorname{cosec}^2\phi/2\,\mathrm{d}\phi = -2\,\mathrm{d}b/a\ .$$

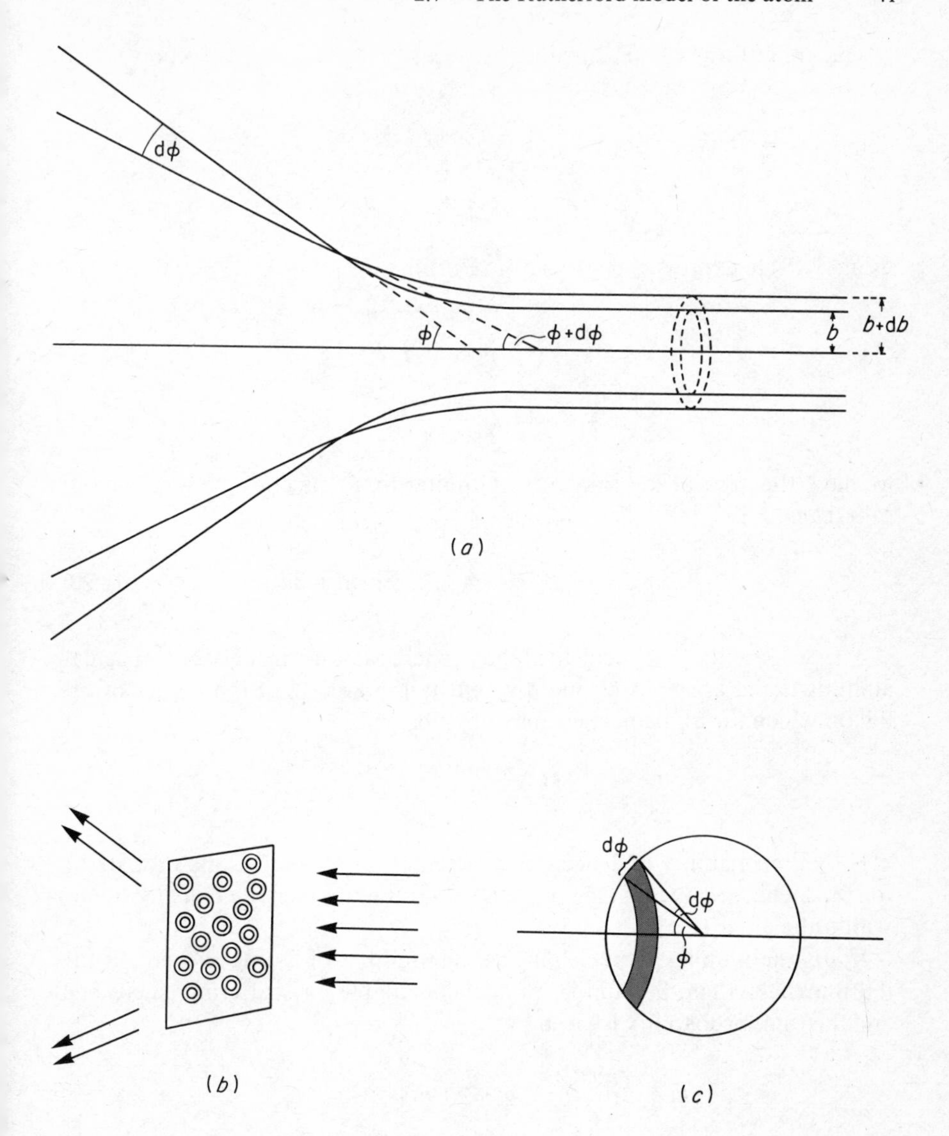

Fig. 2.13. The probability of scattering through an angle between ϕ and $\phi + d\phi$. As shown in (*a*) the alpha particles with impact parameters between b and $b + db$ will be scattered through angles between ϕ and $\phi + d\phi$. Hence the number of alpha particles scattered into the solid angle between ϕ and $\phi + d\phi$ equals the number passing through the annulus $2\pi b\, db$. The total number scattered into the solid angle between ϕ and $\phi + d\phi$ is the number passing through all such annuli, as indicated in (*b*). The solid angle between ϕ and $\phi + d\phi$, shown in (*c*) is the area of the annulus between ϕ and $\phi + d\phi$ on the surface of a unit sphere. The width of the annulus is $d\phi$ (the radius of the sphere being unity) and its circumference is $2\pi \sin \phi$. Hence the solid angle is $2\pi \sin \phi\, d\phi$.

The area of this ring is $2\pi b|\mathrm{d}b|$ (the negative value for $\mathrm{d}b$ merely means b decreases as ϕ increases). Thus the area is given by

$$2\pi b|\mathrm{d}b| = 2\pi b\tfrac{1}{4}a \operatorname{cosec}^2 \phi/2\, \mathrm{d}\phi$$

$$= 2\pi\frac{a}{2}\cot \phi/2 \frac{a}{4}\operatorname{cosec}^2 \phi/2\, \mathrm{d}\phi$$

using (2.19) to express b in terms of ϕ. Putting

$$\cot \phi/2 = \frac{\cos \phi/2 \sin \phi/2}{\sin^2 \phi/2}$$

$$= \frac{\sin \phi}{2 \sin^2 \phi/2}$$

we have the area of the ring corresponding to a deflection between ϕ and $\phi + \mathrm{d}\phi$ is

$$2\pi b|\mathrm{d}b| = \frac{a^2}{16}\operatorname{cosec}^4 \phi/2 \cdot 2\pi \sin \phi\, \mathrm{d}\phi\,. \tag{2.20}$$

The probability of an incident alpha particle passing through this particular annulus is the area of the ring divided by the area A of the region of the foil on which the alpha particle may impinge, viz.

$$\frac{2\pi b|\mathrm{d}b|}{A}\,.$$

The *total* probability of deflection between ϕ and $\phi + \mathrm{d}\phi$ is the probability for this deflection by a single nucleus, times the number of nuclei in the foil within the area A.

If n is the number of nuclei/m^3, i.e. the number of atoms/m^3 and t is the foil thickness, then the number of nuclei available for scattering equals ntA and the total probability is given by:

$$ntA \cdot \frac{2\pi b|\mathrm{d}b|}{A} = nt \cdot 2\pi b|\mathrm{d}b|$$

$$= \frac{nta^2}{16}\operatorname{cosec}^4 \phi/2 \cdot 2\pi \sin \phi\, \mathrm{d}\phi\,.$$

Inserting the value for a, the distance of closest approach in a head-on collision as given by Eq. (2.18), viz.

$$a = \frac{zZe^2}{4\pi\varepsilon_0 E}\,,$$

we find *the probability of an alpha particle of energy E being scattered through an angle between* ϕ *and* $\phi + \mathrm{d}\phi$ *by a foil of thickness t containing n atoms/m*^{3}. *It is given by*

$$P(\phi)\,\mathrm{d}\phi = \frac{nt}{16}\left(\frac{zZe^2}{4\pi\varepsilon_0 E}\right)^2 \operatorname{cosec}^4 \phi/2 \cdot 2\pi \sin\phi\,\mathrm{d}\phi \tag{2.21}$$

where ze is the charge on the alpha particle, Ze is the charge on the nucleus, and E is the energy of the alpha particle.

We have deliberately left the angle dependent factor so that it contains the term $2\pi \sin\phi\,\mathrm{d}\phi$ as this is *the solid angle into which the particles are scattered*, as shown in Fig. 2.13(c). Thus dividing out this term we have that *the probability, per unit solid angle, of scattering through an angle* ϕ, which we express as $\mathrm{d}P(\phi)/\mathrm{d}\Omega$, is given by

$$\boxed{\frac{\mathrm{d}P(\phi)}{\mathrm{d}\Omega} = \frac{nt}{16}\left(\frac{zZe^2}{4\pi\varepsilon_0 E}\right)^2 \operatorname{cosec}^4 \phi/2\ .} \tag{2.22}$$

Equations (2.21) and (2.22) are two versions of the famous Rutherford scattering formula.

2.8 COMPARISON OF THE RUTHERFORD FORMULA WITH EXPERIMENTAL RESULTS

The Rutherford scattering formula (2.22) makes some quite specific predictions which can be tested by experiment:

(i) For a given foil and fixed alpha particle energy, the number of alpha particles scattered, per unit solid angle, through an angle ϕ is proportional to $\operatorname{cosec}^4 \phi/2$.

(ii) For a fixed angle and fixed energy the number of scattered alpha particles is proportional to the foil thickness.

(iii) For a fixed angle and foil thickness, the number of scattered particles is inversely proportional to the square of the alpha particle energy.

(iv) Finally the number of scattered particles is proportional to $(Ze)^2$, the charge on the nucleus. By considering the *absolute* numbers involved one can measure Z, a piece of information that was not otherwise available at the time, apart from the general idea that $Z \sim A/2$ obtained from X-ray scattering.

The predictions of the Rutherford scattering formula were tested in 1913 by Geiger and Marsden using the apparatus illustrated in Figs. 2.14. Alpha particles, from a source placed in the cavity R in a metal block, emerged

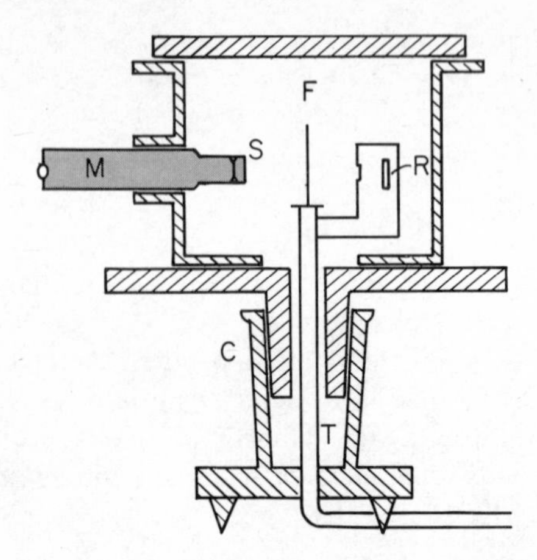

Fig. 2.14. Geiger and Marsden's apparatus for checking the Rutherford scattering law. R—source cavity, F—scattering foil, S—scintillation screen, C—conical joint, T—tube to vacuum pump.

through a small hole to impinge on the foil *F*. After scattering the alpha particles travelling in the appropriate direction struck the zinc sulphide screen S. Zinc sulphide produces a flash of light, or scintillation, when struck by a rapidly moving alpha particle. The scintillations were observed and counted through the microscope M. The microscope with the zinc sulphide screen attached could be rotated about the vertical axis through the foil via the conical joint C. The whole apparatus was evacuated via the tube T.

In Fig. 2.15 we plot some of the results obtained by Geiger and Marsden for a fixed foil thickness and alpha energy, as a function of angle, for gold and silver. The ordinate is $N^{1/4}$ where N is the number of counts per unit interval of time, and the abscissa is cosec $\phi/2$. It will be seen that in both cases the points fall well on a straight line.

In Figs. 2.16 and 2.17 we show the number of counts per unit time at a fixed angle, as a function of foil thickness and of $1/E^2$ respectively, as found by Geiger and Marsden. In both cases they provide impressive support for the Rutherford scattering formula.

They could only obtain the absolute number of scattered alpha particles to about 20% which, of course, is not very precise and only determines Z to about 10%. Later experiments improved a lot on this, but much more accurate methods were forthcoming for the determination of Z, and these will be discussed later in this book.

We have presented some of the initial tests of the Rutherford scattering formula. Since those tests innumerable experiments have proved the validity

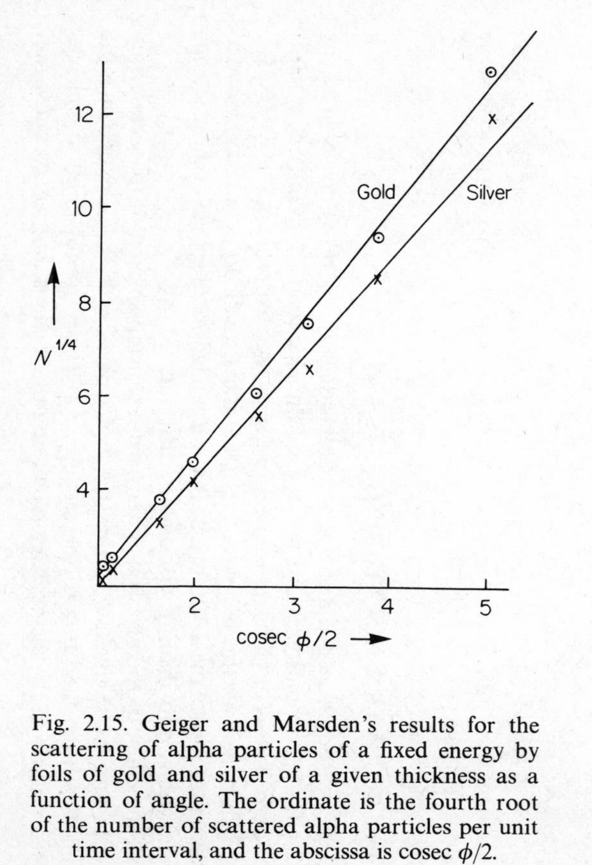

Fig. 2.15. Geiger and Marsden's results for the scattering of alpha particles of a fixed energy by foils of gold and silver of a given thickness as a function of angle. The ordinate is the fourth root of the number of scattered alpha particles per unit time interval, and the abscissa is cosec $\phi/2$.

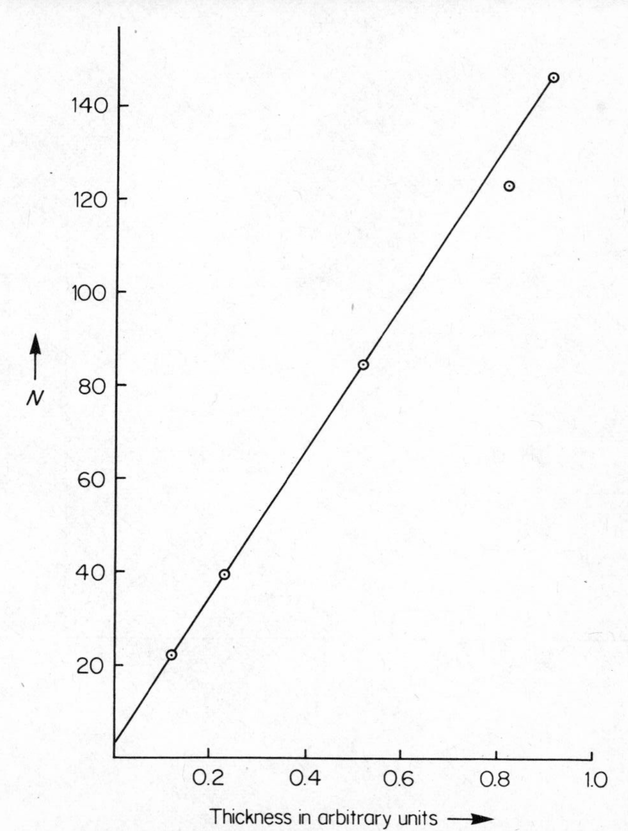

Fig. 2.16. A plot of the number of alpha particles scattered at a fixed angle as a function of foil thickness.

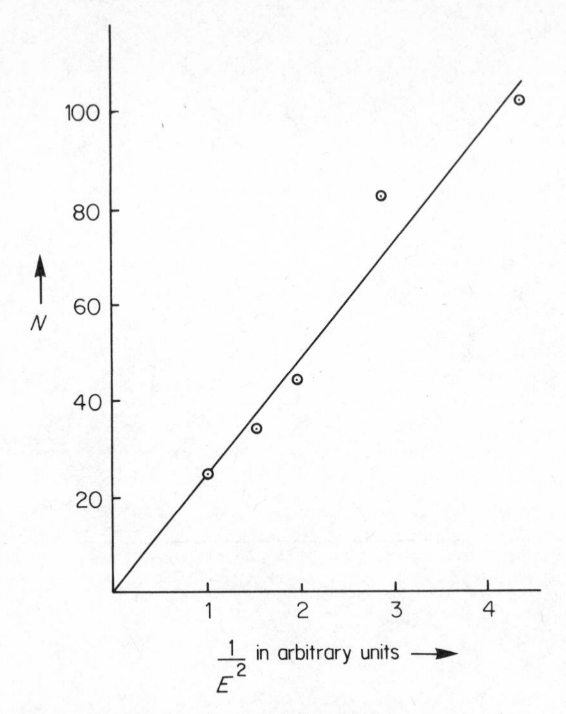

Fig. 2.17. The number of alpha particles scattered at a fixed angle for a given foil as a function of $1/E^2$ where E is the energy.

of the Rutherford formula, and it even survived the advent of quantum mechanics. A full quantum mechanical calculation leads to precisely the same formula.

2.9 FORWARD SCATTERING

It will be noticed that at very small values of ϕ, $dP(\phi)/d\Omega$ as given by Eq. (2.22) becomes greater than one, and indeed tends to infinity as ϕ tends to zero. This is obvious nonsense and contrary to experiment.

The reason for this behaviour is that at these small angles we have violated either or both of the assumptions that we can ignore the effect of the electrons and we are only dealing with single scattering.

The effect of the electrons (apart from causing some multiple scattering) is to screen the nucleus so that the alpha particle does not experience the full Coulomb field due to the nucleus. As yet we do not know the charge distribution of the electrons surrounding the nucleus, but we do know that atomic radii are of the order of 10^{-8} cm and, as atoms are neutral, the screening effect must be complete by this distance so, to at least a first approximation, alpha particles with impact parameters greater than this *will not be scattered at all.* If the number of atoms per unit area of the surface of the scattering target is small, so that $nt\pi b_{\max}^2$ is less than one, where $b_{\max}$ is the largest value of b with incomplete screening, then the total scattering probability is less than one. These circumstances can occur for thin gaseous targets. For example, a gaseous target one centimetre thick at 10^{-2} torr contains approximately 4×10^{14} atoms per square centimeter of surface.

Multiple scattering occurs for impact parameters such that $nt\pi b^2$ is greater than one. When multiple scattering occurs the total scattering probability is *not* equal to the scattering probability per atom times the number of atoms in the target, as once a particle has been scattered, it has been scattered and the probability of scattering is not increased by subsequent encounters. Thus for metal foils the distribution of alpha particles varies smoothly from a Rutherford distribution at large angles to a multiple scattering distribution at small angles, in such a way that the total scattering probability is equal to or less than one.

2.10 THE IDEA OF A CROSS-SECTION

We notice that Eq. (2.22) for the probability of scattering per unit solid angle in the direction ϕ can be broken into two terms; first, the factor nt which arises from the number of scattering centres, and second, the rest of the expression which determines the contribution from an individual scattering centre.

As n is the number of atoms per unit volume and t is the thickness of the target, nt has the dimensions of L^{-2}. As probability is a pure number it follows that the rest of the expression has the dimensions of an area, as will readily be verified in that $(zZe^2/(4\pi\varepsilon_0 E))$ is the distance of closest approach and therefore has the dimensions of length.

This form is not peculiar to Rutherford scattering, but is true of any process where the probability is sufficiently low that we may obtain the total probability by multiplying the probability for scattering by a single centre by the number of centres present. This condition may always be achieved, at least in principle, by considering a case with sufficiently few centres present.

Thus the factor which determines the contribution from an individual centre has the dimensions of an *area* and we call it a *cross-section.*

This idea has more in it than mere dimensionality, for it represents the effective area presented to the beam by each of the centres for the process being considered. In the case of Rutherford scattering this is very obvious in that we arrived at the result by considering the number of particles passing through an annulus, and related the area of this annulus to the angular range of the scattered particles, and different circular rings corresponded to different angles of scattering. This is not the usual case, and occurs only when there is a simple, monotonic dependence of the scattering angle on impact parameter. Nevertheless one can still think of the scattering centre as providing a certain area to the incoming beam for a specific process and refer to the cross-section for this process. Or one may add all possibilities together and refer to a *total* cross-section.

Rutherford scattering is an example of elastic scattering, in which the energy of the scattered particle is the same as that of the incident particle. At sufficiently high energies other processes can occur, too, and we denote the totality of these as absorption. Thus we shall have a cross-section for elastic scattering σ_{el}, and a cross-section for absorption σ_{ab}. Cross-sections are additive, as is indicated in Fig. 2.18, and the elastic scattering and absorption cross-sections can be added together to form a total cross-section σ_T

$$\sigma_T = \sigma_{el} + \sigma_{ab} . \tag{2.23}$$

The absorption cross-section can, in turn, be broken down into partial absorption cross-sections for different processes which we may label 1, 2, 3 etc., thus

$$\sigma_{ab} = \sigma_{ab}(1) + \sigma_{ab}(2) + \sigma_{ab}(3) + \cdots . \tag{2.24}$$

Not only may we quote a cross-section for a particular process, we may quote a cross-section for a particular process to occur with the outgoing particles travelling in a given direction; this is called a *differential* cross-section and is denoted by $d\sigma/d\Omega$. This quantity is the differential cross-section per unit solid angle. For example, the differential cross-section per unit solid angle for Rutherford scattering is

$$\frac{d\sigma}{d\Omega} = \frac{1}{16}\left(\frac{zZe^2}{4\pi\varepsilon_0 E}\right)^2 \operatorname{cosec}^4 \phi/2 . \tag{2.22a}$$

To find the number of particles undergoing a particular process, one multiplies the number of particles in the incident beam by the cross-section for the process and the number of centres per unit area of the target. Thus,

$$Y = nt\sigma I \tag{2.25}$$

where Y is the number of particles that undergo the specified process and I is the number of incident particles. Instead of quoting the thickness t in units of length, it is common to quote the mass per unit area of the target. As one mole of a substance contains N_0 atoms, we get nt is equal to mN_0/A where m is the mass per unit area of the target, in grams, and A is the atomic weight of the substance. Thus,

$$Y = \frac{mN_0}{A}\sigma I . \tag{2.25a}$$

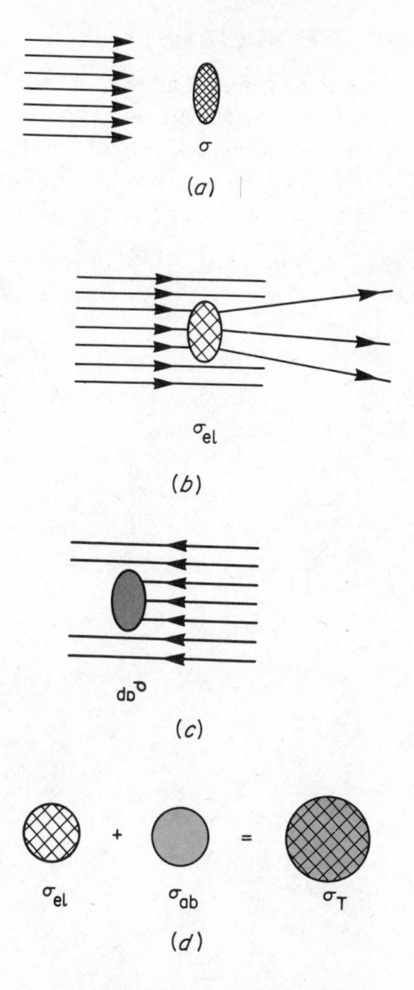

Fig. 2.18. The idea of a cross-section. The probability of an event occurring due to a particular centre (e.g. an atom) may be represented as a cross-section. This is the effective area the centre presents to the incoming beam for the process as indicated in (*a*). The symbol σ is usually used for cross-sections. One can have a variety of cross-sections, such as a scattering cross-section σ_{el} as in (*b*), and a cross-section for absorption as in (*c*). Cross-sections are additive. For example, the scattering cross-section plus the absorption cross-section is equal to the total cross-section which is the cross-section for anything to happen.

2.11 THE SIZE OF THE NUCLEUS

The Rutherford scattering formula has been derived on the assumption that the nucleus and the alpha particle are point charges. This will be a valid approximation so long as the closest distance of approach is larger than the radius of the nucleus, (Fig. 2.19). We may indeed say that the nucleus *must*

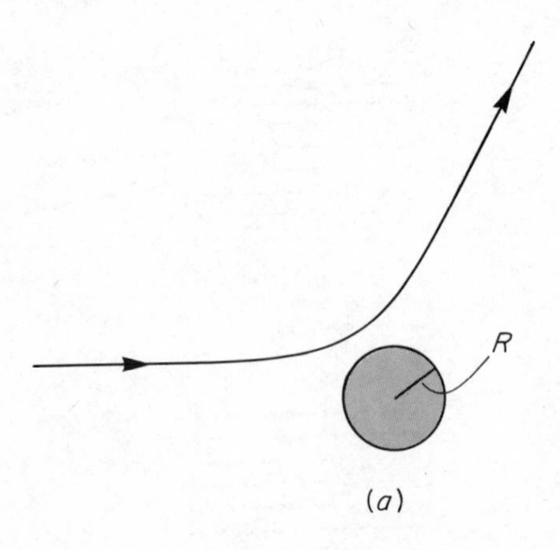

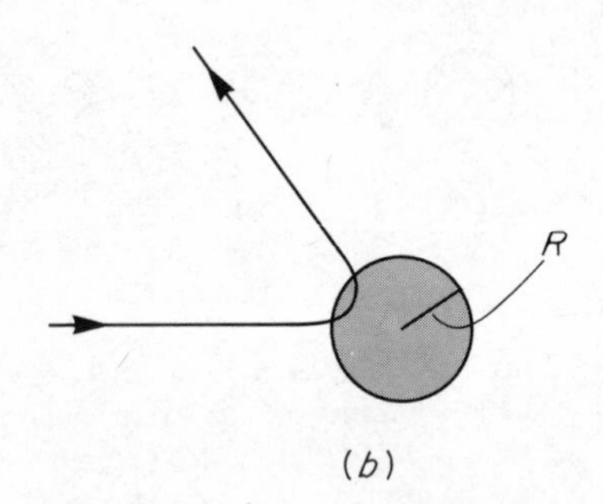

Fig. 2.19. Nuclear sizes. If the computed trajectory for Rutherford scattering is as in (*a*) where the closest distance of approach is greater than the nuclear radius *R*, we expect to observe the Rutherford cross-section. If the closest distance of approach is less than *R*, as in (*b*), we expect deviations from the Rutherford cross-section. In this way we obtain an estimate for *R*.

be smaller than the smallest value of the closest distance of approach for which Rutherford's formula still gives the correct value for the number of scattered alpha particles.

In the scattering by copper of the alpha particles emitted by radium it was observed that the Rutherford law was obeyed out to 180°. The energy of these alpha particles is 5.3 MeV, and from the number of scattered particles it was deduced that the value of Z for copper is 29. Hence applying formula (2.18) we have

$$\frac{2 \times 29 \times (1.6 \times 10^{-19})^2}{4\pi\varepsilon_0 a} = 5.3 \times 1.6 \times 10^{-13} .$$

Hence

$$a = \frac{58 \times 2.56 \times 10^{-38}}{4\pi\varepsilon_0 \times 5.3 \times 1.6 \times 10^{-13}}$$
$$= 1.58 \times 10^{-14}\,\text{m} ,$$

on substituting $1/(4\pi\varepsilon_0) = 9 \times 10^9$. Hence a copper nucleus has a radius that is less than 1.6×10^{-12} cm.

2.12 RESTATEMENT AND CRITICISMS OF THE RUTHERFORD MODEL

Rutherford's experiments had shown that an atom contains a minute central core or nucleus less than about 10^{-12} cm in radius, carrying a positive charge. Further this positive charge, measured in units of the electron charge, was within experimental error equal to the atomic number of the element, i.e. the ordering number of the atoms in the periodic table. Rutherford therefore proposed a model in which an atom consists of a nucleus carrying a positive charge equal to Ze with Z electrons moving round it in such a way that their dynamical motion balanced the forces, and the overall extent of the electron motion accounted for the size of the atoms of about 10^{-8} cm.

On this model hydrogen would have one electron, helium two electrons, lithium three electrons revolving round the nucleus, and so on. There is an appealing completeness about this idea in that it suggests that the chemical properties of an element are determined by the number of electrons contained in an atom of that element, and that all possible elements exist up to the point when it is experimentally observed that the nucleus is no longer stable and it is therefore presumably not possible to hold together any more charge within the nucleus.

However the model is very incomplete in that it gives us no idea in what way the number of electrons determines the chemical properties.

An even more serious objection to the model is that it provides no explanation why all atoms of a particular species appear to have the same size or why atoms of different species all have about the same size. It is true that methods of measuring atomic sizes, such as come from kinetic theory, or the density of solids, involve very many particles, and hence all we get is a mean value, but it is difficult to see how atoms could fall into a regular crystalline structure unless they were all nearly the same size.

Let us study this problem further by considering the simple case of the hydrogen atom. This contains only one electron and so the dynamics are particularly simple. We treat the simplest case and assume that the electron rotates round the nucleus in a circular orbit. Thus we have

$$\frac{mv^2}{r} = \frac{e^2}{4\pi\varepsilon_0 r^2}$$

which gives us the relation that

$$rv^2 = \frac{e^2}{4\pi\varepsilon_0 m}. \tag{2.26}$$

This equation allows any value of r. It merely tells us that if we scale r by some factor k^2, then v must be scaled by the factor k^{-1}. This is simply a particular case of a more general law for a system of particles moving under an inverse square law of force between them. If we find a solution for the motions of the electrons which we express in the form $\mathbf{r}_i(t)$ as some function of t then another solution exists $\mathbf{r}'_i(t)$ where

$$\mathbf{r}'_i(t) = k^2\mathbf{r}_i(t/k^3)$$

which, in fact, is nothing more than Kepler's third law of motion for the planets.

Finally, perhaps the most important objection is that the rotating electrons are continually accelerated, and therefore should radiate energy. They should thus spiral in towards the nucleus.

2.13 THE MISSING CONSTANT

In the classical Rutherford picture of the atom the only fundamental constants that enter are the mass and charge of the electron. There are other fundamental constants in nature (e.g. Avogadro's number, the velocity of light), but they do not appear to enter into the description of the atom. Therefore we do not have a sufficient set of quantities with which to construct a fundamental unit of length appropriate to an atom.

If we artificially import the velocity of light, we can construct a length

$$\frac{e^2}{4\pi\varepsilon_0 mc^2} = 2.8 \times 10^{-15}\ \mathrm{m}$$

but this seems far too small to have anything to do with atomic dimensions. Also, the only way the velocity of light could enter would be through relativistic effects. We have carefully kept everything non-relativistic, which is equivalent to letting the velocity of light go to infinity. As the reader will be aware, the dilemma was solved with the introduction of Planck's quantum of action, which we discuss in the next chapter.

PROBLEMS 2

2.1 Use Eq. (2.1) to find an expression for the mean rate of energy emission from an oscillating charge e. A gaseous source emits light of wavelength 6×10^{-7} m. Assuming each atom acts as an oscillator of charge e and amplitude 10^{-10} m, calculate the average rate of energy emission per atom.

2.2 A 60 W fluorescent lamp is 1 m long and 4 cm in diameter and contains a monatomic vapour at a pressure of 1 cm Hg. Assuming the results of question 1 estimate the fraction of the atoms in the tube emitting at any one time.

2.3 The reduction in intensity of a beam of X-rays passing through graphite is given by $I = I_0 \exp(-x/x_0)$. Calculate x_0 assuming the reduction in intensity is solely due to scattering by electrons, given that the Thompson cross-section of the electron is $6.66 \times 10^{-29}\ \mathrm{m}^2$. (Each carbon atom contains 6 electrons, and carbon has an atomic weight of 12 and a density of $2.3 \times 10^3\ \mathrm{kg\,m^{-3}}$.)

2.4 1 g of radium emits 4.7×10^{10} alpha particles per second. A piece of radium weighing 200 mg is placed in vacuo 5 cm from a circular plate 2 cm in diameter, on a line perpendicular to the plate and passing through its centre. The plate is insulated and has a capacity to earth of 5 pF. Calculate the rate of rise in the potential of the plate due to the alpha particles hitting it.

2.5 A 5 MeV beam of alpha particles passes between two plates 1 cm apart and 10 cm long. Calculate the potential difference between the plates required to deflect the beam of alpha particles through 1 cm on a screen placed 40 cm from the end of the plates.

What value of the magnetic field over the length of the plates would compensate for this deflection?

2.6 A 5 MeV beam of alpha particles from an accelerator falls perpendicularly on a gold foil with surface density of 1 mg per cm^2. If the beam current is 1 μA calculate the number of alpha particles per second scattered into a circular detector 1 cm in diameter placed, facing the gold foil, at a distance of 10 cm and at an angle of 30° to the beam.

2.7 A detector is placed so as to detect alpha particles that have been scattered through 90° by a copper foil ($Z = 29$). At low alpha particle energies the number scattered is found to agree with the predictions of the Rutherford scattering law but to deviate from these predictions for energies above 17.5 MeV. Calculate the radius of the copper nucleus. You may neglect the effect of the recoil of the copper nuclei. (Hint: use conservation of energy and angular momentum to determine the closest distance of approach in a 90° scatter in terms of the energy and impact parameters.)

2.8 Show that the cross-section for Rutherford scattering through angles *greater* than ϕ is given by

$$\sigma(\phi) = \frac{\pi\delta}{4}\left(\frac{zZe^2}{4\pi\varepsilon_0 E}\right)^2 \cot^2 \phi/2 \ .$$

2.9 Two gold foils separately give identical Rutherford scattering. When put together they are found to give 2.1 times the scattering of either of them at the same angle. If the initial energy of the alpha particles is 5 MeV calculate the energy loss in passing through one of the foils.

2.10 We may describe a scattering process either in the laboratory frame of reference, where the target is stationary, or in the centre of mass frame where the centre of mass of the projectile and target particles is stationary. Show that if the projectile and target particles have the same mass the laboratory scattering angle is half the corresponding centre of mass scattering angle.

The Rutherford formula Eq. (2.22) strictly applies to the centre of mass frame, and the energy E is the energy of the two particles in the centre of mass. Show that if the projectile and target particles have the same mass the formula for Coulomb scattering in the laboratory frame is

$$\frac{\mathrm{d}P(\theta)}{\mathrm{d}\Omega} = nt\left(\frac{zZe^2}{4\pi\varepsilon_0 E}\right)^2 \operatorname{cosec}^3 \theta \cos \theta$$

where θ is the laboratory scattering angle and E is the incident energy in the laboratory frame of reference.

2.11 An electron rotating in a circle round a nucleus with a single positive charge has an inward acceleration $e^2/4\pi\varepsilon_0 mr^2$. Use Eq. (2.1) to calculate the time required for the electron to spiral into the nucleus, starting from an initial radius of 10^{-8} cm.

CHAPTER 3

Planck's constant, radiation and photons

3.1 INTRODUCTION

We saw in the last chapter that although Rutherford's model of the atom was spectacularly successful in accounting for the scattering of alpha particles, there was nothing in classical physics that could give us any idea at all why atoms are of the size they are. Further, the model is unstable as the circulating electrons in the atom should radiate. We summarized this by saying that it looked as if there was some hitherto undiscovered constant in nature which determined these matters.

The reader will probably be aware that at the time Rutherford proposed his model (1911) just such a constant had been discovered, namely Planck's constant of action (or angular momentum, the two quantities have the same dimensions). This constant was introduced by Planck when he read a paper on black body radiation to the German Physical Society on December 14, 1900. This chapter will be concerned with some of the evidence for the universal nature of this constant, particularly with regard to electromagnetic radiation.

3.2 BLACK BODY RADIATION

Black body radiation is the name given to the radiation that occurs in a constant temperature enclosure. It is called 'black body' radiation because it is identical in all respects with the radiation emitted by a body at the same temperature, and which absorbs all the radiation falling on it. This can be proved with the aid of the laws of thermodynamics. It is more convenient to think of black body radiation as the radiation within a constant temperature enclosure, as then one can treat it as the working substance in a heat engine and apply the laws of thermodynamics to it. Also, it suggests that one can construct a very close approximation to a black body source by making a small hole in such an enclosure and thereby make measurements of the intensity of the emitted radiation as a function of wavelength and temperature.

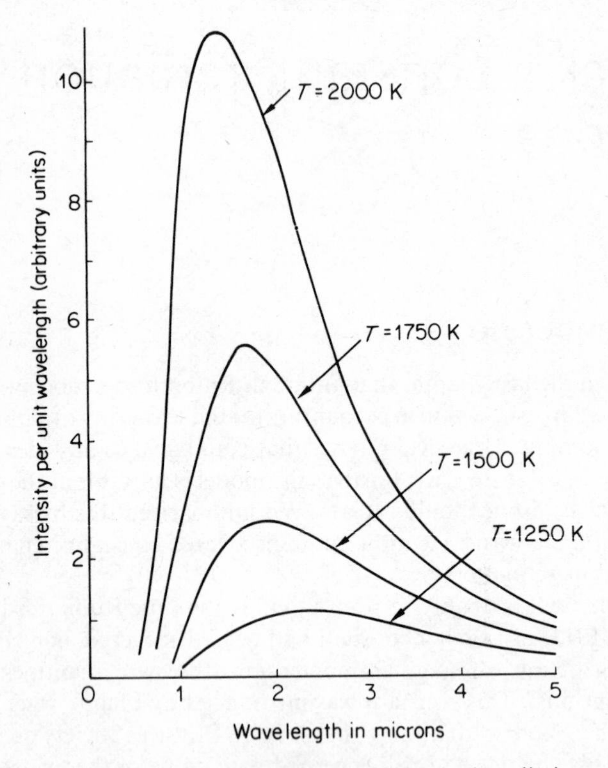

Fig. 3.1. The spectral distribution of black body radiation at several temperatures.

The results of such measurements are shown in Fig. 3.1 which shows the intensity of radiation from a black body as a function of wavelength, for various temperatures of the body. The principal features of this are as follows:

(*a*) The curve obtained for any particular temperature lies above that obtained for a lower temperature (and below that for a higher temperature) at all wavelengths.
(*b*) The curve has a peak in it, and the position of this peak moves to shorter wavelengths the higher the temperature.

Experimentally we find

$$\lambda_{max} T = 2.90 \times 10^{-3} \text{ m K} . \qquad (3.1)$$

Eq. (3.1) is known as Wien's displacement law. The form of this equation can be derived from thermodynamics, but not the value of the constant.

A considerable effort was made at the turn of the century to derive an expression which would fit the observed data, but all attempts using classical physics failed to produce a formula which would give satisfactory agreement with experiment. This effort however led to a deeper understanding of the problem, and finally culminated in a demonstration, using the most general arguments, that classical physics would always give an expression *without* a maximum, i.e. the constant in Eq. (3.1) would be zero, and the intensity of radiation would increase to infinity as the wavelength goes to zero.

A basic assumption of the classical argument was that, at any particular wavelength or frequency, the energy of the radiation could assume any value. In his paper to the German Physical Society, Planck made the radical assumption that, at any frequency ν, the energy of the radiation at that frequency could only change by an integral multiple of the quantity $h\nu$ where h is a new fundamental constant of nature. With this assumption Planck was able to devise a formula for the spectral distribution of black body radiation in excellent agreement with experiment. He was also able to give a value for the constant in formula (3.1) in the form

$$\lambda_{max} \frac{kT}{c} = \frac{h}{4.965} .$$

Therefore

$$\frac{hc}{k} = 4.965 \times 2.90 \times 10^{-3} \text{ m K} .$$

Inserting

$$c = 2.998 \times 10^{8} \text{ m s}^{-1} \text{ and } k = 1.380 \times 10^{-23} \text{ J K}^{-1}$$

we get

$$h = 6.67 \times 10^{-34} \text{ J s} .$$

A modern value is

$$\boxed{h = (6.626196 \pm 0.000050) \times 10^{-34} \text{ J s} .}$$

3.3 ATOMIC SIZES

Is Planck's constant of universal significance or is it just peculiar to radiation? In Chapter 2 we could not find a fundamental length to associate with the size of the atom, as we only had two constants to play with, namely the electron mass and the electron charge. We tried arbitrarily importing the velocity of light, but found the resultant length too small to be of any significance. Let us try importing Planck's constant into the situation.

In black body radiation we have used $h\nu$ as a unit of energy. However, we normally find in physical problems that if we express things in terms of a frequency ν, we get awkward factors of 2π turning up, whereas if we use the angular frequency ω the equations are usually much tidier. Therefore let us define $\hbar = h/2\pi = 1.054 \times 10^{-34}$ J s so that our unit of energy is $\hbar\omega$. We now try to make a unit of length from e, m and $\hbar$. In S.I. units the force between two charges is given by

$$F = \frac{q_1 q_2}{4\pi\varepsilon_0 r^2}$$

hence $e^2/(4\pi\varepsilon_0 r^2)$ has the dimensions of force, MLT^{-2}, and $e^2/(4\pi\varepsilon_0)$ therefore has dimensions $\mathrm{ML}^3\mathrm{T}^{-2}$. m and $\hbar$ have the dimensions M and $\mathrm{ML}^2\mathrm{T}^{-1}$ respectively. We now form the quantity

$$\left(\frac{e^2}{4\pi\varepsilon_0}\right)^{\alpha} m^{\beta}\hbar^{\gamma} \tag{3.2}$$

and choose α, β and γ so that Eq. (3.2) has the dimensions of length. Inserting the dimensions of m, $\hbar$ and $e^2/4\pi\varepsilon_0$ the dimensions of Eq. (3.2) are

$$(\mathrm{ML}^3\mathrm{T}^{-2})^{\alpha}\mathrm{M}^{\beta}(\mathrm{ML}^2\mathrm{T}^{-1})^{\gamma} = \mathrm{M}^{\alpha+\beta+\gamma}\mathrm{L}^{3\alpha+2\gamma}\mathrm{T}^{-2\alpha-\gamma} .$$

Eq. (3.2) will have the dimensions of length, therefore, provided we choose α, β and γ such that

$$\alpha + \beta + \gamma = 0 \tag{3.3a}$$

$$3\alpha + 2\gamma = 1 \tag{3.3b}$$

$$2\alpha + \gamma = 0 . \tag{3.3c}$$

From (3.3b) and (3.3c) we obtain $\alpha = -1$, $\gamma = 2$, which when inserted into (3.3a) gives us $\beta = -1$. Hence our unit of length is

$$a_0 = \frac{4\pi\varepsilon_0\hbar^2}{me^2} . \tag{3.4}$$

Inserting the values

$$\hbar = 1.0545 \times 10^{-34}\ \mathrm{J\,s}$$

$$m = 9.109 \times 10^{-31}\ \mathrm{kg}$$

$$e = 1.602 \times 10^{-19}\ \mathrm{C}$$

$$4\pi\varepsilon_0 = 1.113 \times 10^{-10}\ \mathrm{Fd\,m^{-1}}$$

we calculate $a_0 = 0.529 \times 10^{-10}$ m. a_0 is called the *Bohr radius*. In Gaussian units Eq. (3.4) becomes

$$a_0 = \frac{\hbar^2}{me^2} \tag{3.4a}$$

with

$$\hbar = 1.0545 \times 10^{-27}\ \mathrm{erg\,s}$$

$$m = 9.109 \times 10^{-28}\ \mathrm{g}$$

$$e = 4.803 \times 10^{-10}\ \mathrm{e.s.u.}$$

and we obtain

$$a_0 = 0.529 \times 10^{-8}\ \mathrm{cm}\ .$$

a_0 is clearly of the order of magnitude of atomic dimensions and it would thus appear that Planck's constant plays some part in determining the size of atoms, but at the moment we have no idea why.

Of course in an approach like this we do not expect to get exact answers. What we are trying to do is to set a scale, and when we do a proper calculation we expect to get an answer which is our unit times a small number, of order 1. Thus, if we had taken h, instead of $\hbar$ we should have got a unit some forty times bigger.

In fact, as the reader will probably know, in 1913 Niels Bohr amended Rutherford's model of the hydrogen atom by asserting that only those circular orbits are allowed whose angular momentum is an integral multiple of $\hbar$, i.e.

$$mvr = n\hbar\ ,$$

where n is an integer. Inserting this condition into Eq. (2.26) we get

$$r = n^2\left(\frac{4\pi\varepsilon_0\hbar^2}{me^2}\right) = n^2 a_0\ .$$

a_0 is thus the radius of the first orbit allowed by Bohr's condition and does truly represent the size of the hydrogen atom.

3.4 THE PHOTOELECTRIC EFFECT

The photoelectric effect was discovered by Hertz in 1887 when experimenting on electromagnetic waves. He used an oscillating circuit containing a spark gap as a generator, and a similar circuit as a detector. He noted that the size of the gap in the detecting circuit which allowed sparks to pass was smaller if a glass plate was interposed between the two gaps. As glass does not allow the passage of ultraviolet light he concluded that the illumination of the detector spark gap by the ultraviolet light from the generator facilitated the passage of a spark at the detector. He also observed that the light from another spark was equally effective, provided the detector terminals were smooth and clean, and the light actually fell on the terminals. Hertz's experiments are illustrated in Fig. 3.2.

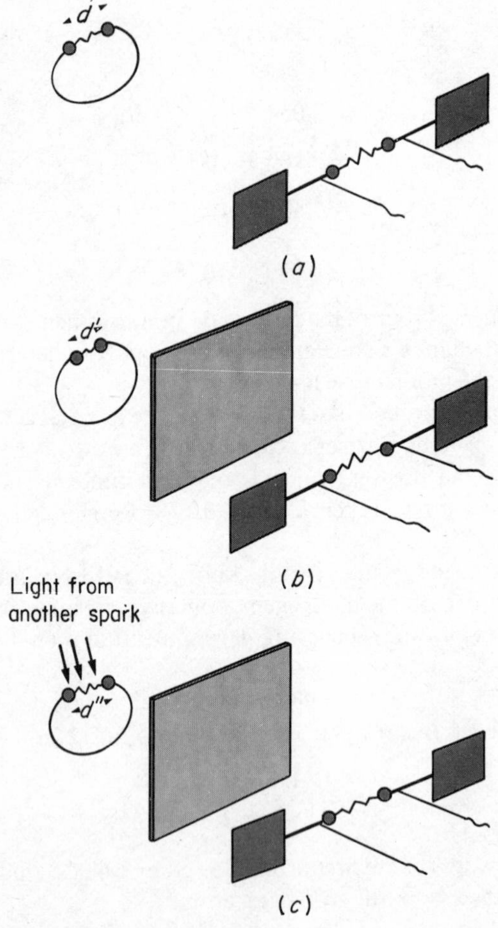

Fig. 3.2. The photoelectric effect. Hertz noticed that d was noticeably greater than the distance d' that occurred when a glass plate was inserted between the two circuits as in (*b*). He also noticed that when the gap on the left was illuminated from another spark as in (*c*) d'' was greater than d' and about equal to d.

The nature of the phenomenon was further elucidated by Hallwachs (1888) and Stoletow (1890) whose experiments are illustrated in Figs. 3.3 and 3.4 respectively. Hallwachs discovered that ultraviolet light would

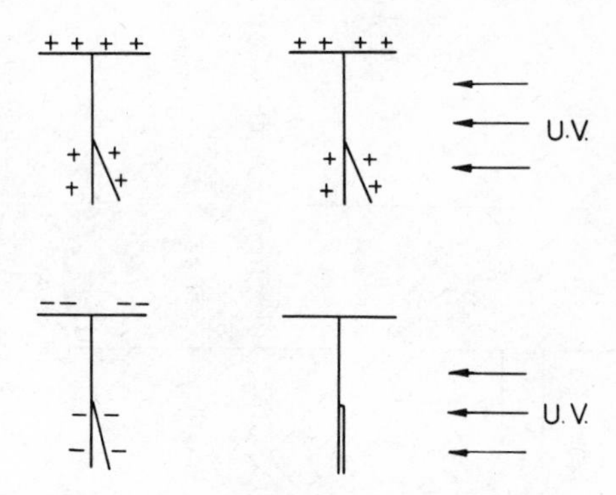

Fig. 3.3. The photoelectric effect. Ultraviolet light discharges a negatively charged electroscope, but not a positively charged one.

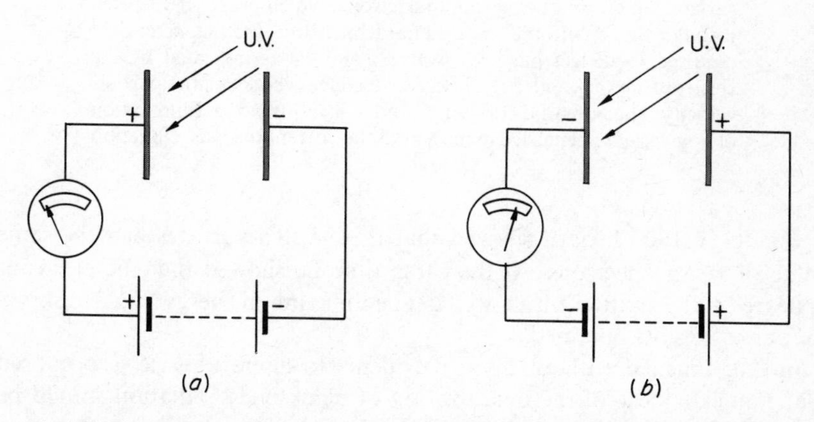

Fig. 3.4. The photoelectric effect. When ultraviolet light illuminates a plate at a positive potential, as in (*a*) no current flows, but when the illuminated plate is negative, as in (*b*), current flows, indicating that ultraviolet light causes the emission of negatively charged particles.

discharge a negatively charged electroscope but not one that was positively charged. Stoletow essentially constructed the first photoelectric cell and showed that the current must be due to the emission of negatively charged particles from the illuminated surface, as a current passed between the surface and a nearby electrode if the latter was positively charged, but not if it was negatively charged as illustrated in Fig. 3.4. The fact that the current continued to flow even when the highest attainable vacuum was produced strongly supported the view that the current was due to the emission of charged particles.

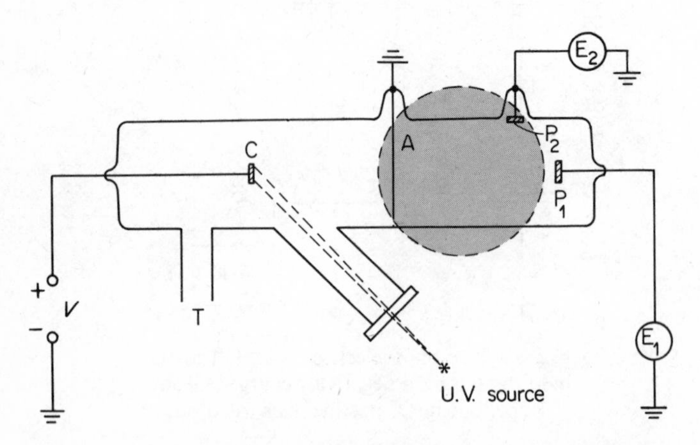

Fig. 3.5. Lenard's apparatus for e/m for photoelectrons. The apparatus was evacuated through T. An ultraviolet source illuminated the surface of C producing photoelectrons which were accelerated through the small hole in A. Lenard measured the magnetic field required to deflect particles from P_1 and P_2 as indicated by the electrometers E_1 and E_2. This gives us mv/e where v is the particle velocity. The potential between C and A gives us mv^2/e. Elimination of v^2 gives e/m. Lenard also measured the current to P_1 as a function of V.

Finally in 1900 Lenard showed that the emitted particles had the same value of e/m as electrons. At the same time he showed that the electrons appeared to be emitted with a well defined maximum energy. Fig. 3.6 shows the sort of results obtained.

In 1905 Einstein, without any real evidence to support his views, proposed that Planck's ideas of the quantization of black body radiation should be extended. Planck had proposed that inside a constant temperature enclosure radiation of frequency ν could only change by multiples of $h\nu$. Einstein proposed that we should actually think of the light as consisting of packets of energy, each of magnitude $h\nu$, and that each of these packets in some way retained its identity so that the whole of its energy could be concentrated on a single electron.

When such a 'photon', to give it its modern name, fell on an electron and ejected it, the electron would in general lose energy on the way out.

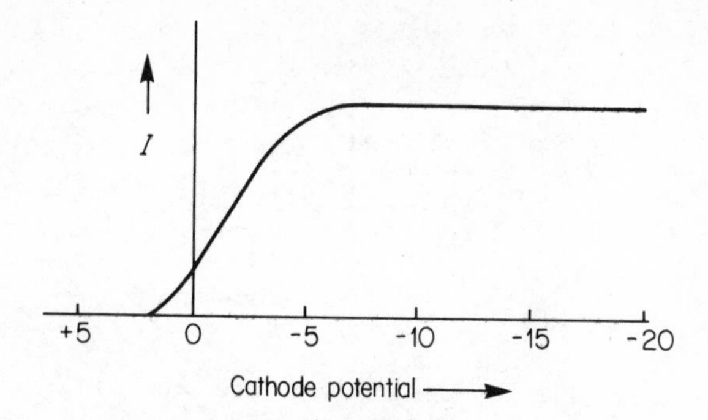

Fig. 3.6. The current I to P_1 of Fig. 3.5 in arbitrary units as a function of the voltage V of Fig. 3.5. The results show a definite cut-off when the cathode is a few volts positive, indicating a maximum energy of emission.

Those ejected from near the surface would lose the least energy. Einstein proposed that the maximum energy should be given by the relation

$$eV = h\nu - W \tag{3.5}$$

where W, known as the work function, is the least energy required to remove an electron from a metal. Einstein's hypothesis is illustrated in Fig. 3.7.

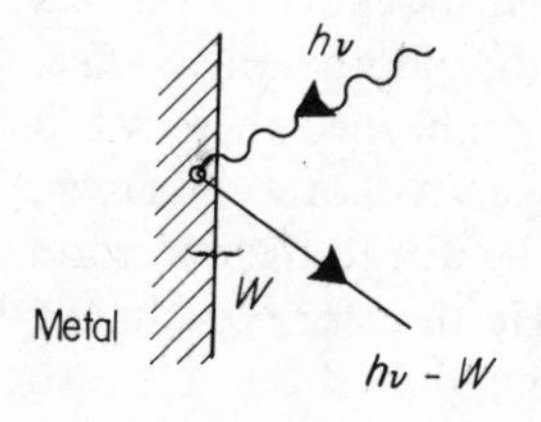

Fig. 3.7. Einstein's hypothesis concerning the photoelectric process. An energy W is required to move an electron from just inside to just outside a metal. A photon of energy $h\nu$ therefore ejects an electron of energy $h\nu - W$.

The concept of the work function had already been introduced by Richardson in accounting for thermionic emission.

There must be a minimum energy to remove an electron from a solid otherwise they would just pour out of the material. A part, at least, of this energy in the case of a metal comes from the work which must be done against the force of attraction between the electron, when it is outside the metal, and the image charge it induces in the metal, as is shown in Fig. 3.8.

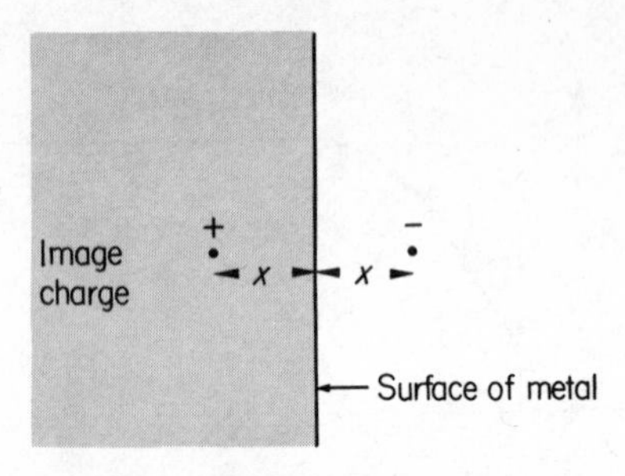

Fig. 3.8. Image charge and work function. Part at least of the work function comes from the force between an electron outside a metal surface and its image charge in the metal. From

$$F = \frac{e^2}{4\pi\varepsilon_0 4x^2}$$

we get

$$W_I = \int_{x_0}^{\infty} F\,\mathrm{d}x = \frac{e^2}{4\pi\varepsilon_0 4x_0}.$$

The work required to remove a charged particle from a distance x_0 from a metal surface out to infinity is $e^2/(4\pi\varepsilon_0 \cdot 4x_0)$. If we take x_0 to be about an atomic diameter, i.e. 2 Å, and put in the value for e we get an answer of 1.8 eV. Work functions are typically from about 2.0 to 7 eV, so the image effect forms a substantial portion of the work function.

Einstein's ideas were brilliantly confirmed by a classic series of experiments performed by Millikan in 1916. Early results on the photoelectric effect had not produced consistent results. It was clearly established there was a maximum energy to the emitted electrons but its value was not well known. Millikan conjectured that these discrepancies might be due to the uncertain nature of the surfaces of the metals being studied. He therefore conducted experiments on surfaces that had been freshly prepared *in vacuo*. He also took great care, when measuring the potential between the surface under study and the collecting electrode, to measure the contact potential difference between the two. The results he obtained were as follows:

(i) Below a certain threshold frequency no electron emission took place.
(ii) For frequencies above the threshold the current rose from zero for a small negative potential to a maximum around zero volts, as shown in Fig. 3.9, and stayed constant thereafter.
(iii) The magnitude of the maximum photoelectric current, for a fixed frequency, was strictly proportional to the light intensity over a range of 10^4 to 1.

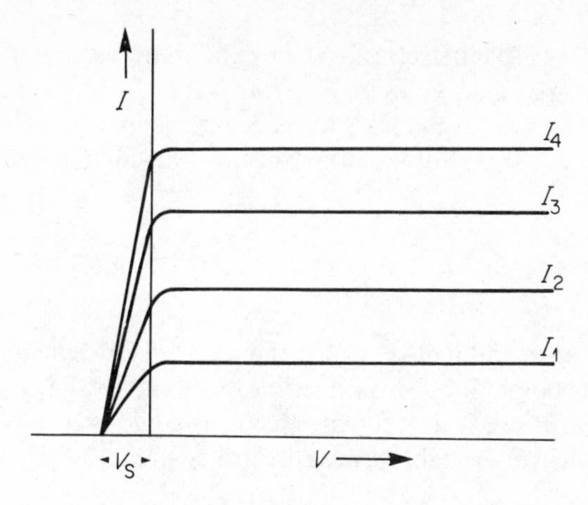

Fig. 3.9. Millikan's results I. The photoelectric current as a function of the anode voltage for various intensities of light of a fixed frequency.

(iv) The magnitude of the potential which just stopped the flow of current for a fixed frequency was independent of the intensity.

(v) When the stopping potential V_S was plotted against frequency ν for a given metal the results fell on a straight line. The slope of this line was the same for *all* metals and was equal to Planck's constant divided by e, as shown in Fig. 3.10.

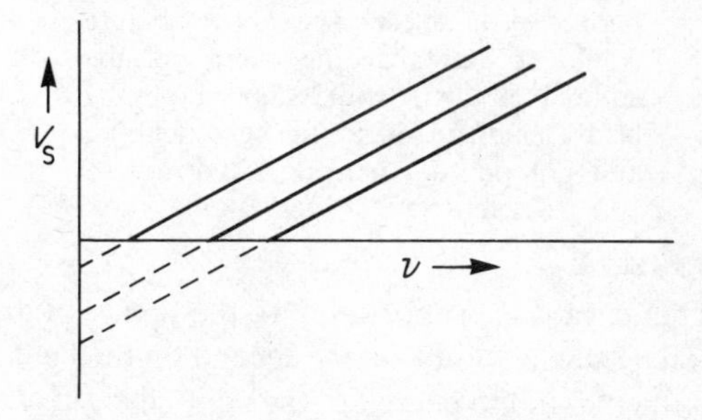

Fig. 3.10. Millikan's results II. The stopping potential V_S as a function of frequency ν for various surfaces. All the lines are parallel and have a slope h/e. The intercept on the V_S axis gives the work function.

Thus we have very strong evidence that, at least in the photoelectric effect, light behaves as if it were corpuscular, each photon carrying an energy equal to $h\nu$, and in its interaction with an atom it is capable of giving up the whole of its energy to an electron. We have no information on whether light can give less energy than $h\nu$ to an electron, but it cannot give more.

This is an extension of Planck's ideas, for in his theory there was nothing to prevent there being photons of frequency ν having energies of $2h\nu$, $3h\nu$, etc.

3.5 THE DIFFICULTIES OF THE PHOTOELECTRIC EFFECT FROM A CLASSICAL VIEW

Let us look at the difficulties which the photoelectric effect presents from a classical viewpoint. First, it is possible to visualize that a photoelectric effect would occur classically, for the electrons could presumably be ejected by the electric field of the light wave as illustrated in Fig. 3.11. But the intensity

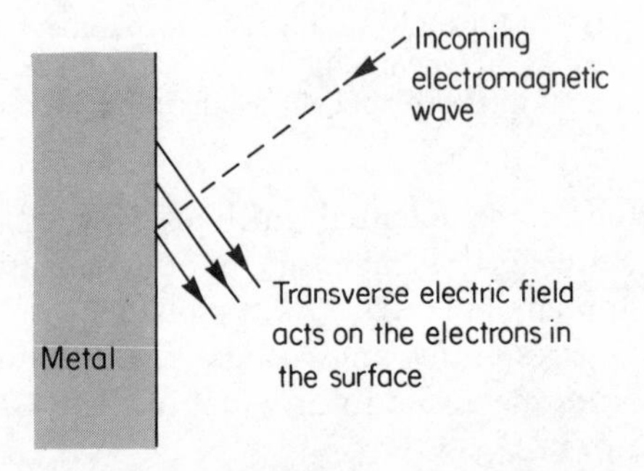

Fig. 3.11. A possible classical view of the photoelectric effect. The above picture would suggest that the energy of the emitted electron would depend on the field strength and hence the light intensity, and not on the frequency, contrary to experimental observation.

of an electromagnetic wave is proportional to the square of the field, and the energy given to the electron would surely depend on the field strength. Thus we would expect the maximum energy of the emitted electrons to depend on the intensity of the light, but not on the frequency, i.e. results quite the contrary of what is observed.

An equally glaring difficulty is raised by energy considerations. Classically the energy in an electromagnetic field is distributed uniformly over the wave front. A photoelectron might be expected to absorb energy from an area roughly equal to that of an atom. It is true that if the electron in the atom has a natural frequency of vibration exactly equal to that of the incoming wave then it could collect energy from an area about the size of the wavelength squared, but if this were the case we should observe the photoelectric effect as a resonance phenomenon, and this is not the case. We can hardly suppose that there exists an *exact* resonance for *all* frequencies above the threshold. In fact, the smooth variation of the effect above threshold strongly supports the view that there are *no* resonances that are effective.

Let us work out the time that would be taken for an atom to absorb the energy required to produce a photoelectron. Clearly this will depend on the intensity of light at the photoelectric surface. We may choose our intensity as follows. We find that on a good clean surface about 10% of the light energy is converted into energy of the photoelectrons (including the energy required to remove them from the surface). Now these days it is quite easy to measure currents down to 10^{-14} A, but let us not be as extreme as that and assume a current of 10^{-10} A from a sensitive surface of 10 cm^2. If the production of each electron uses up 3 eV of energy (a typical value) we have

$$10^{-10}\ \text{A} = 6 \times 10^{8}\ \text{electrons/sec}\ .$$

As each electron released uses up 3 eV, the total photoelectric energy released per second equals

$$1.8 \times 10^{9}\ \text{eV sec}^{-1}\ .$$

This energy is released from an area of 10 cm^2. If we assume an efficiency of 10% the light intensity must equal

$$1.8 \times 10^{9}\ \text{eV cm}^{-2}\ \text{sec}^{-1}\ .$$

Hence the energy per second falling on an atom of area 10^{-16} cm^2 is

$$1.8 \times 10^{9} \times 10^{-16} = 1.8 \times 10^{-7}\ \text{eV sec}^{-1}\ .$$

Thus the time required to acquire 3 eV in the atom is approximately equal to

$$1.7 \times 10^{7}\ \text{sec}$$

$$\approx 200\ \text{days}\ .$$

Yet in these circumstances one finds that photoelectrons appear as soon as the light is switched on.

★3.6 TIME DELAY WITH VERY LOW LIGHT INTENSITIES

Suppose we switch on a light intensity which corresponds to the production of less than 10^{9} photoelectrons per second. This figure is chosen as it is easily possible to measure time intervals of the order of 10^{-9} sec electronically.

We may then study what interval if any occurs between switching on the light and the release of the first photoelectron. To do this we need a fast light switch. Such a switch is provided by a Kerr cell, which consists of two metal plates with nitrobenzene between them. When a potential difference is applied between the plates, the electric field causes the nitrobenzene to become doubly refracting, with the result that it can be used to change the state of polarization of an initially polarized beam of light. If a Kerr cell is placed between two polarizers which have polarization directions at right angles, so that no light passes when there is no electric field in the nitrobenzene, light will be passed through the system when a potential difference is applied between the plates of the cell.

If we measure the time interval between the switching on of the light and the appearance of the first photoelectron many times we find that a whole range of time intervals occurs, and that we may represent the distribution with which these time delays occur by the expression

$$P(t)\,\mathrm{d}t = \mathrm{N}\,\mathrm{e}^{-Nt}\,\mathrm{d}t\ . \tag{3.6}$$

Here N is the average number of electrons emitted per second and $P(t)\,\mathrm{d}t$ is the probability of the first photoelectron appearing at a time between t and $t + \mathrm{d}t$ after switching on the light. This distribution is illustrated in Fig. 3.12. This is exactly the distribution expected for a completely random

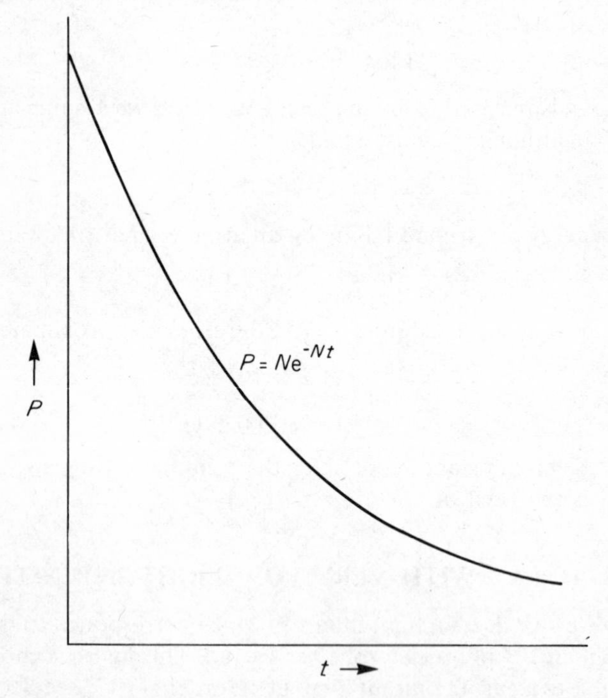

Fig. 3.12. The time delay in photoelectric emission. t is the time delay between the switching on of a light source and the appearance of the first photoelectron. $P\,\mathrm{d}t$ is the probability of a delay between t and $t + \mathrm{d}t$ occurring as found experimentally. It is found that the results can be fitted by $P = N\mathrm{e}^{-Nt}$ where N is the number of photoelectrons emitted per second when the light is on.

process, with an average rate N (see H. D. Young, *Statistical Treatment of Experimental Data*, McGraw-Hill, New York, 1962, p. 57) and is just what would be expected if the photoelectric effect were caused by the random arrival of photons at some mean rate, each photon having a certain probability p of producing a photoelectron (p being the same for each photon).

3.7 DIFFICULTIES OF A PHOTON VIEW OF THE PHOTOELECTRIC EFFECT

The photoelectric effect receives a ready explanation in terms of corpuscles of light, which we call photons, but quite aside from the difficulties of explaining interference and diffraction phenomena on a corpuscular view, we need the wave picture of light to give us our frequency, for it is this that determines the size of our packets of energy. We do not measure frequency directly in this region of the electromagnetic spectrum; what we do is to measure the velocity of light c and its wavelength λ by some interferometric method on the assumption that light is a wave motion, and deduce the frequency from the relation

$$\lambda \nu = c\ .$$

Thus we appear to have a clear conflict of behaviour between interference phenomena and the photoelectric effect and yet for the photoelectric effect we seem to need both pictures. Before we take the first steps in resolving this problem we will look at another experimental phenomenon which gives support to the photon picture of light.

3.8 THE COMPTON EFFECT

We mentioned in section 2.3 that the X-ray absorption coefficient fell below the classically derived value for wavelengths less than 0.2 Å. This region was investigated by A. H. Compton in 1923. He discovered that when monochromatic X-rays are scattered by a suitable scatterer, the scattered radiation contains two components, one at the original wavelength, and one at a longer wavelength. He found that the wavelength *difference* was a function of the angle of scatter *only* and was independent of the wavelength of the incident radiation and of the scattering material, though both of these affected the relative intensities of the two lines, as can be seen in Fig. 3.13.

A general feature of these experiments was that, very close to 0°, only the unshifted line occurred. As the angle was increased the shifted line moved away from the unshifted line and grew stronger. The rate at which it grew depended on the wavelength, being more rapid the shorter the wavelength.

In this wavelength region we ought to be able to regard the electrons as free according to the criteria in Section 2.3. Compton therefore proposed to treat the problem as an elastic collision between a photon, treated as a

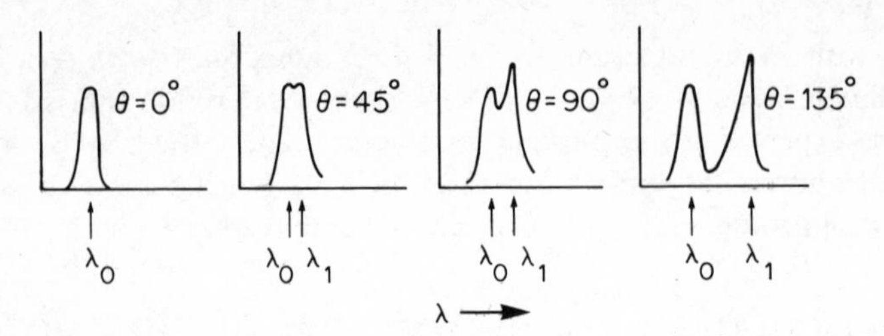

Fig. 3.13. Wavelength spectra of quanta scattered at various angles from a carbon foil. (From A. H. Compton. *Phys. Rev.*, **22**, 409, 1923).

particle, and a stationary electron. In such a collision the scattered particle (the photon) has less energy after the collision than before and, being a photon, will have a longer wavelength.

We must use relativistic mechanics as the energies involved are comparable with mc^2, but we only need the very simplest relations, namely:

(i) The *total* energy of a particle moving with velocity v is given by

$$E = mc^2/\sqrt{(1 - \beta^2)} \tag{3.7a}$$

where m is the rest mass of the particle and $\beta = v/c$.

(ii) The momentum at velocity v is given by

$$\begin{aligned} p &= mv/\sqrt{(1 - \beta^2)} \\ &= mc\beta/\sqrt{(1 - \beta^2)} \ . \end{aligned} \tag{3.7b}$$

By squaring Eqs. (3.7a) and (3.7b) we obtain the relation

$$\begin{aligned} E^2 - p^2c^2 &= m^2c^4/(1 - \beta^2) - m^2c^4\beta^2/(1 - \beta^2) \\ &= m^2c^4 \ . \end{aligned} \tag{3.8}$$

We note that if m is zero we must have $\beta = 1$, i.e. $v = c$, and that $p = E/c$. Hence for a photon we have

$$p = \frac{h\nu}{c} = \frac{h}{\lambda} \ . \tag{3.9}$$

Consider the situation shown in Fig. 3.14 where an incoming photon of energy E_0 and momentum p_0 is scattered by an electron with rest energy

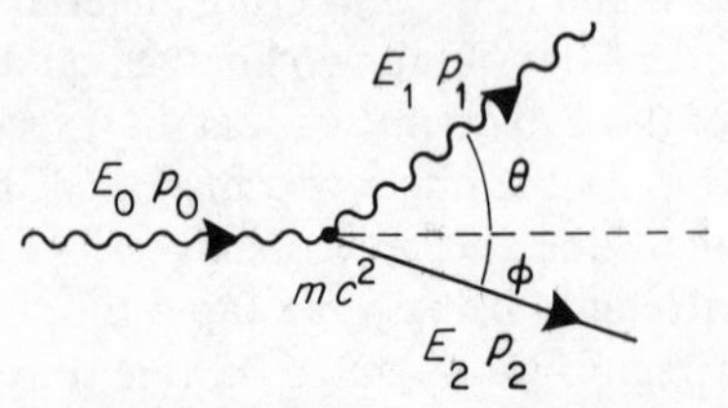

Fig. 3.14. Compton scattering.

mc^2. After the scattering we have a photon of energy E_1 and momentum p_1 at an angle θ to the original direction, and the electron has energy and momentum E_2 and p_2 at an angle ϕ to the original direction.

From conservation of momentum we have

$$p_0 = p_1 \cos\theta + p_2 \cos\phi$$

$$p_1 \sin\theta = p_2 \sin\phi \ .$$

We square and add to eliminate ϕ and get

$$p_2^2 = p_0^2 + p_1^2 - 2p_0p_1\cos\theta$$

$$= (p_0 - p_1)^2 + 2p_0p_1(1 - \cos\theta) \ . \tag{3.10}$$

From Eq. (3.8) we have

$$p_2^2 = \frac{1}{c^2}(E_2^2 - m^2c^4)$$

$$= \frac{1}{c^2}(E_2 - mc^2)(E_2 + mc^2) \ . \tag{3.11}$$

From conservation of energy we have

$$E_2 = E_0 + mc^2 - E_1$$

$$= c(p_0 - p_1) + mc^2 \ .$$

If we substitute this expression for E_2 in Eq. (3.11) we obtain

$$p_2^2 = (p_0 - p_1)(p_0 - p_1 + 2mc)$$

$$= (p_0 - p_1)^2 + 2mc(p_0 - p_1) \ . \tag{3.12}$$

We equate the right hand side of Eq. (3.10) to the right hand side of Eq. (3.12) and cancel out common terms to obtain

$$mc(p_0 - p_1) = p_0p_1(1 - \cos\theta)$$

or

$$\frac{1}{p_1} - \frac{1}{p_0} = \frac{1}{mc}(1 - \cos\theta)$$

which, after substitution from Eq. (3.9) for p_0 and p_1, becomes

$$\lambda_1 - \lambda_0 = \frac{h}{mc}(1 - \cos\theta) . \qquad (3.13)$$

This formula was found to give excellent agreement with the observed experimental results.

The quantity $h/mc = 0.0243$ Å is known as *the Compton wavelength of the electron* and plays a fundamental role in quantum electrodynamics.

Thus, once again we have an experimental result which is well accounted for by treating light as consisting of photons of energy $h\nu$ and momentum $h\nu/c$, and which cannot be accounted for on a wave picture. On a wave picture we expect that the electron will oscillate under the influence of the electric field of the incoming wave and thus radiate a wave of the *same* frequency.

3.9 THE SOURCES OF PHOTONS

We now have evidence from the photoelectric effect and the Compton effect that in its interaction with electrons light behaves as if it were made up of corpuscles, which we have called photons, each photon having an energy $h\nu$. Further, the observed results for black body radiation could only be reproduced theoretically if it was assumed that radiation of frequency ν changed its energy in units of $h\nu$.

If light then exists in packets of energy $h\nu$, it is reasonable to suppose it is produced in packets of $h\nu$, and we ought to be able to find some evidence for this. It will obviously be much easier to look for the evidence in a source which produces light of one or more well defined wavelengths, rather than in one producing a continuous spectrum. Such a source is provided by any atomic vapour in which an electrical discharge is running. Fig. 3.15 shows a

Fig. 3.15. A portion of the emission spectrum of mercury vapour.

portion of the emission spectrum of mercury vapour, which was used by Franck and Hertz in experiments similar to those described in this section.

As we wish to investigate how light is produced in photons we wish to control the energy of excitation, and to measure it. We may do this with the apparatus in Fig. 3.16. Electrons from the filament F are accelerated by a

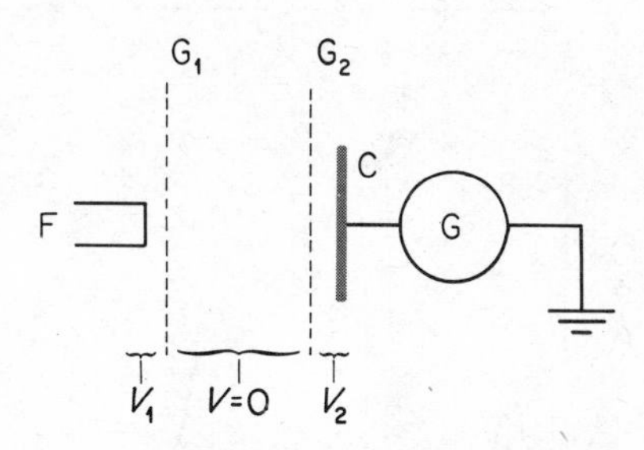

Fig. 3.16. Apparatus for determining the nature of electron energy losses in monatomic vapours and their relation to the frequencies of emitted radiation.

voltage V_1 between the filament and the grid G_1. The region between G_1 and G_2 is field free and contains a suitable monoatomic vapour (e.g. sodium, mercury etc.) at a pressure of about 0.01 mm of mercury. At this pressure it is found that the mean free path for electrons of a few volts energy is of the order of 1 cm. If the pressure is too high an electron will suffer many collisions in its path, and the results will be difficult to analyse. If the pressure is too low, many electrons will suffer no collisions at all, and the atomic excitations we are looking for will be too feeble to observe. We need a pressure such that the mean free path is of the same order as the distance between G_1 and G_2.

Between G_2 and C we have a voltage V_2 opposing V_1, and G measures the current arriving at C. By varying V_2 we may analyse the energies of the electrons passing through G_2.

We now slowly raise V_1 and watch what happens. Up to a voltage V_1 such that $eV_1 = E_1$ we find *nothing* happens, and all the electrons have the full energy eV_1. At $V_1 = E_1/e$ we find that a small value of V_2 causes a substantial reduction of the current and at the same time light is emitted from the space between G_1 and G_2. An analysis of the energy of the electrons arriving at G_2 shows either they have a full energy, or essentially zero energy. Raising V_1 slightly raises both groups of electrons in energy, the energy difference remaining constant at E_1. For sodium vapour $E_1 = 2.1$ eV and for mercury $E_1 = 4.9$ eV. The sort of results obtained are shown in Fig. 3.17.

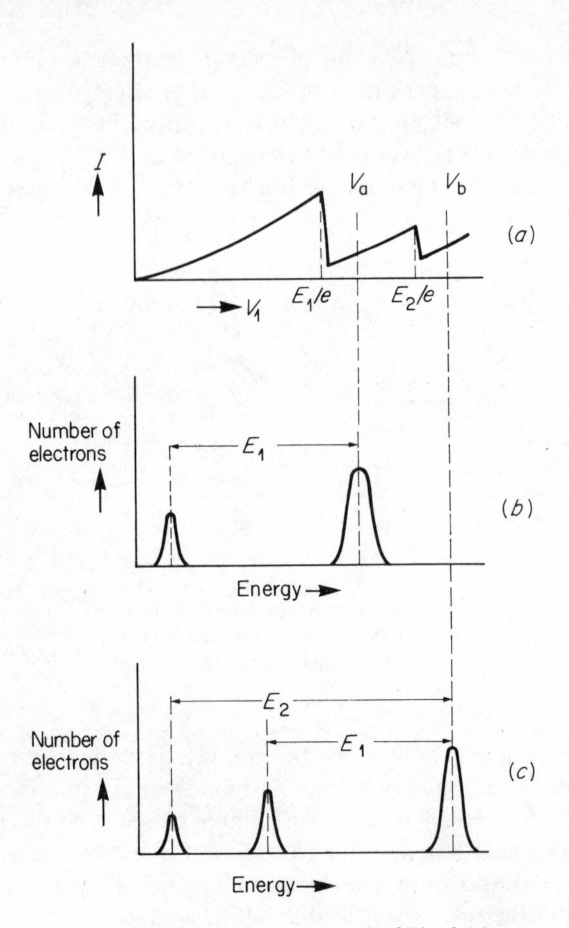

Fig. 3.17. (*a*) The total current to C of Fig. 3.16 as a function of V_1.
(*b*) The energy analysis of electrons arriving at C when $V_1 = V_a$.
(*c*) As (*b*) but for $V_1 = V_b$.

If we now measure the wavelength of the light we find that it corresponds to a frequency ν such that

$$h\nu = E_1 \quad . \tag{3.14}$$

On raising V_1 further we find a second energy E_2 at which further spectral lines appear. Again for an energy E slightly greater than E_2 we find electrons of energy E, $E - E_1$ and $E - E_2$ but not in between. Again for sodium this occurs at 3.2 eV.

Analysis of the light emitted shows that, at most, we now have *three* frequencies given by

$$h\nu_1 = E_1$$

$$h\nu_2 = E_2$$

$$h\nu_{21} = E_2 - E_1 \ .$$

In the case of sodium only the frequencies ν_1 and ν_{21} appear.

If we increase V_1 still further we find yet more energies at which new excitations appear. The energies of the electrons reaching G_2 are discrete, corresponding to energy losses $E_1, E_2 \ldots E_n$ and the frequencies of the emitted light are always given by the relation

$$h\nu_{rs} = E_r - E_s \tag{3.15}$$

but *not all possible combinations* ν_{rs} *actually appear.*

Above a certain energy E_I we find that energy losses greater than E_I are continuous. By a rearrangement of the voltages in Fig. 3.16 it is possible to show that this is because *ionization* is now possible. Up to this energy essentially *all* the energy lost by an electron must have been used in internally exciting the atom; the electron is so much lighter than any atom that conservation of momentum forbids any but the most minute amount of the electron's energy being converted into kinetic energy of atomic motion. However, once it is possible to knock out another electron, this second electron may carry away any amount of energy consistent with the conservation of energy, and so the original electron can be left with any energy less than $E - E_I$. The voltage V_I, such that $eV_I = E_I$, is known as the *ionization potential.* The difference between these two cases is illustrated in Fig. 3.18.

We may summarize our findings as follows:

(i) An electron can only give up certain specific amounts of energy E_1, $E_I, \ldots E_I$ to an atom up to the ionization energy E_I, above this it can deliver any amount of energy.

(ii) When an electron gives up a certain amount of energy E_r to an atom, the atom emits light of various *discrete* frequencies ν_{rs} such that

$$h\nu_{rs} = E_r - E_s \ .$$

(iii) Not all possible frequencies ν_{rs} actually appear.

From these findings we may make the following deductions about atoms.

(*a*) An atom cannot possess *any* internal energy, but only certain discrete energies $E_1, E_2, \ldots$ up to the ionization energy E_I. We call these 'energy states' or 'energy levels'.

(*b*) Atoms emit light by going from an energy state E_r to another energy state E_s. The frequency of the emitted light is given by $h\nu_{rs} = E_r - E_s$. This is known as the Bohr frequency condition.

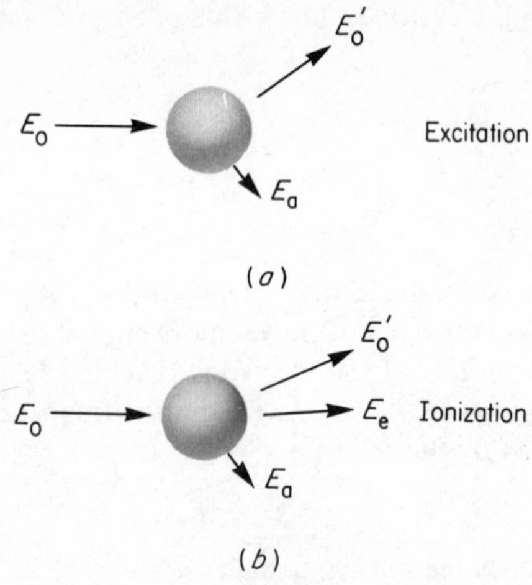

Fig. 3.18. Excitation and ionization by electrons: (*a*) Excitation. An incident electron of energy E_0 excites an atom to a level of energy E_1, and goes off with a reduced energy E_0'. By conservation of energy $E_0 = E_0' + E_1 + E_a$ where E_a is the recoil energy of the atom. Because the atom is so heavy compared with the electron conservation of momentum makes $E_a = P_a^2/2M$ negligible, where P_a is the recoil momentum of the atom. Hence $E_0' = E_0 - E_1$. (*b*) Ionization. An incident electron of energy E_0 ionizes an atom causing the emission of an electron with energy E_e. The original electron goes off with a reduced energy E_0. By conservation of energy $E_0 = E_0' + E_e + E_I + E_a$ where E_I is the energy required to ionize the atom and E_a is again the recoil energy of the atom. E_a is negligible for the same reason as before, but the large mass of the atom means that its momentum is able to balance the total momentum equation without affecting the energy equation. Hence we have $E_0' = E_0 - E_I - E_e$ with no restriction other than that E_0' and E_e both be positive.

(*c*) Not all transitions $E_r \rightarrow E_s$ occur with the emission of light, i.e. *selection rules* exist for radiative transitions between states.

Figure 3.19 illustrates the levels and transitions in sodium.

Spectroscopic measurements are, of course, much more accurate than the experiments of the type just described. They show, for example, that what

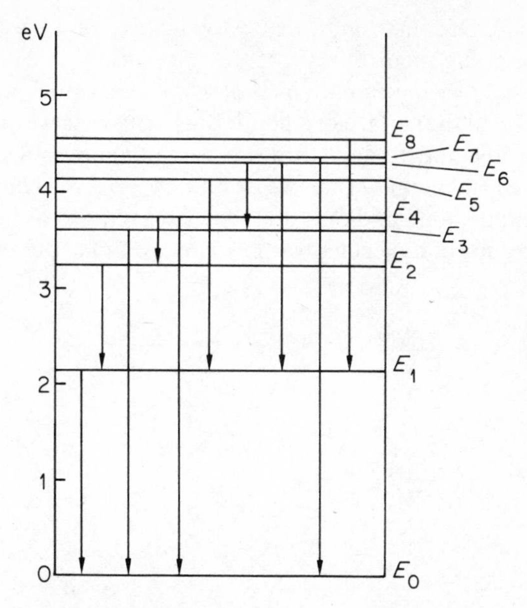

Fig. 3.19. The energy levels of the sodium atom and the stronger radiative transitions. The frequencies of the emitted lines are given by

$$\nu_{rs} = (E_r - E_s)/h \ .$$

appears as one level in an electron bombardment experiment may in fact be a group of levels, and it is through spectroscopic studies that the energy levels of atoms have been accurately determined.

3.10 RADIATION—WAVES OR PARTICLES?

Before we can discuss this question we have to be sure what we mean by a 'wave' and a 'particle'. Both of these ideas, and particularly that of a particle, are extrapolations from macroscopic experience. For example, from the point of view of celestial dynamics the earth is a particle. That is to say that, in discussing its motion, we may regard it as a point object, as its dimensions are small compared with its distance from any other of the heavenly bodies (except the moon, and in this case we cannot regard it as a particle), and we do not have to worry about its internal constitution and degrees of freedom. Again an orbiting satellite is a particle from the point of view of its motion round the earth, and we can go on down the scale, at least as far as macroscopic bodies are concerned. We develop faith in this idea because we find we can correctly predict their paths using Newtonian mechanics (or relativistic mechanics if the velocities are very high). But this gives us no reason to suppose these ideas can be extrapolated indefinitely, down to the atomic scale and beyond. Of course, until we have evidence that they do *not* work we naturally go on using them.

In the same way we develop an idea of what we mean by a wave. We can see a water wave on the surface of a pond. But in this case we know we cannot extrapolate the idea indefinitely, for if we were of about molecular size, then even on a classical picture what we would experience would be a lot of molecules rushing about, and only after we had averaged over the behaviour of very large groups of molecules would we recover our wave. (See Fig. 3.20.)

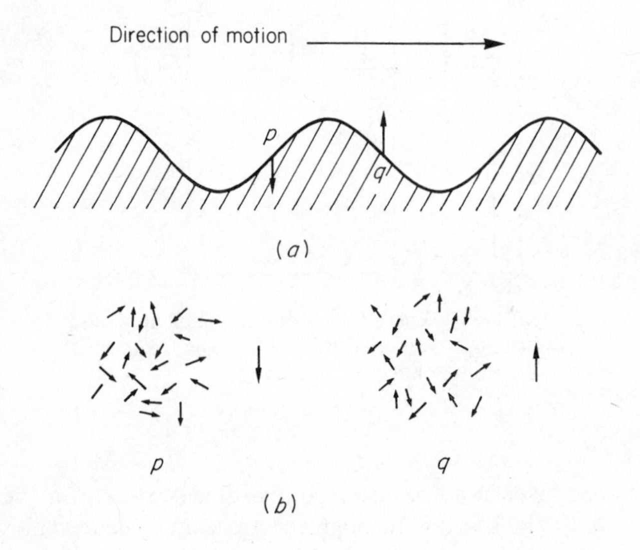

Fig. 3.20. Water waves: (*a*) A macroscopic view. The waves are travelling to the right, and the velocity of the surface at the points p and q are indicated by the attached arrows. (b) A microscopic view at the points p and q. In each case one observes a large number of molecules with apparently random motion, but after summing over a very large number one recovers a net motion downwards at p, and upwards at q.

All this is just to emphasize the point that our dilemma may only be apparent and occurs only because we insist on extrapolating the macroscopic ideas of our daily experience down to a level where they may no longer be valid, and that the distinction should not be between particles and waves, as classically understood, but between some other sorts of properties (as indeed we shall find is the case, but not in this chapter).

What, then, about light waves? We cannot *see them.* Let us summarize the reasons why we believe they are a wave motion.

(i) We can observe interference and diffraction effects, which we associate with wave phenomena.
(ii) The distribution of energy in space and time is correctly predicted by Maxwell's wave equation for a range of wavelengths extending from the indefinitely long down to ~ 0.02 Å.
(iii) For radio waves at least not only can we measure their wavelength, but we can measure the amplitude and phase (and hence the frequency) by a device that responds to the field strength.

Let us look at these points more closely. We take the last one first. There is a limit to the smallest amount of energy we can measure which is set by the thermal agitation in the detecting aerial. For all practical measurements this means that we must detect at least 10^{10} photons/sec so all the last point tells us is that a lot of photons acting together on the electrons in an aerial look like an electric field.

We may estimate the limit set by thermal agitation in the detecting aerial as follows. If we wished to detect radiation at some frequency f we would connect the aerial to an amplifier, tuned to the frequency f, and detect the amplified signal as shown in Fig. 3.21(*a*). The amplifier should only give

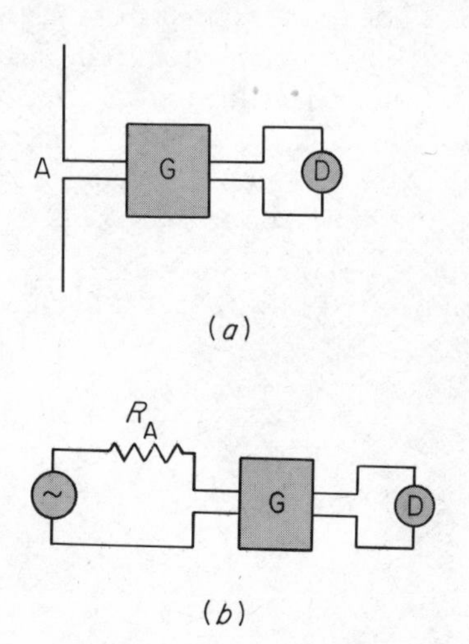

Fig. 3.21. (*a*) An aerial A, tuned amplifier G and detector D for detecting radio waves. (*b*) The equivalent circuit of (*a*): The aerial is represented by a generator in series with a resistance R_A which is due to the fact that the aerial can radiate. This resistance generates 'noise' in the input circuit of G.

substantial amplification over a certain band of frequencies Δf, and the ratio $f/\Delta f$ is called the selectivity of the amplifier and is denoted by the symbol Q. Clearly the value of Q is for us to choose over fairly wide limits, and we shall discuss the factors which affect our choice in a moment.

From the point of view of the amplifier, the aerial looks like a generator, due to the signals induced in it by the incoming radiation, in series with the impedance of the aerial. The aerial impedance has a resistive component R_A due to the fact that it can re-radiate, i.e. it is coupled to space. It is this resistance we are interested in, and which we show in Fig. 3.21(b) in the equivalent circuit.

If we have a resistor, R, the thermal motion of the electrons in the resistor causes a random fluctuating potential difference to occur between its ends which we call noise. It was shown by Nyqvist that the mean square value of this noise voltage is given by

$$\overline{V^2} = 4kTR\Delta f \tag{3.16}$$

where k is Boltzman's constant, T is the absolute temperature of the resistor R, and Δf the frequency range over which the detector will respond. It can further be shown that Eq. (3.16) applies however the resistance arises. In our case the resistance is due to the coupling of the aerial to space, and the noise in fact is due to the random radio frequency fields in space due to the thermal motions that determine its temperature. Thus we may use Eq. (3.16) to determine the noise voltage in our aerial provided that we insert for T the temperature of the space to which it is coupled. Radio astronomy observations have shown this is never less than 2.7 K.

The noise *power* in the aerial is

$$\frac{\overline{V^2}}{R_A} = 4kT\Delta f$$

$$= 4kTf/Q$$

where Q is the selectivity of the amplifier as previously defined. The number of photons per second, of frequency f, which will give the same power is

$$\frac{\overline{V^2}}{R_A hf} = \frac{4kT}{Qh}$$

$$\approx 10^{11}\frac{T}{Q}$$

on putting in numerical values for k and h. We have already stated that T cannot be less than a few degrees absolute. Q is also a measure of the number of cycles over which the detecting device averages, so if we wish to see whether the original field varies smoothly we must take Q to be fairly low. Thus we may put T/Q to be of order of magnitude unity, and say that the noise in the aerial is equivalent to about 10^{11} photons per second. Clearly we shall not be able to measure the amplitude and phase of a radio wave if it corresponds to many fewer photons per second than this, and so our statement that there is a limit of about 10^{10} photons per second for these measurements to be achieved is fairly conservative.

On the other hand it is easy to show that in an ordinary optical interferometer quite average intensities correspond to having only one photon in the instrument at a time. Fig. 3.22 shows an example of this in a Young's slit experiment. An extreme case was studied by G. I. Taylor in 1909 who showed that one still got the usual interference pattern with a light source so weak that it took three months to register the photograph.

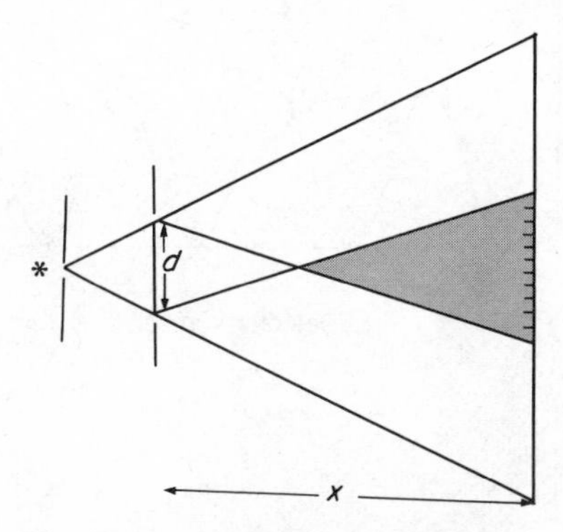

Fig. 3.22. A schematic representation of a Young's slits interferometer. Coherent light obtained from the first slit is diffracted at the two later slits and interference occurs in the shaded region. If $x = 20$ cm and $d = 0.1$ mm and each of the pair of slits is 0.02 mm wide the fringe spacing will be 1 mm and about 10 of them will be readily visible. Let the entry slit be narrowed until about 10^8 photons per second pass through the interferometer. The flight time between the double slit and the screen is 0.67×10^{-9} s so there will be only one photon in the apparatus at a time. 10^8 photons per second corresponds to 5×10^{-12} J s^{-1} in each fringe. This is the same energy as would enter the eye from a 100 watt lamp at a distance of 1 km, which can be readily seen at night.

We can look at this from a photon point of view by replacing the photographic plate by a lot of narrow photoelectric surfaces, as illustrated in Fig. 3.23, so, by detecting the emitted electron we can tell where the photon

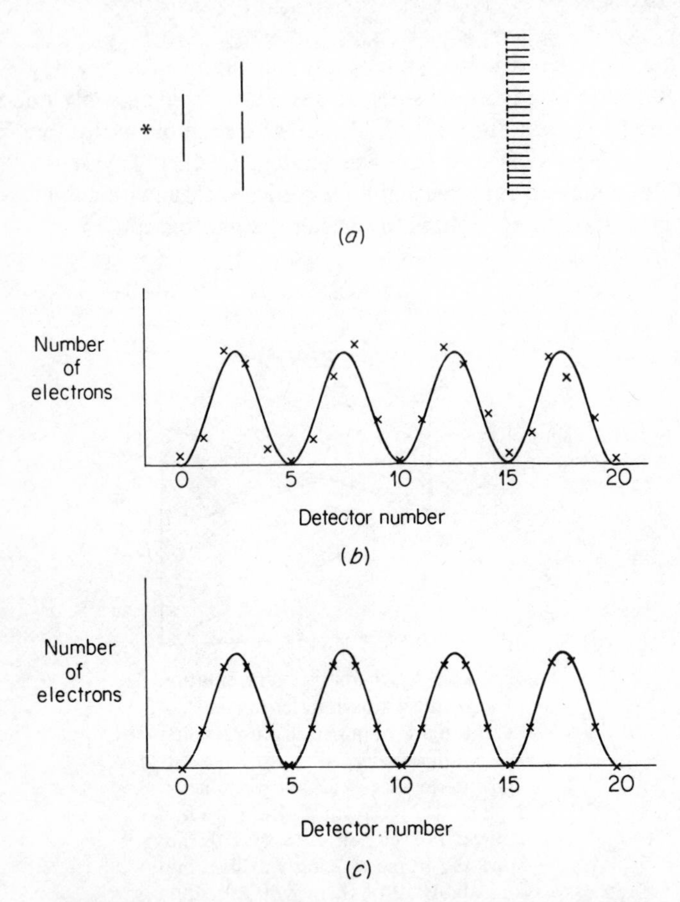

Fig. 3.23. (*a*) An interferometer with a large number of photoelectric detectors in place of a screen. The results obtained are shown in (*b*) for small numbers of electrons detected and in (*c*) for large numbers. In each case the crosses are the observed values and the solid line the expected curve.

landed. The total light energy that has fallen on any one detector at any moment is proportional to the number of photons that have arrived at that detector. When we perform this experiment we find that, when only a few photons have been detected, the results correspond roughly to the expected interference pattern but the numbers of detected photons vary quite widely from the intensity we expect on the wave theory. As we build up bigger and bigger numbers the fractional fluctuations get smaller and smaller.

So it appears that the number of photons arriving at any point is proportional to the intensity predicted at that point by wave theory *on average*, but only on average, and that wide fluctuations can occur when the number of photons involved is small.

Our interference pattern depends on the light going through *both* slits, as can be checked by covering up one of the slits. The fact that when light goes through two slits we do not get the sum of the intensities from the slits separately, but see interference, is what convinced people in the first place that light is a wave phenomenon. Does this mean that photons can be split in two? Remember we only have one photon at a time in our apparatus. We can check this by setting up the experiment in Fig. 3.24. We arrange to split

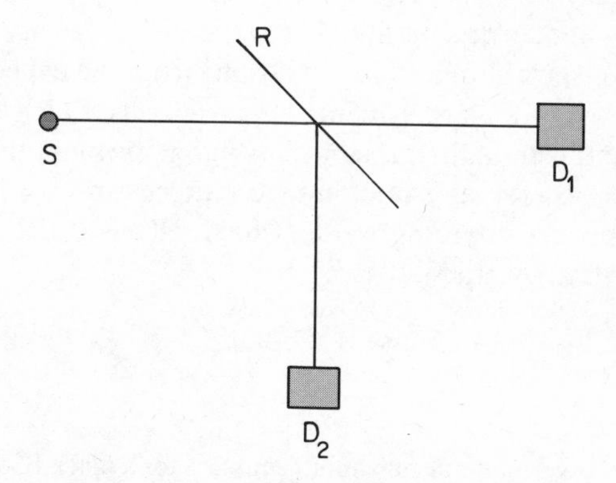

Fig. 3.24. Apparatus for detecting split photons. D_1 and D_2 are two detectors, S is a monochromatic source and R a half silvered mirror.

a beam in two by a half-silvered mirror. Prior to putting in the mirror we adjust our detectors so that they will respond to electrons ejected by the photons from our source (we choose an appropriate monochromatic source) but not to photons of half the energy. We count the number of photons arriving at D_1 without the mirror. We put the mirror in position, and if the photons are split we would expect to detect no photons at either detector. But, in fact, we detect half the number of photons at each detector. But perhaps the photons do not split equally and so some bits are above our threshold and some below. To check this we turn down the threshold level.

It makes no difference to the count rate. Finally we arrange so that we only count events when photons arrive at our two detectors simultaneously. We find very very few of these, and those we do detect we find we can account for by assuming the photons are entering our apparatus randomly and working out the chance that two photons will enter the apparatus together.

Thus we are led to the conclusion that it is impossible to split photons, and yet a photon appears to be able to go through two slits simultaneously and interfere with itself.

Well, why not? It is true that a classical particle cannot behave this way, but in fact we shall find that on the atomic scale there are *no* classical particles. But if we assert that a photon is a non-classical particle which can behave in this way, and the electromagnetic wave tells us something about the probability of finding the photon in a particular place, in fact that the probability is proportional to the intensity of the wave at that point, then our difficulty is solved. It also would lead us to expect the result observed in the experiment shown in Fig. 3.23, for if the intensity only tells us the *probability* of finding a photon, then it is not surprising that, when the total intensity corresponds to only a *few* photons, we should find deviations from the expected number.

Of course if we do an experiment which tells us which slit of a pair a photon has gone through, then it can't have gone through the other slit and we ought not to observe any interference pattern, and we find that this is exactly what happens. Anything we do which tells us which slit the photon passed through destroys the pattern.

PROBLEMS 3

3.1 Calculate the wavelength for maximum emission for a black body at 300 K (room temperature) and 6000 K (the temperature of the sun's surface).

3.2 Light of 4000 Å wavelength falls on a surface with a work function of 2.1 eV. What is the maximum energy of the emitted electrons?

3.3 In the above example photoelectric current to the collecting electrode ceases when a potential of 1.5 eV is applied. What is the work function of the material of the collecting electrode?

3.4 A photon of 20 keV energy collides with a stationary electron and is scattered through 45°. Find the energy of the scattered photon and the energy, momentum, and direction of motion of the recoiling electron.

3.5 Calculate the energy of a 1 MeV photon scattered through 180° from an electron (*a*) at rest (*b*) moving towards the photon with an energy of 100 keV.

3.6 The minimum light intensity that can be perceived by the eye is about 10^{-10} W m^{-2} Estimate the number of photons per second entering the eye that this corresponds to.

3.7 When sodium vapour is bombarded by electrons a strong yellow light of wavelength 5893 Å is observed provided the electron energy is in excess of 2.11 eV. Show the light is due to a transition from an excited state at 2.11 eV to the ground state.

3.8 Under good resolution the yellow sodium light is observed to consist of two components, one at 5890 Å and one at 5896 Å. Calculate the positions of the corresponding energy levels.

3.9 When a beam of electrons of energy 10.0 eV enters a monatomic gas the gas emits light of the following wavelengths 1402 Å, 2536 Å and 3132 Å. The light of wavelength 2536 Å is appreciably stronger than the other two components. What energies would you expect the emerging electrons to have?

3.10 Assume to an electron an atom looks like a sphere of radius 10^{-10} m. Calculate the pressure of a monatomic vapour at room temperature for which the mean free path of electrons is 10^{-2} m. (You may assume the electron velocity is very much greater than the thermal velocity of the atoms.)

CHAPTER 

Electrons and waves

4.1 THE ELECTRON AS A PARTICLE

So far we have implicitly treated the electron as if it were a classical particle. We have had very good reasons for doing so. Electrons are the basic carriers of electric charge. When a body is charged it either has a few more or a few less electrons than when in the neutral state, and Millikan's experiment showed that minute droplets could carry single charges. Experimentally one finds that negatively charged atoms can exist and these almost always carry a single charge (only two cases of doubly charged negative ions are known, viz. oxygen and fluorine). All this shows that an electron can be localized to a considerable degree, and this is just what one expects of a particle.

Also we found the ratio e/m for the electron by deflecting it in electric and magnetic fields, and using classical mechanics to calculate the result. In these experiments we found no phenomena which suggested that our approach was in any way faulty, and thus it would appear that the electron is at least very close to being a classical particle.

The only failure we have had with this picture of the electron is in the atom itself. We saw in Chapter 2 that on a classical model we could find no reason for atoms being the size they are; we could construct no natural unit of length from the constants involved. In section 3.3 we showed that if we introduced Planck's constant into the situation we could indeed construct a unit of length of the right size, but we have no idea why this should be so.

In section 3.9 we showed that atoms can only exist in certain preferred energy states, and could not take up just any energy. This is certainly contrary to classical mechanics as we pointed out in section 2.12. For classical mechanics once we have one solution we can apply an arbitrary scaling factor to get another solution.

Thus we are in the situation where the dynamics of the free electron are in agreement with classical dynamics, but the bound electron does not seem to obey classical dynamics, and that Planck's constant has something to do with atoms.

4.2 ELECTRONS AND WAVES

Could the electron have wave-like properties? So far Planck's constant has always been associated with what we regard classically as a wave motion, though as we saw in the last chapter, a deeper view shows that light in some of its aspects is very like a classical particle. If the electron did have wave-like properties we can see straight away this might offer a solution to the problem of why only certain energy states appear in an atom. A characteristic of wave-motion within a confined region is that it has to satisfy certain boundary conditions. A simple example of this is a vibrating string fixed at each end, in which only certain modes are allowed which correspond to the fact that the ends of the string cannot move (see Fig. 4.1). As is well-known the condition is that the wavelength λ must be related to the length l of the string by

$$\lambda = 2l/n \tag{4.1}$$

where n is an integer.

If an electron could behave like a wave of wavelength λ and there were some relation between λ and its energy (via Planck's constant?) for an electron in an atom then it is possible that only certain energy states would be allowed. This would, of course, depend on the existence of appropriate boundary conditions.

At the same time such a behaviour need not be contrary to the fact that classical mechanics has worked well for free electrons. If the associated wavelength were sufficiently small we could be in the same state as we are in geometrical optics, where, in spite of the wave nature of light, we talk of light travelling in straight lines etc. The classical path would then correspond to a ray as in optics.

Thus we have good reason for performing experiments specifically designed to look for wave properties associated with electrons.

4.3 ELECTRON AND X-RAY DIFFRACTION

Clearly the mere fact that no diffraction phenomena were observed with electrons fairly early on implies that any wavelength which may be associated with a moving electron must be small. Further for an electron in an atom the

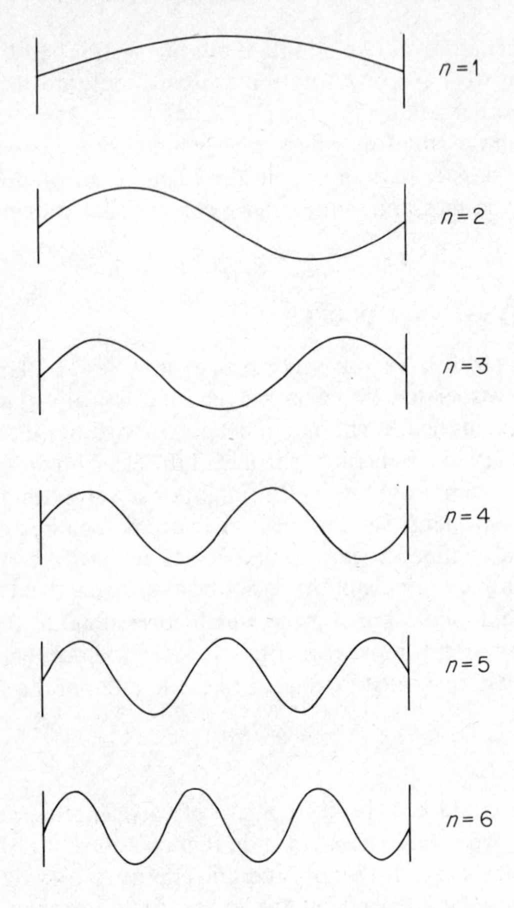

Fig. 4.1. Some of the allowed modes of a vibrating string.

wavelength can hardly be bigger than atomic dimensions, or the atoms would surely appear bigger than they do. This means that if we are to look for diffraction effects we cannot use ordinary slits, but must use the methods of X-ray diffraction from crystal lattices.

Consider first a crystal lattice. Through such a lattice we can draw a set of equidistant parallel planes which pass through all the atoms which compose the crystal. There are many such sets as is shown in Fig. 4.2, each of which has a characteristic spacing known as the Bragg spacing.

Consider now a monochromatic beam falling on a crystal. Each atom in the crystal will scatter some of the incident radiation, giving rise to a spherical wavelet. In order to observe a strong diffracted beam in any particular direction all the secondary wavelets must add up in phase in that direction.

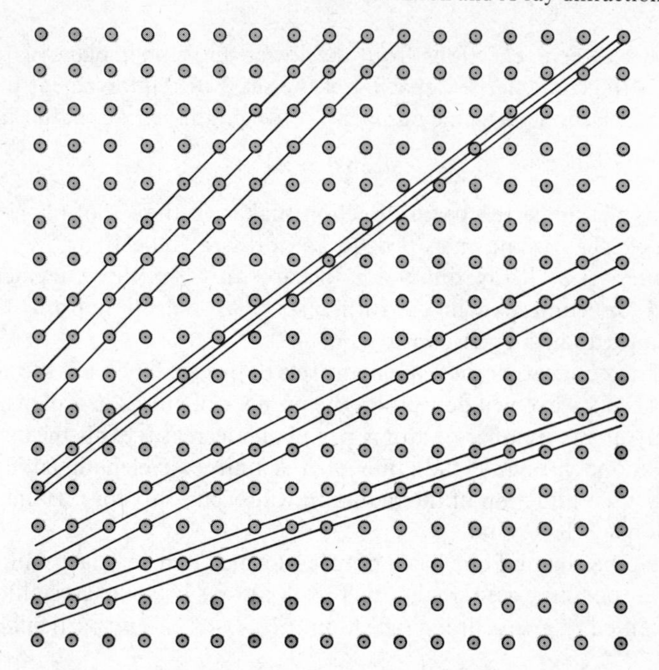

Fig. 4.2. Some of the sets of Bragg planes that can be drawn in a cubic lattice. Note that the number of atoms per plane varies quite widely.

To see whether this will occur we split the crystal up into sets of Bragg planes. If there exists a set of Bragg planes for which the diffracted direction is that for specular reflection of the incident beam from this set of planes, then the wavelets from all the atoms from any one plane will be in phase. If, in addition, the wavelets from successive planes in the set are also in phase, then we have the required condition for a diffraction maximum to exist.

Let us take two successive members from the set of Bragg planes which satisfy the specular reflection condition. From Fig. 4.3 we see that the path

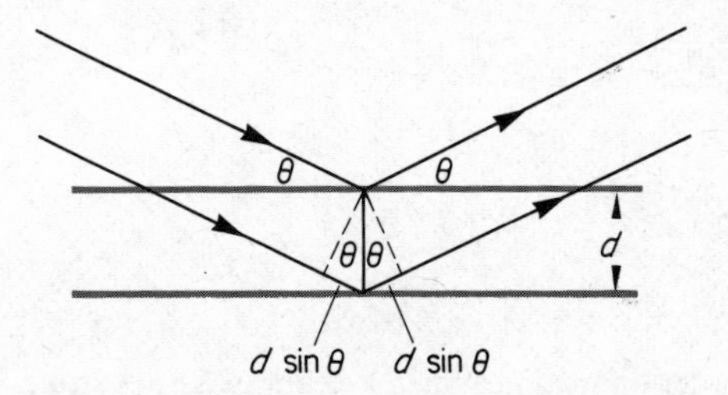

Fig. 4.3. Bragg's law. The path difference of the two rays shown is $2d \sin \theta$. For reinforcement this must equal a whole number of wavelengths.

difference between reflections from the lower and upper planes is $2d \sin \theta$ where d is the characteristic spacing of the set. For reinforcement this path difference must equal a whole number of wavelengths, so we get the law

$$2d \sin \theta = n\lambda \tag{4.2}$$

where θ is the angle the beam direction makes to the set of planes and n is an integer. This is known as *Bragg's law* for X-ray reflection.

To summarize, X-ray diffraction occurs only for those incident and diffracted directions which constitute specular reflection from a set of Bragg planes and which also satisfy Eq. (4.2).

As pointed out earlier there are many sets of Bragg planes, but the intensity of the reflected wave will depend on the number of atoms in a plane, falling off rapidly as the number of atoms per plane decreases. This means that if one has monochromatic radiation of a definite wavelength falling on a single crystal, diffraction of the radiation will occur only for certain specific orientations of the crystal.

On the other hand if one has a powder, formed of a very large number of crystallites, one may assume that there will always be some crystallites correctly oriented to give a Bragg reflection (Fig. 4.4). For any particular Bragg

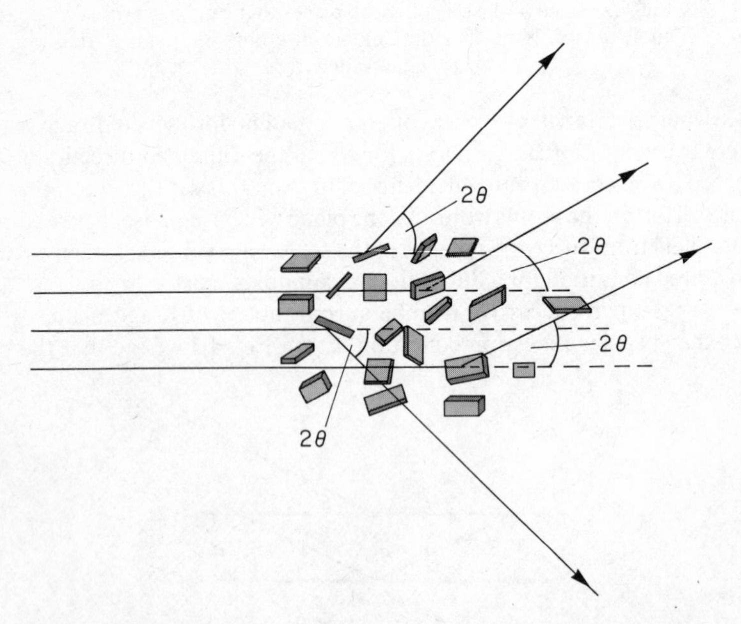

Fig. 4.4. Powder diffraction. When a beam of X-rays strikes a powder only those crystalites oriented to the beam at a Bragg angle θ will reflect some of the beam. The totality of these reflections produce a cone of reflected X-rays about the incident direction of semi-angle 2θ.

reflection it is only the relative orientation of the plane and the incoming beam that counts, and when this is correct Bragg reflection will take place, whatever the azimuthal orientation of the plane about the incoming beam axis. Thus, if a monochromatic beam is incident on a powder a particular Bragg reflection will produce a *cone* of outgoing radiation with the incident direction as axis and a semi-angle equal to 2θ, where θ is the Bragg angle. Thus if the scattered radiation falls on a photographic plate placed normal to the beam direction we shall see a series of *rings*, each ring corresponding to a particular Bragg reflection. This method of studying X-ray diffraction was devised by Debye and Scherrer. An example of a powder diffraction photograph is shown in Fig. 4.5.

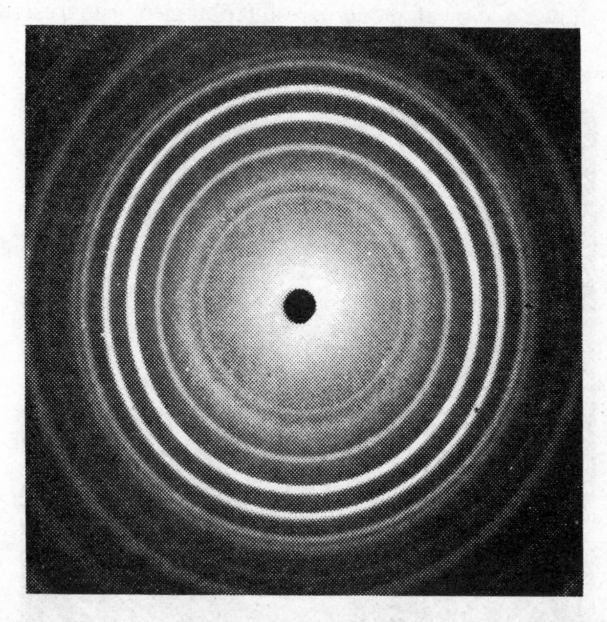

Fig. 4.5. Powder X-ray diffraction. The X-ray diffraction pattern from zirconium oxide powder (From U. Fano and L. Fano, *Basic Physics of Atoms and Molecules*, Wiley, New York, 1959).

Following the enunciation of Bragg's law in 1912, much X-ray work was done which led to the determination of many crystal structures. A careful comparison of the diffraction of X-rays by crystals and gratings established the *absolute* spacings of the crystal planes and has led to the most accurate method of determining Avogadro's number.

The first suggestion that electrons might have wavelike properties was made by L. de Broglie in 1924. In analogy with the photon case de Broglie suggested that the wavelength to be associated with the motion of an electron should be given by

$$\lambda = h/p \tag{4.3}$$

where h is Planck's constant and p is the momentum of the electron. These ideas were followed up in 1928 by Davisson and Germer, and also by G. P. Thomson. Thomson used the Debye and Scherrer method to investigate the possibility of electron diffraction. He fired a beam of 10 kilovolt electrons at a polycrystalline gold foil. The foil was prepared by evaporating gold in vacuum on to an organic backing and then dissolving away the backing leaving a very thin self-supporting foil of gold.

Fig. 4.6 shows a typical modern result obtained using essentially the

Fig. 4.6. Electron diffraction. The electron diffraction pattern from gold (From U. Fano and L. Fano, *Basic Physics of Atoms and Molecules*, Wiley, New York, 1959).

same methods. The similarity with Fig. 4.5 is obvious. Many such experiments on the diffraction of electrons have been performed, clearly showing that electrons can exhibit a wavelike behaviour.

4.4 THE WAVELENGTH OF ELECTRONS AND OTHER PARTICLES

Our knowledge of crystal dimensions allows us to determine the wavelength associated with electrons through Eq. (4.2). Experiments performed with electrons of various energies have shown the wavelength is always given by Eq. (4.3), as suggested by de Broglie, and for this reason Eq. (4.3) is known as the de Broglie wavelength of the particle.

Is this relation peculiar to electrons, or has it more universal validity? Later experiments have shown that all particles can be made to show wave-like behaviour under suitable circumstances and the wavelength is *always* given by Eq. (4.3). For example, in 1929 Estermann and Stern showed that helium atoms and hydrogen molecules are diffracted and the wavelength is given by Eq. (4.3). This equation therefore is of universal validity and must express something very fundamental.

A consideration of the magnitudes involved will show why such phenomena had not been observed before. For a wavelength of 1 Å we find the energies involved are

for photons	12 keV
for electrons	150 eV
for neutrons	0.081 eV
for helium atoms	0.02 eV.

As the *thermal* energy of a particle at room temperature is 0.025 eV it is clear that very special circumstances will be required to observe the effects of the wave nature of particles more massive than the helium atom. Thus looking at things from a different standpoint, we find a particle of mass 10 micrograms, moving at 1 cm per second has a wavelength associated with it of *6.6×10^{-22} cm!*

4.5 ELECTRONS—WAVES OR PARTICLES?

This discussion follows very much the same lines as in section 3.10. We saw in that section that when we detect a photon we either get the whole photon or nothing at all; so that photons are indivisible. In the same way we find when we detect an electron we get the whole electron or nothing at all. Attempts to split an electron always fail. We may perform an experiment similar to that illustrated in Fig. 3.24, and in the same way we find we split the *number* of electrons.

What happens when we pass a beam of electrons through a screen containing two slits, so that we have an electron interferometer? We find, just as we did in the optical case, that we get an interference pattern. Once again we may use a very weak source of electrons so that there is only one electron at a time in the interferometer, and once again we find electrons fall randomly and unpredictably on the detector as far as individual events are concerned, but that as the number of electrons detected increases a pattern builds up which is the usual two-slit interference pattern.

Let us try to envisage an experiment which would tell us through which slit the electron has passed. We set up a double slit electron interferometer, as shown in Fig. 4.7, which includes a light source, L, which projects a beam

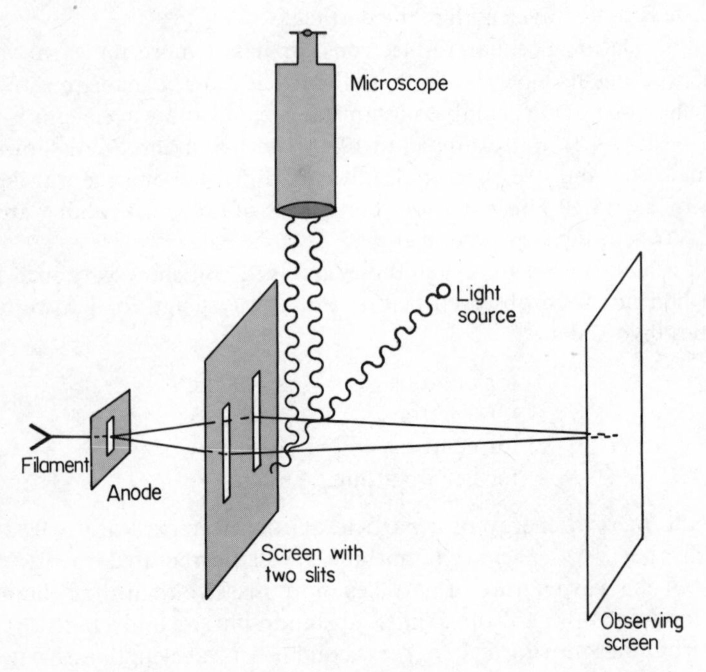

Fig. 4.7. An electron interference experiment in which an attempt is made to determine through which of the two slits each electron passes.

of light across the path of the electrons close to the slits, as shown. When an electron passes through the light beam it will scatter some of the light, and by noting the location of the source of scattered light with the aid of a microscope we could hope to determine which slit the electron has passed through. When we do this we find our interference pattern has been completely destroyed. This is not surprising, though, because we know from the Compton effect that when the electron scatters a photon it receives an impulse from the photon which will change its momentum. We also know that lowering the light intensity will not help, as photons of a given wavelength have the same energy and momentum. All that will happen is that the *chance* of the electron scattering a photon will be reduced, so some electrons will arrive at the detector without having given a light signal. If, however, we do the experiment and keep count of electrons which did and did not give a light signal, we find that those that give a light signal do not give an interference pattern, whereas those that did *not* give a light signal do indeed still show an interference pattern, i.e. they have behaved as if the light were not there. But as they did not interact with a photon there is no way for anything to be changed.

Clearly the only way to minimize the disturbance is to use photons which will give less momentum to the electron when they are scattered, i.e. use a longer wavelength. When we do this we find that due to diffraction effects the position from which the light flash comes is less well defined. The results we observe depend on the wavelength of light we use. If it is still short enough to discriminate between the two slits then the interference pattern does not appear, but when we use wavelengths such that diffraction effects make it impossible for us to distinguish between the slits the interference pattern reappears.

It might be thought that, when using light of a short wavelength, although the electron receives an impulse when it scatters the light, we could determine the magnitude and direction of this impulse from the equations leading up to Eq. (3.13); and thus we could choose electrons disturbed in a *known* way. Then surely our observation could not wash out the interference pattern. However we find that when we measure the angle through which the light has been scattered with sufficient accuracy to give us the necessary accuracy in our knowledge of the transverse momentum of the electron, diffraction effects prevent us from knowing the source of the light flash with sufficient accuracy to determine through which slit the electron has passed. An example is shown in Fig. 4.8 where an electron moving in the z-direction scatters a

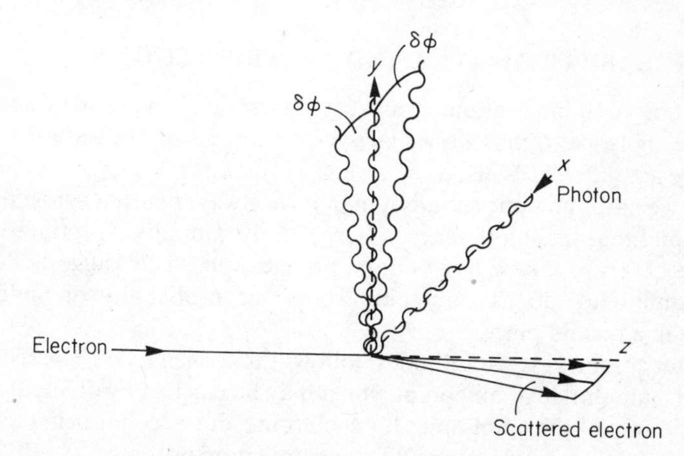

Fig. 4.8. The photon and electron paths in the experiment of Fig. 4.7. A photon incident from the x-direction is scattered through $(90 \pm \delta\phi)^\circ$ by an electron incident in the z-direction. The uncertainty in the knowledge of the final direction of the photon leads to a corresponding uncertainty in the direction of the scattered electron.

low energy photon, initially travelling in the x-direction, through $90° \pm \delta\phi$. If $h\nu$ is very much less than mc^2 the electron receives momentum in the x-direction of

$$\frac{h\nu}{c}(1 \pm \delta\phi) .$$

The uncertainty in the x-momentum of the electron is

$$\pm\frac{h\nu}{c}\delta\phi = \pm\frac{h}{\lambda}\delta\phi .$$

If the electron fringe pattern is to persist the above uncertainty in the x-momentum of the electron must produce an angular spread in the beam which is less than the angular spacing between the fringes. The angular spread in the beam is $2\,\Delta p_x/p_z$ i.e. $2h\,\delta\phi/\lambda p_z$ and the angular spacing between fringes is λ_e/d where λ_e is the electron wavelength h/p_z. Thus for the fringes to still be observable we must have

$$2h\,\delta\phi/\lambda p_z < h/p_z d$$

or

$$\lambda/2\,\delta\phi > d .$$

But $\lambda/\delta\phi$ is the closest distance at which it is possible to resolve two objects using light of wavelength λ confined within an angle of $\pm\delta\phi$. Hence we cannot tell which slit the electron passed through.

4.6 PROBABILITY WAVES AND SUPERPOSITION

How are we to intepret our results? The situation obviously bears a very close resemblance to the one we faced when discussing the nature of light, but there are differences also.

First, we have the same situation in that we always detect a whole particle and we find that the interference pattern is only faithfully reproduced when we have observed a *very large* number of electrons. This suggests that the wave amplitudes tell us something about the probability of finding an electron in a certain place.

Secondly, the wave amplitudes follow the *principle of superposition*; that is the amplitude at any point which may be reached by different routes is just the sum of the amplitudes for each route, due account being taken of any phase differences which result from different path lengths.

The evidence we have for this is that the observed interference patterns are the same as those observed in X-ray diffraction using X-rays of the same wavelength. Now the whole of the electromagnetic wave theory is built up on the principle of superposition, which in this case is a consequence of the fact that the wave equation is a homogeneous linear differential equation.* There is accurate numerical agreement between the predictions of electromagnetic theory and the observed phenomena. As the two cases give identical results we believe that superposition holds for electron waves also. This in turn means that the wave equation governing these waves must also be a homogeneous linear differential equation.

We can see how this sort of picture also explains the fact that we destroy our interference pattern when we detect which slit an electron passes through. To find where we expect to find the electron we must solve our wave equation subject to certain boundary conditions. The boundary conditions say that the electron came from the source and might pass through either slit. It was the possibility that the electron could pass through either slit that caused the interference. By observing which slit the electron passed through we have changed the boundary conditions. We now *know* the electron is in one region and it cannot therefore be in the other region, and our new solution has no interference wiggles in it.

In the case of light waves the intensity is given by the mean square amplitude, and we associated this with the probability of finding a photon. As the number of electrons observed follows the same pattern as the intensity of light, we assume that probability is also related to the square of the wavefunction.

However there are notable differences between light waves and those associated with electrons. In the first place we not only know the wavelength of light waves but also their frequency, through the relation $\lambda\nu = c$ for short wavelengths and by direct observation for long wavelengths. For electron waves we have, as yet, no knowledge of their frequency or velocity, or whether these concepts mean anything.

Also in the case of electromagnetic radiation we can observe the actual wave pattern vary as a function of position and time when we have a sufficient number of photons present, as in the case of radio waves, but no similar phenomenon has been found for the waves associated with electrons (or neutrons or protons for that matter).

Thus we have concluded, on grounds of plausibility, that the waves associated with electrons give us information on the probability of finding an electron at a given place, and these waves are governed by a homogeneous linear differential equation and hence obey the principle of superposition. We know also that the wavelength is given by $\lambda = h/p$ but we know nothing else. In particular we do not know the equation governing their motion.

* A homogeneous linear differential equation has the property that, if A and B are two solutions, then $A + B$ is a solution also. This in fact is just the principle of superposition.

4.7 REPRESENTATION OF A UNIFORMLY MOVING PARTICLE

In the last section we pointed out that the waves associated with electrons must be governed in their behaviour by a homogeneous linear differential equation. Our main purpose is to discover this equation and apply it to a wide variety of physical problems. We cannot *derive* this equation; we shall have to postulate it. Derivation implies the existence of a more fundamental law from which the derivation is made; but we are trying to find the fundamental equation which governs the behaviour of electrons (and of other atomic particles).

We are thus involved in a guessing game, and in such circumstances one is more likely to guess correctly the more one knows; at the moment we know very little. It would clearly be a great help if we knew more about the nature of the waves which form some of the solutions of the equation we wish to postulate. To this end, in this and the following section, we discuss one of the simplest solutions, namely that of a uniformly moving particle.

Consider a particle travelling in a given direction, say the x-direction. This is presumably to be represented by a plane wave. If it were an electromagnetic wave it would have a wavefunction ψ of the form

$$\psi = A_0 \sin (kx - \omega t + \delta) , \tag{4.4}$$

where

$$k = 2\pi/\lambda , \qquad \omega = 2\pi\nu ,$$

and the phase δ is determined by the origin of x and t. A_0 is a constant amplitude.

There is however no need for an electron wavefunction to be of precisely the same form as Eq. (4.4). What we have observed so far is an agreement between electron diffraction patterns and X-ray diffraction patterns. The observed patterns are the time averaged intensities at certain positions in space; for example over the surface of a photographic plate. An electron wavefunction for a given wavelength must give the same time averaged intensities, as a function of position, as the corresponding electromagnetic wavefunction. The time averaged intensity of Eq. (4.4) is $\frac{1}{2}A_0^2$, i.e. it is uniform and independent of x and so must be the time averaged intensity for the corresponding electron wavefunction. Therefore any generalization of Eq. (4.4) which satisfies this condition is a candidate for our electron wavefunction.

An alternative way of writing Eq. (4.4) is

$$\psi = A' \sin (kx - \omega t) + B' \cos (kx - \omega t) \tag{4.5}$$

where

$$A'^2 + B'^2 = A_0^2 , \qquad B'/A' = \tan \delta .$$

There is only one way to generalize Eq. (4.5), and that is to allow the constants A' and B' to be complex. If we allow this, then the wavefunction ψ will also be complex. We defer a full discussion of the implications of this until it proves necessary, but we must consider now what we mean by the intensity of such a wave. As we require the intensity in any situation to give the observed electron diffraction pattern it must be a real quantity. We therefore take the intensity to be $|\psi|^2$, the square modulus of the wavefunction.

With this definition we have

$$|\psi|^2 = |A'|^2 \sin^2(kx - \omega t) + |B'|^2 \cos^2(kx - \omega t) + (A'^*B' + B'^*A') \sin(kx - \omega t) \cos(kx - \omega t)$$

and the time averaged intensity is

$$\overline{|\psi|^2} = \tfrac{1}{2}(|A'|^2 + |B'|^2) \ . \tag{4.6}$$

An alternative way of writing Eq. (4.5) which will prove more convenient is

$$\psi = A\,\mathrm{e}^{\mathrm{i}(kx-\omega t)} + B\,\mathrm{e}^{-\mathrm{i}(kx-\omega t)} \tag{4.7}$$

where

$$A = (B' - \mathrm{i}A')/2, \quad B = (B' + \mathrm{i}A')/2, \quad |A|^2 + |B|^2 = |A'|^2 + |B'|^2 \ .$$

We therefore take Eq. (4.7) with A and B complex as our representation of an electron moving with a uniform velocity in the x-direction. The reader should be prepared for the possibility that the equation we eventually postulate for governing the motion of electrons may impose some restrictions on the constants A and B.

4.8 FREQUENCY AND VELOCITY

The wavefunction Eq. (4.7) we have written down for a particle moving in the x-direction has two parameters in it, the wavenumber $k = 2\pi/\lambda$ and the frequency ω. We know how to determine k through electron diffraction but we have no information concerning ω. If Eq. (4.7) is to be a solution of a differential equation we need a relation between k and ω.

Let us consider for a moment, how we establish, experimentally, the relation between k and ω for electromagnetic waves. At the low frequency end of the spectrum we would find we could measure the wavelength (by diffraction), and also the frequency, and we should find we had the relation

$$\omega/k = c \tag{4.8}$$

where c is a constant independent of frequency and equal to the velocity of propagation.

At the high frequency end of the spectrum it is no longer feasible to measure ω directly, and we must look for other methods or checking whether Eq. (4.8) is true over the whole frequency range, or is merely a low frequency approximation. One way is to measure the velocity of propagation. But to do this we have to chop our light beam into short packets in order to measure their transit time, and hence find the velocity. It is well known that the velocity with which such packets travel is not the phase velocity ω/k but the group velocity v_g given by

$$v_g = \frac{d\omega}{dk} \,. \tag{4.9}$$

(See F. G. Smith and J. H. Thomson, *Optics* (Manchester Physics Series), Wiley, London 1971, chap. 1.) The result of such an experiment would be to discover that

$$\frac{d\omega}{dk} = c \tag{4.10}$$

and hence

$$\omega = kc + \text{constant} \,. \tag{4.11}$$

In view of Eq. (4.8) one would feel justified in putting the constant equal to zero and saying that Eq. (4.8) holds over the whole frequency range.

Let us now return to electron waves. We have no idea how to measure ω or even whether it is a directly measurable quantity. We can, however, measure the velocity of an electron as illustrated in Fig. 4.9 for example.

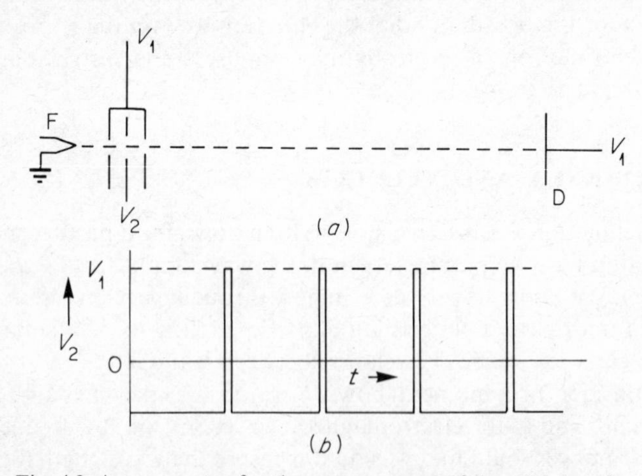

Fig. 4.9. An apparatus for the measurement of the velocity of electrons. F is a filament and D a detector. The potential V_2 of the central electrode is shown as a function of time in (*b*). When V_2 is below earth potential no electrons pass through. Bunches of electrons are produced when V_2 is suddenly switched to the value V_1, enabling their transit time to D to be measured.

To do this we have to determine the position of the electron in order to measure its transit time to some other position. As a result the electron cannot be represented by the same function as in Eq. (4.7) which represents a wave extending uniformly from $x = -\infty$ to $x = +\infty$. Rather we must take a wave packet, a superposition of waves of the type of Eq. (4.7) but with a range of k (and hence of ω) centred around a mean value k_0.

This wave packet will propagate with the group velocity $\mathrm{d}\omega/\mathrm{d}k$ and our scheme of things will only make sense if this group velocity is equal to the velocity of the particle, i.e.

$$\frac{\mathrm{d}\omega}{\mathrm{d}k} = v \; . \tag{4.12}$$

Now

$$k = p/\hbar$$

$$v = \hbar \frac{\mathrm{d}\omega}{\mathrm{d}p} \; .$$

In the non-relativistic region we have $p = mv$ where m is the rest mass, so that Eq. (4.12) becomes

$$\frac{\mathrm{d}\omega}{\mathrm{d}v} = \frac{mv}{\hbar} \; . \tag{4.13}$$

Integration of Eq. (4.13) gives us

$$\hbar\omega = \tfrac{1}{2}mv^2 = E \tag{4.14}$$

where we have left out any constant of integration. We do this as we then obtain the simplest relation between k and ω. For a freely moving particle the energy E is just the kinetic energy $p^2/2m$. With $k = p/\hbar$ Eq. (4.14) becomes

$$\hbar\omega = \frac{\hbar^2 k^2}{2m} \tag{4.15}$$

giving us our desired relationship between k and ω.

4.9 A DIFFERENTIAL EQUATION FOR A PARTICLE IN FREE SPACE AND THE COMPLEX NATURE OF THE WAVEFUNCTION

The simplest differential equation which has solutions of the form of Eq. (4.7) which behave in the desired way, and has k and ω related as in Eq. (4.15), is the equation

$$\mathrm{i}\hbar \frac{\partial \psi}{\partial t} = -\frac{\hbar^2}{2m}\frac{\partial^2 \psi}{\partial x^2} \; . \tag{4.16}$$

The simplest solutions of Eq. (4.16) are

$$\psi_+ = A\,\mathrm{e}^{\mathrm{i}(kx-\omega t)} \tag{4.17a}$$

$$\psi_- = B\,\mathrm{e}^{-\mathrm{i}(kx+\omega t)} \tag{4.17b}$$

where k and ω are constants related by Eq. (4.15). We note that Eq. (4.17a) is just Eq. (4.7) with B equal to zero. Eq. (4.17a) represents a plane wave with wavenumber k travelling in the positive x-direction, and Eq. (4.17b) a plane wave, again with wavenumber k, travelling in the negative x-direction. We may verify that Eq. (4.17a), for example, is a solution of Eq. (4.16) by direct substitution; thus

$$\mathrm{i}\hbar\frac{\partial\psi}{\partial t} = \hbar\omega A\,\mathrm{e}^{\mathrm{i}(kx-\omega t)}$$

and

$$-\frac{\hbar^2}{2m}\frac{\partial^2\psi}{\partial x^2} = \frac{\hbar^2k^2}{2m}A\,\mathrm{e}^{\mathrm{i}(kx-\omega t)}$$

and therefore

$$\mathrm{i}\hbar\frac{\partial\psi}{\partial t} = -\frac{\hbar^2}{2m}\frac{\partial^2\psi}{\partial x^2}$$

provided $\hbar\omega = \hbar^2k^2/2m$ as postulated in Eq. (4.15).

It is of *extreme importance* to note that the solutions ψ_+ and ψ_- given in Eqs. (4.17) are *complex*. This, at once, means that the *wavefunctions themselves are not observable quantities*, and indeed neither is the frequency ω. However, this need not surprise us too much. As we mentioned in section 4.6 and elsewhere the parallel is between X-ray diffraction intensities and the numbers of electrons observed at various places in an electron diffraction experiment, provided we observe a large number of electrons in total. We pointed out that this suggests that there is a relation between the electron wavefunction amplitude and the probability of finding an electron.

The intensity of an electromagnetic wave is determined by the square of the amplitude, and hence we assume that the probability of finding an electron at a particular place is related in a somewhat similar way to the electron wavefunction. Put this way one might find it surprising, rather, if one could directly observe a quantity which depends on the square root of the probability of finding an electron!

We will have to determine how the probability of finding an electron in a particular place, which is a real measurable quantity, is related to the electron wavefunction which, in the case we have been discussing, is complex. We postulate that the probability of finding an electron in a particular place is proportional to the square modules of the wavefunction at that place, i.e.

$$P(x)\,dx \propto |\psi(x)|^2\,dx \tag{4.18}$$

where $|\psi(x)|^2 = \psi^*(x)\psi(x)$ and $\psi^*(x)$ is the complex conjugate of $\psi(x)$.

As Eq. (4.16) is homogeneous the solutions are defined only up to a multiplicative constant, i.e. if $\phi(x)$ is a solution, so is $\psi(x) = a\phi(x)$ where a is any arbitrary complex number. In cases where the wavefunction is localized so that the integral $\int_{-\infty}^{+\infty} |\psi(x)|^2\,dx$ is finite it is convenient to choose the constant a so that

$$\int_{-\infty}^{+\infty} |\psi(x)|^2\,dx = 1 \tag{4.19}$$

in which case Eq. (4.18) becomes

$$P(x)\,dx = |\psi(x)|^2\,dx \tag{4.20}$$

i.e. the probability of finding the particle between x and $x + dx$ is just $|\psi(x)|^2\,dx$. This procedure is called *normalization*. This cannot be done for Eqs. (4.17) as these solutions are not square integrable, i.e. the integral $\int_{-\infty}^{+\infty} |\psi(x)|^2\,dx$ diverges. In such circumstances one is usually interested in the flux of particles, e.g. the number going from left to right as compared with the number going from right to left, in which case one is only interested in the ratios of the components of the wavefunction and one may choose the constant to suit the problem.

Using Eq. (4.18) as our definition of intensity, we see that Eqs. (4.17) both have an intensity which is independent of t and x. They thus more than satisfy the condition we laid down in section 4.7 that the time averaged intensity should be independent of x, for the intensity itself is independent of x and *no time averaging is needed*. If we superpose waves of the same wavelenth travelling in different directions we again find that the intensity directly gives us the diffraction pattern without time averaging. It is gratifying that this is the case, for it is difficult to see where a time varying component in the intensity of electron diffraction patterns could come from.

We have shown that Eqs. (4.17) are solutions to Eq. (4.16). As Eq. (4.16) is a homogeneous linear differential equation, any linear combination of solutions of the type of Eq. (4.17) will also be a solution, provided only that Eq. (4.15) is obeyed for each pair of k and ω that occurs. Thus the most general solution to Eq. (4.16) will be

$$\psi(x, t) = \frac{1}{\sqrt{(2\pi)}} \int_{-\infty}^{+\infty} A(k)\, e^{i(kx - \omega t)}\, dk \tag{4.21}$$

where $A(k)$ is the amplitude of the wave with wavenumber k and k and ω are related as in Eq. (4.15). The factor of $1/\sqrt{(2\pi)}$ is included for later convenience.

So far we have taken our plane waves to be travelling in the x-direction, and hence our differential equation (4.16) is one involving x and t only. If we take a plane wave travelling in an arbitrary direction in free space, this will be given by

$$\psi(\mathbf{r}, t) = A\,\mathrm{e}^{\mathrm{i}(\mathbf{k}\cdot\mathbf{r} - \omega t)} \tag{4.22}$$

where the direction of $\mathbf{k}$ specifies the direction of travel, and $\hbar\omega = \hbar^2 k^2/2m$ as before. The equation corresponding to (4.16) which has solutions of the form (4.22) is

$$\boxed{i\hbar\frac{\partial\psi}{\partial t} = -\frac{\hbar^2}{2m}\nabla^2\psi\ .} \tag{4.23}$$

We postulate that Eq. (4.23) is the equation governing the motion of free particles of mass m.

We note that Eq. (4.23) has the following properties which we have been careful to build in.

(i) It has plane wave solutions.

(ii) It is a homogeneous linear differential equation, so a sum of solutions is also a solution; i.e. its solutions satisfy the principle of superposition, as indicated for the one-dimensional case by Eq. (4.21).

(iii) The relation between the parameters k and ω which appear in the solutions is such as to ensure that the group velocity of a wavepacket is equal to the particle velocity.

It must be emphasized that Eq. (4.16) and its three dimensional extension Eq. (4.23) are merely the *simplest* differential equations which have plane wave solutions in which k and ω are related as in Eq. (4.15). In the flexual vibrations of a uniform bar, for example, k and ω are related by

$$\omega = (EI/m)^{1/2}k^2$$

where E is Young's modulus for the material of the bar, I the second moment of the cross-section of the bar about the centre line at right angles to the direction of vibration, and m the mass per unit length of the bar; a precisely similar relation to Eq. (4.15). However, the differential equation governing these vibrations is

$$\frac{\mathrm{d}^2 y}{\mathrm{d}t^2} = -\frac{EI}{m}\frac{\mathrm{d}^4 y}{\mathrm{d}x^4}\ .$$

4.10 THE UNCERTAINTY PRINCIPLE—A CONSEQUENCE OF THE WAVE PICTURE

We have represented a uniformly moving particle by a wave function

$$\psi = A\,e^{i(kx - \omega t)} \quad . \tag{4.17a}$$

Let us consider this wavefunction at some fixed instant of time, say $t = 0$. Then we have

$$\psi = A\,e^{ikx} \quad . \tag{4.24}$$

The probability of finding the particle between x and $x + dx$ is, according to the discussion of section 4.9,

$$P(x)\,dx \propto \psi^*(x)\psi(x)\,dx \quad .$$

This expression is independent of x, i.e. the particle is equally likely to be anywhere along the x-axis and we have no knowledge of its position.

In many experiments we do have a knowledge of the position of a particle to some specified degree of accuracy.

Such a situation could arise in the following way, illustrated in Fig. 4.10.

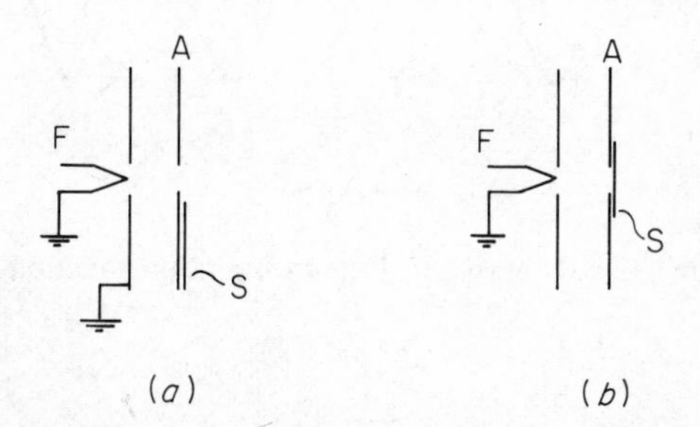

Fig. 4.10. Apparatus consisting of a filament F, an anode A, and a fast acting shutter S.

Electrons from a heated filament F are attracted to an anode A, at a potential V, which has a small hole in it. This hole can be covered by a fast operating shutter. We now do an experiment in which the shutter is opened for a very small time interval Δt. (We assume that the electron emission rate, and the time interval Δt, are such that only one electron at a time gets through the hole.) If the shutter is opened at $t - \Delta t/2$ and closed at $t + \Delta t/2$, then at time t, any electron which gets through is within a distance of $\pm\Delta x$ of the plane of the shutter, where $\Delta x = v\Delta t/2$ and v is the velocity of the electron. We have thus determined the position of the electron to within a region of $\pm\Delta x$.

We would expect therefore, to represent this electron by a wave*packet* of the form

$$\psi(x, t) = A(x, t)\, e^{i(kx - \omega t)} \tag{4.25}$$

where

$$k = \frac{2\pi}{\lambda} = \frac{mv}{\hbar}$$

and $A(x, t)$ is a modulating function which ensures that $\psi(x, t)$ only exists over a region of length $2\Delta x$ as determined above and the centre of the packet travels with velocity v. We choose the origin of x and of the time t so that at $t = 0$ the centre of the packet is at $x = 0$. We illustrate the situation at $t = 0$ in Fig. 4.11 by plotting the real part of Eq. (4.25) for $t = 0$ and the function $A(x, 0)$ as functions of x.

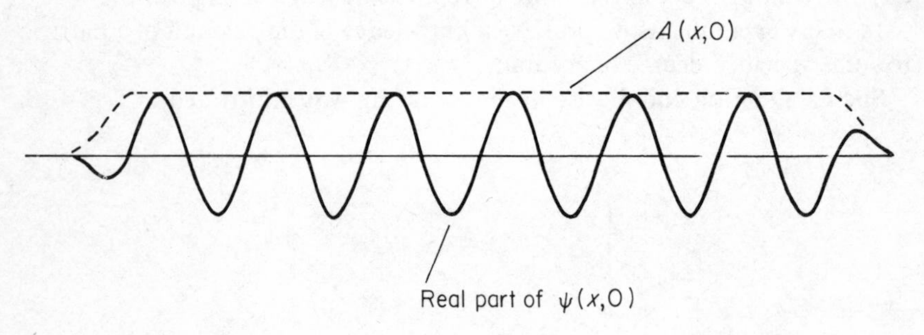

Fig. 4.11. The real part of $\psi(x, 0)$, Eq. (4.25), and the function $A(x, 0)$.

Now such a wavepacket, as is well known, does not consist of a single wavelength λ. This is because wavelength is *defined* in terms of sine waves (or exponential waves such as e^{ikx}) which necessarily go from $-\infty$ to $+\infty$. Our wavepacket at $t = 0$

$$\psi(x, 0) = A(x, 0)\, e^{ikx} \tag{4.25a}$$

does not consist of a single wavelength $\lambda = 2\pi/k$ but rather consists of a superposition of waves of different wavelengths.

It is intuitively obvious that if we have a wavepacket of the form of Eq. (4.25a) in which the wavelength is constant within the envelope, then the longer the packet, the closer we get to the ideal case, and the smaller will be the spread of wavelengths involved (see Fig. 4.12). In section 4.14 we illustrate this by an example which allows us to state precisely, for a given wavepacket,

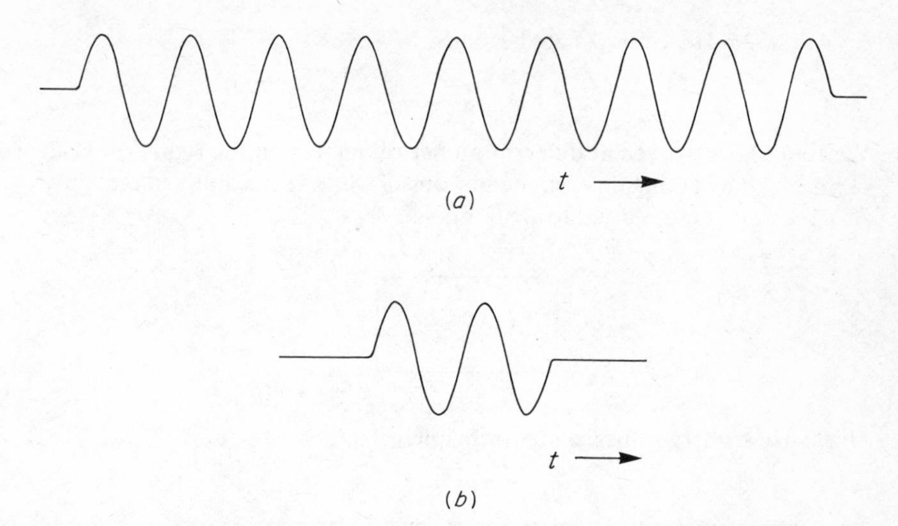

Fig. 4.12. A long wave train with many oscillations as in (*a*) has a small frequency spread. A short wave train with few oscillations as in (*b*) has a large frequency spread.

the relative amplitudes of the wavelengths involved. For the present we shall proceed more qualitatively.

Let us suppose that we have a wavepacket of the form (4.25a), as illustrated in Fig. 4.11, and let us suppose it consists of a superposition of wavelengths, with wavenumbers ranging from $k - \Delta k$ to $k + \Delta k$. We wish to estimate the range of wavenumbers $\pm\Delta k$ involved, given the uncertainty in position of the particle $\pm\Delta x$. If we take $\Delta k/k \approx \lambda/(2\Delta x)$, then the extreme wavelengths will just be cancelling the centre wavelength at either end of the wavepacket, assuming they are all in phase at the centre of the packet. This is clearly the right order of magnitude. Employing this criterion then, we have

$$\frac{\Delta k}{k} \approx \frac{\lambda}{2\Delta x}$$

$$\approx \frac{2\pi}{k \cdot 2\Delta x}$$

i.e.

$$\Delta k \Delta x \approx \pi \ . \tag{4.26}$$

The significance of this relation is clearer if instead of the spread in k we consider the corresponding spread in momentum p. As $p = hk/2\pi$,

$\Delta p = h\Delta k/2\pi$. Hence Eq. (4.26) becomes

$$\Delta p \Delta x \approx h/2 \,. \tag{4.27}$$

We could have arrived at different numerical factors on the right-hand side of Eq. (4.27) by taking different definitions of what we meant by uncertainty. As a result it is more usual to write Eq. (4.27) as

$$\boxed{\Delta p \Delta x \approx h \,.} \tag{4.27a}$$

This is Heisenberg's uncertainty principle.

★ 4.11 FURTHER DISCUSSION OF THE UNCERTAINTY PRINCIPLE

The importance and meaning of Eq. (4.27a) can be seen if we consider how the uncertainties involved have arisen, and for this purpose we return to our starting point and consider what happens in the apparatus of Fig. 4.10. The uncertainty in position Δx arises from the finite time for which the shutter is open, and this is simple enough. But how does the uncertainty in momentum arise, and how is this affected by the *time* for which the shutter is open? At first sight it might seem that all electrons which pass through the hole will have the same energy. This would be true if the hole were open all the time, but the act of opening or closing the shutter slightly changes the distribution of equipotentials and consequently affects the energy of the electrons. In Fig. 4.13 we draw the equipotentials in the region of the anode when the shutter is closed, and when it is open. Consider now an electron which is at x_1 when the shutter is opened, and is at y_1 when the shutter is closed. It has been travelling in the 'shutter open' condition and at y_1 has not yet acquired its full potential energy eV. The act of closing the shutter destroys the field on the right hand side of the anode, and so the electron *never* acquires the full energy. Similarly an electron which has got to x_2 before the shutter opens has already acquired essentially the full energy eV; the act of opening the shutter now creates a field on the right hand side of the anode giving the electron an *extra* acceleration, so it ends up with *more* than the full energy eV. Finally an electron which travels from x_3 to y_3 while the shutter is open is negligibly affected, as the change in potential at x_3 due to moving the shutter is negligible, and at y_3 it has passed through essentially the whole of the potential. Thus we see the final energy is dependent on the exact history of the electron, or to put it another way, the act of

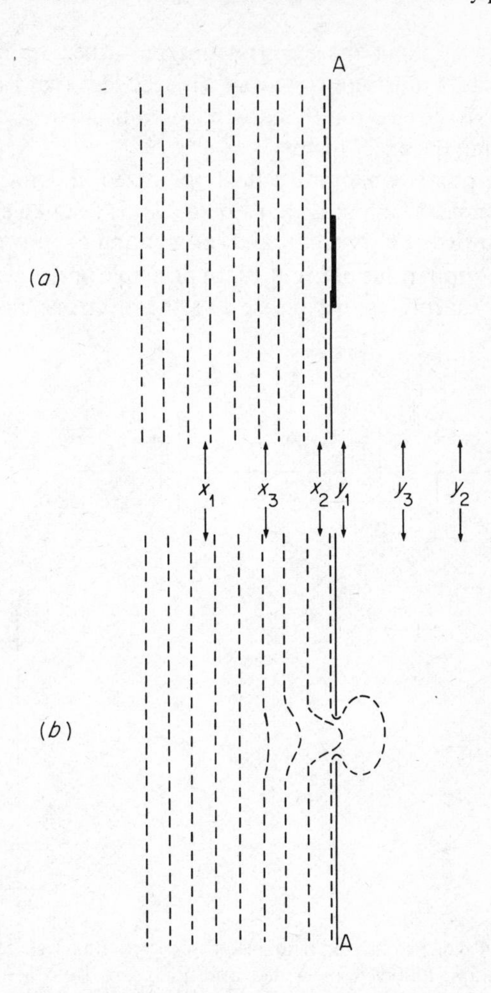

Fig. 4.13. The equipotentials close to the anode A of Fig. 4.10 (*a*) when the shutter is closed, and (*b*) when it is open. An electron which is at x_1 when the shutter opens has reached y_1 by the time it closes again. Similarly electrons which are at x_2 and x_3 when the shutter opens have reached y_2 and y_3 respectively when it closes. An electron which is at x_1 when the shutter opens never acquires the full potential as the field to the right of y_1 is destroyed by closing the shutter. An electron which is at x_2 when the shutter opens experiences an additional field beyond the aperture and hence acquires more than the full potential.

measuring the position of the electron has perturbed its momentum. In saying this the reader must not run away with the idea that if only one knew the exact history of the electron then there would be no uncertainty. There is no way of knowing the exact history.

There are two points which must be emphasized in connection with the uncertainty principle. The first is that Eq. (4.27a) is a statement about one's *simultaneous* knowledge of position and momentum for *predictive purposes.* Thus consider the apparatus of Fig. 4.14. In principle one can set the momentum analyser as sharply as one pleases, so the electrons emerging from it

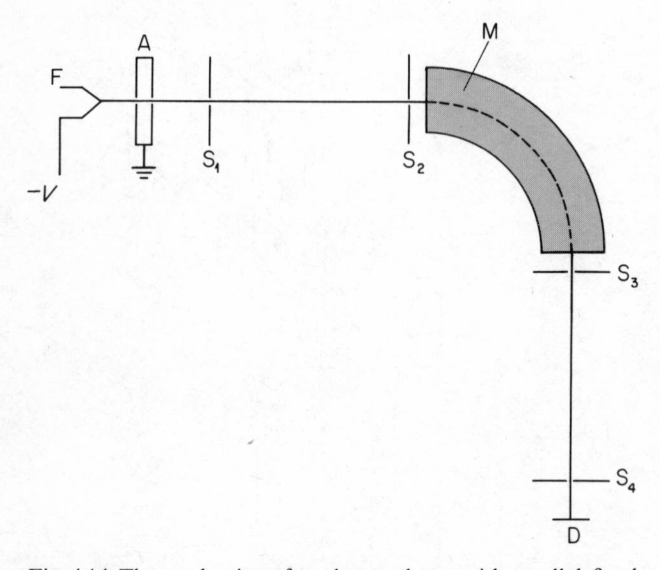

Fig. 4.14. The production of an electron beam with a well defined momentum, followed by a detector. Electrons from a heated filament F are accelerated to an earthed anode A. S_1 and S_2 define the direction of the electrons passing through A and entering a 90° uniform field magnet M. S_3 and S_4 define the direction of the emergent beam detected by the detector D.

have a very well defined momentum. The electrons are now detected in a detector, thus determining their position, and their time of arrival. Thus we know their momentum prior to their arrival at the detector, and their instant of arrival. Hence we can determine their position at any time prior to their arrival at the detector with great accuracy, so that $\Delta p \Delta x \ll h$. However, this is all hindsight and is quite useless for predicting the future behaviour of the electrons, because the operation of detecting them disturbs their momentum.

The useless nature of the information on position and momentum we have obtained in this way is emphasized by the fact that it is impossible to write down a wavefunction describing the above situation. Any wavefunction will always lead to an uncertainty in position and momentum whose product is at least of order $\hbar$, because wavefunctions are what we use to predict the results of measurement.

The second point is that Eq. (4.27a) represents a lower limit on the uncertainty in position and momentum one can obtain. It is very easy to have circumstances where ones knowledge is much worse than this, and in fact this is normally the case.

The classic example of the disturbance produced in a measuring process is Heisenberg's gamma ray microscope, illustrated in Fig. 4.15. Here one

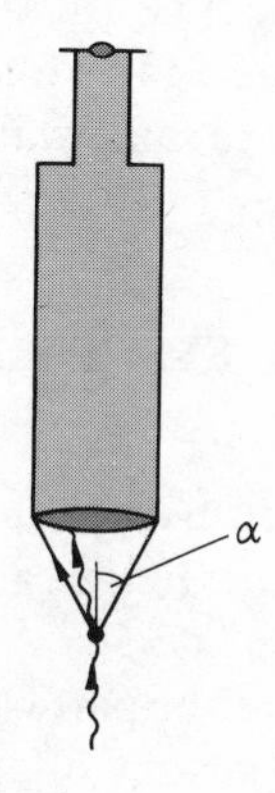

Fig. 4.15. Heisenberg's γ-ray microscope.

attempts to determine the position of a particle using very short electromagnetic waves. The positional accuracy is determined by the resolution of the microscope which is given by the usual formula $\Delta x = \lambda \sin \alpha$ where α is the semi-angle taken in by the microscope. For the light to enter the microscope at least one photon must have been scattered by the electron into the microscope objective. The direction of travel of this photon is not known. We only know that it is within the cone taken in by the objective. The transverse momentum of the photon is then uncertain by $\pm h \sin \alpha/\lambda$ and this uncertainty exists in the recoil of the particle. Thus we have

$$\begin{aligned}&\text{positional uncertainty } \Delta x \approx \lambda/\sin \alpha\\&\text{momentum uncertainty } \Delta p \approx h \sin \alpha/\lambda\ .\end{aligned}$$

Therefore

$$\Delta p \Delta x \approx h\ . \tag{4.27a}$$

From an experimental view therefore we see that the uncertainty arises from an irreducible minimum interaction between the particle and the detector.

We have mentioned and analysed two situations and seen how the uncertainty arises, though in only one of these cases, the Heisenberg gamma-ray microscope, have we actually produced a semi-quantitative solution. It is easy to visualize more sophisticated methods of measuring position and momentum in which the problem of analysing the uncertainties that exist becomes truly formidable. This however, is unnecessary, for, as we showed in the previous section, the mere introduction of a wave description brings uncertainty in its train. This is true in classical physics as well as quantum physics and uncertainty relations are not peculiar to quantum phenomena. Thus, if one wants to measure a purely classical oscillation frequency, the accuracy with which this can be done depends on the time one is prepared to spend on the measurement. One may of course *quote* a more accurate result on the *assumption* that the oscillation continues in the same way outside the period of measurement but one does not *know* this.

4.12 THE UNCERTAINTY RELATION BETWEEN ENERGY AND TIME

In addition to there being an uncertainty relation between position and momentum, there is also one between energy and time. In the case of a wave-packet we can deduce this from our earlier results.

In section 4.10 we assumed that, at the instant $t = 0$, the particle was within $\pm\Delta x$ of the plane of the shutter. Thus the actual instant of arrival of the particle at the plane of the shutter is uncertain to an amount $\pm\Delta t$ given by

$$\Delta t = \frac{\Delta x}{v} \tag{4.28}$$

where v is the velocity of the particle.

As the particle possesses an uncertainty in its momentum, it also possesses an uncertainty in its energy; thus from

$$E = \frac{p^2}{2m}$$

we get

$$\Delta E = \frac{p\Delta p}{m}$$

$$= v\Delta p\,. \tag{4.29}$$

Multiplying Eq. (4.28) and Eq. (4.29) together we get

$$\Delta E \Delta t = \Delta x \Delta p \approx h \,. \tag{4.30}$$

We have arrived at Eq. (4.30) through considering the motion of a free particle, but we make the assumption that this relation nevertheless holds in all circumstances. Undoubtedly the most important circumstances in which Eq. (4.30) is used occur in the consideration of the properties of the bound states of atoms and other systems. In section 3.9 we presented experimental evidence that atoms could only exist in certain discrete energy states. If we investigate these states more closely we find that a particular bound state can occur over a small range of energies ΔE; that is, its energy is uncertain to a degree ΔE. ΔE is called the width of the level.

We also find that an atom does not normally exist in a particular state indefinitely, but decays usually by the emission of electromagnetic radiation. The length of time spent by an atom in a particular state is found to vary, but if a sufficiently large sample of identical atoms is taken it is found that there is a definite mean life time τ associated with the state. The distribution of actual life times which occurs is found to be such that the uncertainty in the life time, defined as the root mean square variation, is actually equal to the mean life time τ. The life time, and hence its uncertainty, is not independent of ΔE but is related to it through Eq. (4.30), and thus we have, through the uncertainty principle, a relation between the width of a level and its life time

$$\tau \Delta E \approx h \,. \tag{4.31}$$

4.13 FURTHER DISCUSSION OF THE ENERGY–TIME UNCERTAINTY RELATION

Let us consider an experiment in which we study the light emitted when an atom in a state with energy E_1, makes a transition to a lower state of energy E_0. The frequency of the emitted light will be given by $h\nu = E_1 - E_0$.

The emitted light is to be studied in a spectrometer of the highest possible resolution. For simplicity's sake we will assume that the state E_0 is the state with the lowest energy allowed in that particular atomic species, the *ground state* as it is called. A suitable experimental set up is shown schematically in Fig. 4.16. The source is a beam of atoms all moving with the same low velocity at right angles to the line of sight of the spectroscope. By using a source of this nature we ensure that, for all practical purposes, each atom is isolated, and its energy levels are those of the isolated atom, and further, there is no Doppler spread of the observed spectral line due to different atoms in the source having different components of velocity along the line of sight. Nevertheless, we find that, as we increase the resolving power of our spectrometer,

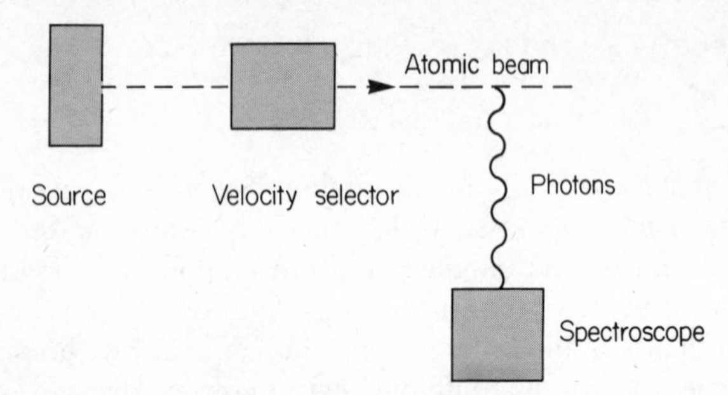

Fig. 4.16. Apparatus for determining the natural width of a spectral line. Excited atoms from a suitable source (e.g. a discharge tube) pass through a velocity selector and the light they emit at right angles to their direction of motion is studied with a spectroscope.

the observed spectral line does not continue to get narrower indefinitely, but settles down to a small but finite width. Thus we must assume that the photons coming from the source do not all have exactly the same frequency but cover a range of frequencies $\Delta\nu$. If this is the case, and we assume that the relation $E = h\nu$ *holds exactly*, then it follows that the excited atoms themselves did not all have the same energy, but covered a small range of energies $\Delta E = h\Delta\nu$ i.e. the excited state E_1 has a certain 'width' ΔE. The reason for ascribing the whole of the frequency spread to the upper level, and none to the ground state, will appear in a moment.

We may now perform another experiment with the same apparatus. This is to measure the intensity of the emitted light as a function of distance along the beam. When we do this we find that the intensity decreases exponentially as we move along the beam; as shown in Fig. 4.17.

We can explain this result by assuming that each atom in its excited state E_1 has a certain probability λ per unit time of decaying to the ground state. If the number of excited atoms at time t is N, then the number decaying between t and $t + \Delta t$ will be $\lambda N \Delta t$. Therefore the rate of change of N is given by

$$\frac{\mathrm{d}N}{\mathrm{d}t} = -\lambda N; \tag{4.32}$$

from this we get

$$\begin{aligned} N &= N_0\,\mathrm{e}^{-\lambda t} \\ &= N_0\,\mathrm{e}^{-t/\tau} \end{aligned} \tag{4.33}$$

where $\tau \equiv 1/\lambda$ has the dimensions of time, and is known as the 'lifetime' of the state.

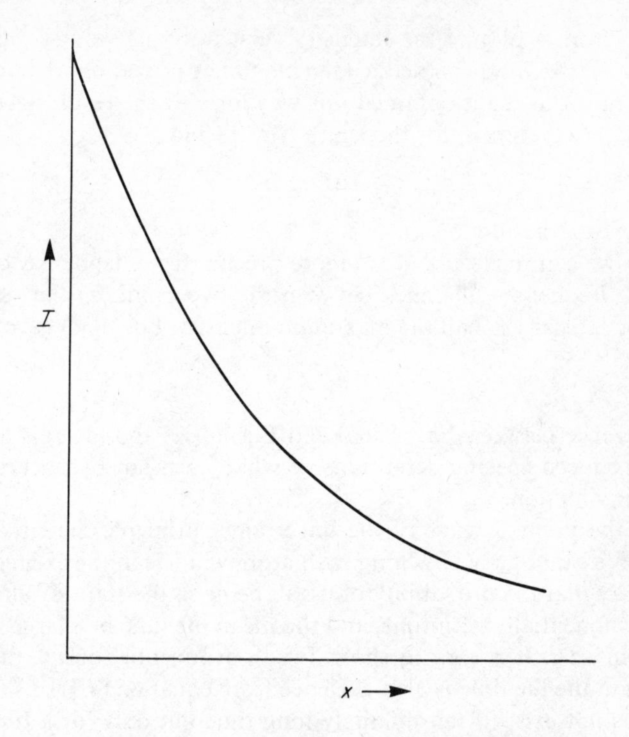

Fig. 4.17. Plot of the light intensity versus distance along the beam in the experiment illustrated in Fig. 4.16.

If we assume there are N_0 excited atoms per unit length of the atomic beam at the point $x = 0$, then, as it takes a time $t = x/v$ to arrive at the point x the number of excited atoms per unit length at x will be

$$N = N_0\,\mathrm{e}^{-x/v\tau}\ . \tag{4.34}$$

The intensity, I, seen by the spectrometer at the point x is proportional to the number λN decaying per second at that point. Therefore as

$$\lambda = 1/\tau$$

$$\begin{aligned} I &\propto N/\tau \\ &\propto \frac{N_0}{\tau}\,\mathrm{e}^{-x/v\tau}\ . \end{aligned} \tag{4.35}$$

Hence, from a plot of the intensity versus distance along the beam, as shown in Fig. 4.17, we can deduce the life time τ providing we know v. If we now multiply the result obtained this way for τ by the result obtained from the previous experiment for the width ΔE, we find

$$\tau \Delta E \approx h \ . \tag{4.36}$$

This is an experimental result.

In fact we can make Eq. (4.36) more precise. If we define ΔE by defining $\Delta \nu$ as the frequency difference between the two points in the spectral line where the intensity is half the maximum intensity, Eq. (4.36) becomes

$$\tau \Delta E = \hbar \ . \tag{4.36a}$$

The difference between Eq. (4.36a) and Eqs. (4.31) and (4.36) is because we have introduced specific definitions of what we mean by uncertainty in a particular situation.

From the point of view of the uncertainty principle the situation is as follows. We cannot say how long each atom will live in the excited state E_1. We can say that the probability of it still being in the state E_1 after a time t decays exponentially with time, and the mean life time of a large number of such atoms is τ. It is easy to show, for an exponential decay, that the uncertainty in the life time is also τ. Hence from equation (4.31), as the excited state does not exist for an infinitely long time but only for a time with an uncertainty τ, its energy must be uncertain to an amount

$$\Delta E = \hbar/\tau$$

corresponding to the experimentally measured width.

It is now clear why we ascribed the whole width to the spectral line to the level E_1, and none to the ground state E_0. Because E_0 is the lowest available state there is no lower level for it to decay to, and hence, under isolated conditions it will stay in that state indefinitely, and therefore has a precise value E_0, with no width.

★ 4.14 AN ANALYTICAL EXAMPLE OF THE UNCERTAINTY PRINCIPLE

In section 4.10 we considered, qualitatively, the situation where a beam of electrons, of well-defined momentum initially, was allowed to pass through a shutter which was open only for a brief moment of time. We now consider this case more quantitatively. We assume that the shutter is open for a definite time Δt so that only those electrons will get through the shutter which, at time $t = 0$, are within the region $\pm x_0$ (see Fig. 4.18(a)).

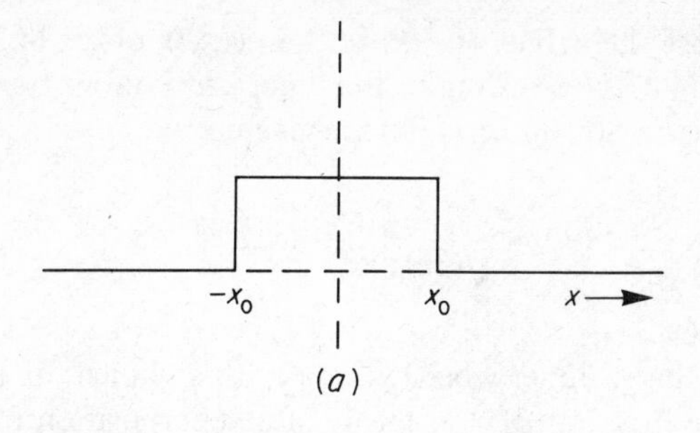

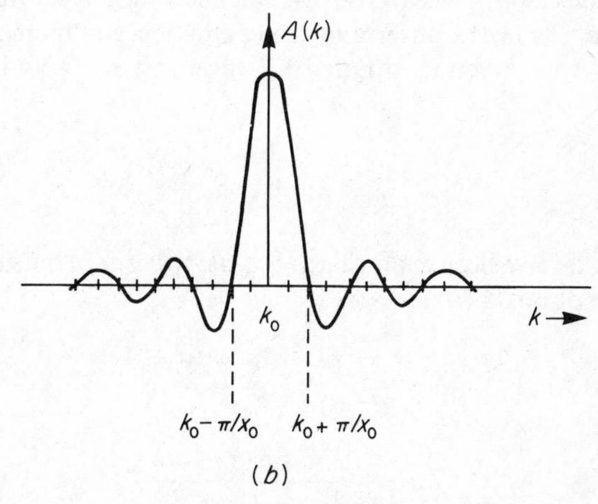

Fig. 4.18. (a) The envelope of the wavefunction at $t = 0$.
(b) The amplitude $A(k)$ as a function of k at $t = 0$.

Our wavefunction at $t = 0$ then, will be

$$\phi(x) = \begin{cases} \dfrac{1}{\sqrt{(2x_0)}} \mathrm{e}^{ik_0x} , & |x| \leqslant x_0 , \\ 0 , & |x| > x_0 . \end{cases} \tag{4.37}$$

(The factor of $1/\sqrt{(2x_0)}$ ensures $\int_{-\infty}^{\infty} \phi^* \phi \, \mathrm{d}x = 1$.)

According to Eq. (4.21) the most general solution to Eq. (4.16) is

$$\psi(x, t) = \frac{1}{\sqrt{(2\pi)}} \int_{-\infty}^{\infty} A(k) \, \mathrm{e}^{\mathrm{i}(kx - \omega t)} \, \mathrm{d}k \tag{4.21}$$

with arbitrary values of the coefficients $A(k)$. At $t = 0$ this becomes

$$\psi(x, 0) = \frac{1}{\sqrt{(2\pi)}} \int_{-\infty}^{\infty} A(k) \, \mathrm{e}^{ikx} \, \mathrm{d}k \ . \tag{4.38}$$

We wish to choose the values of $A(k)$ so that $\psi(x, 0)$ of Eq. (4.38) is identical with $\phi(x)$ of Eq. (4.37). According to the theory of Fourier transforms, given in Appendix A, if the $A(k)$ in Eq. (4.38) are given by

$$A(k) = \frac{1}{\sqrt{(2\pi)}} \int_{-\infty}^{\infty} \phi(x)\, \mathrm{e}^{-ikx}\, \mathrm{d}x \tag{4.39}$$

then $\psi(x, 0) \equiv \phi(x)$.

What we will have done, when we make this choice for the $A(k)$, is to express the wave function $\phi(x)$ in terms of a superposition of plane waves, with amplitudes $A(k)$ given by Eq. (4.39). The range of k over which $A(k)$ is appreciable tells us the uncertainty in k, and hence in momentum, associated with the wave function of Eq. (4.37). Inserting Eq. (4.37) into Eq. (4.39) gives us

$$A(k) = \frac{1}{2(\pi x_0)^{1/2}} \int_{-x_0}^{x_0} \mathrm{e}^{\mathrm{i}(k_0 - k)x}\, \mathrm{d}x \tag{4.40}$$

where the limits have been changed to $\pm x_0$ as $\phi(x)$ is zero outside this region.

Evaluation of Eq. (4.40) gives us

$$\begin{aligned} A(k) &= \frac{1}{2(\pi x_0)^{1/2}} \left[\frac{\mathrm{e}^{\mathrm{i}(k_0 - k)x}}{\mathrm{i}(k_0 - k)} \right]_{-x_0}^{x_0} \\ &= \frac{1}{(\pi x_0)^{1/2}} \frac{\sin (k_0 - k)x_0}{(k_0 - k)} . \end{aligned} \tag{4.41}$$

We sketch this function in Fig. (4.18(b)). We see that it has a maximum at $k = k_0$, and crosses the axis at $(k_0 - k) = \pm\pi/x_0$. We therefore take our uncertainty in k to be $2\pi/x_0$, and $\Delta p = \hbar\, \Delta k = 2\pi\hbar/x_0$. Therefore, with $\Delta x = 2x_0$, we get

$$\Delta p\, \Delta x = 4\pi\hbar .$$

PROBLEMS 4

4.1 The layers of atoms parallel to the surface of a single crystal of silver are spaced 2.0388 Å apart. Calculate the Bragg angles at which diffraction occurs for X-rays of wavelength 1.7892 Å. If this was an experiment to measure the wavelength of the X-rays, to what accuracy would you have to measure the angles to be sure of the significance of the last figure in the wavelength?

4.2 X-rays of unknown wavelength are incident on the same silver surface as in question 1 and diffraction is observed at angles to the surface of 25° 27′ 51″ and 40° 9′ 32″. What is the wavelength of the X-rays and at what other angles would you expect to observe diffraction maxima?

4.3 The angular resolution of a Bragg single crystal spectrometer is 18″ of arc and each of the arms can move over angles to the crystal surface from 0° to 60°. What is the shortest wavelength for which adjacent orders could be resolved if the crystal spacing is 2 Å? Assume the angular spacing between successive orders must be twice the angular resolution. What orders would one be resolving?

4.4 Calculate the maximum energy photons, electrons and neutrons for which adjacent orders could be separated in question 3.

4.5 Calculate the approximate wavelengths for electrons of energy 1 keV, 100 keV, 1 MeV and 10 MeV. (You should take the rest energy of an electron, mc^2, as 500 keV.)

4.6 Show the group velocity $v = \mathrm{d}\omega/\mathrm{d}k$ can also be expressed as $v = u - \lambda(\mathrm{d}u/\mathrm{d}\lambda)$ where u is the phase velocity.

4.7 The position of a nominally stationary electron is determined to 0.1 Å. Estimate the order of magnitude of the electron energy after the observation. (Express your answer in eV.)

4.8 A parallel beam of waves of wavelength λ pass through a slit of width a. Calculate the angular width of the beam due to diffraction after passing through the slit and compare this with the angular width given by your estimate from the uncertainty principle assuming $\lambda = h/p$.

4.9 A beam of electrons of energy 1 eV has its direction defined by two equal slits placed 1 m apart. What should be the width of the slits for the best definition of the beam direction?

4.10 Estimate the width of a level with a lifetime of

$$(a)\ 10^{-8}\,\mathrm{s} \quad (b)\ 10^{-16}\,\mathrm{s} \quad (c)\ 10^{-22}\,\mathrm{s}\ .$$

4.11 How accurately can you measure the position of an electron moving with a kinetic energy of 1 eV if you do not wish to change its energy by more than one part in ten thousand?

CHAPTER 5

Quantum mechanics and the Schrödinger equation

5.1 INTRODUCTION

In the first three chapters, and in the first half of Chapter 4, we have been setting the background for the development of quantum mechanics. We have found that the phenomena of atomic physics cannot all be described by the equations of classical physics. Classical physics gives us no idea why atoms are the size they are, or why they are dynamically stable. Further, classical physics prescribes a definite path for any given particle, once we know the interactions it suffers and its initial conditions. Yet in the phenomenon of electron diffraction we have encountered a situation where it appears impossible to do this, and that all one can talk about are the probabilities of various different final results. Thus we must look for a new equation to replace those of classical physics.

In the non-relativistic region, this new equation is the Schrödinger equation. This, indeed, does not describe the path of an electron, or any other particle, as classically defined, but only tells us about probabilities, and this is the most that one can know.

It is most important to emphasize that we *cannot deduce* the Schrödinger equation, we have to *postulate* it. In postulating any equation, of course, one is guided by experimental results, and what one considers to be reasonable assumptions. For example, in postulating Eq. (4.23) for the behaviour of free particles, we were guided by the experimental results on electron diffraction, which implied an equation possessing wavelike solutions together with the reasonable assumption that the group velocity should equal the particle velocity as classically defined. We then postulated the simplest equation satisfying these conditions. We could have been wrong. There are other, more complicated equations which satisfy these conditions, and it is not true that the group velocity always represents the average rate of propagation of energy; in light, for example, in regions of heavy absorption the group velocity $d\omega/dk$ can go negative.

The assumptions that have been employed, therefore, in postulating a particular equation may not be valid in circumstances other than those which one set out to explain in the first case. If one is lucky, this may merely mean that the postulated equation is a special case of a more general equation.

The fact that a particular equation or hypothesis explains the results it was devised to explain is not therefore greatly to its credit; Ptolemy's epicycles, for example, explained the apparent motions of the planets extremely well. If, however, one applies the postulated equation in other circumstances, and finds it makes accurate predictions one's faith in the correctness of the equation starts to grow. It usually occurs that when one starts to apply the new equation to wider fields additional postulates are necessary in order to make sense of the phenomena and make accurate predictions. Provided the additional phenomena explained are much greater in number than the new postulates, this is not a source of worry. On the contrary, it can mean that one acquires a sense of really understanding what is happening over a very wide field in terms of a very few ideas. This indeed is the situation in quantum mechanics. The Schrödinger equation, applied to the motion of a single particle, was able to explain electron diffraction, give general reasons why the bound states of atoms should be discrete, correctly predict the energy levels of the hydrogen atom, and explain the penetration of potential barriers as exemplified by alpha-particle decay. In order to explain the angular momentum properties of atoms and the energy levels of the alkali metal atoms it was necessary to introduce the concept of electron spin, and when many particle problems were tackled it was found necessary to introduce the Pauli exclusion principle in its quantum mechanical form.

The Schrödinger equation is a non-relativistic equation and can only be expected to give the right answers in situations where all velocities are a small fraction of the velocity of light. This covers almost the whole of extranuclear physics and most of nuclear physics, too. Within this field, the Schrödinger equation coupled with the concepts of intrinsic spin and the exclusion principle has been applied to an enormously wide range of phenomena, and no case of disagreement with predictions has occurred which cannot be traced to the influence of relativity. It is this great range of successes that convinces us of the correctness of the Schrödinger equation within the non-relativistic realm.

No new developments in physics are now likely to make us discard the Schrödinger equation, any more than the development of quantum mechanics has made us discard Newtonian mechanics for ordinary macroscopic phenomena. What it has done is define the limits of applicability of classical mechanics. In the same way one now has relativistic quantum field theories which at least go some way towards defining the limits of applicability of the Schrödinger equation.

5.2 THE SCHRÖDINGER EQUATION

We must now try to postulate the Schrödinger equation, guided by the principles enunciated in the last section. In Eq. (4.23) we postulated an equation for the motion of particles in field-free space which was specifically designed to account for the conclusions one can draw from the phenomenon of electron diffraction. We cannot really test whether this equation is right or not, as there are no other phenomena against which one can test it without introducing an interaction of some sort, and no provision is made for this in Eq. (4.23). Nevertheless, we assume Eq. (4.23) is correct for field-free space, and so our more general equation must reduce to Eq. (4.23) in the absence of any fields.

It is clear that some modification of Eq. (4.23) is required, otherwise, once an initial wavefunction for a particle had been prescribed, its subsequent development would be independent of the potential field in which it found itself. The simplest thing we can do to change this situation is to add a term depending linearly on the potential in which the particle finds itself. Thus Eq. (4.23) becomes

$$\boxed{-\frac{\hbar^2}{2m}\nabla^2\psi + V(\mathbf{r}, t)\psi = i\hbar\frac{\partial\psi}{\partial t}\ .} \tag{5.1}$$

This is the Schrödinger equation, and its real justification depends on the enormous success it has had when applied to an extremely wide range of problems, as mentioned in section 5.1. Note that we have not assumed the applied potential is independent of time, and it therefore covers situations in which the applied field varies as a function of time as well as position.

★ 5.3 SOME PLAUSIBILITY ARGUMENTS

We have repeatedly emphasized that it is not possible to derive the Schrödinger equation, but some readers may wish to have some plausible reasons for choosing the form given in Eq. (5.1) other than its great success in providing solutions over an enormous range of physical problems. That is, they would like some prior reason for choosing this form.

All such arguments essentially centre around reasons for postulating that, in the presence of a potential, the expression for ω given in Eq. (4.14) should be replaced by

$$\hbar\omega = E = T + V \tag{5.2}$$

where T and V are respectively the kinetic and potential energies of the particle. As ω is not an observable quantity, such arguments can only be ones of plausibility.

The start of the whole of the discussion on the wave-like properties of particles was based on the observation of the phenomenon of electron diffraction and its similarity to X-ray diffraction provided one ascribed a wavelength to the electron given by $\lambda = h/p$. One can pursue this similarity even further. If one makes a very careful study of the angles at which X-ray diffraction occurs one finds that the wavelength that must be inserted into the Bragg law

$$2d \sin\theta = n\lambda \tag{4.2}$$

is the wavelength in the medium, and not the free space wavelength, and that the angle θ is related to the angle of observation θ_0 by Snell's law of refraction. A precisely similar refinement is found to occur in electron diffraction. On entering a metal, for example, the electron acquires additional energy due to the work function of the metal. This changes the momentum and hence the wavelength of the electrons and at the same time alters its direction of travel. The analogy with the X-ray case is found to be exact if one assumes that the metal has a different refractive index from free space and that the ratio of the refractive index in the metal μ to the refractive index in free space μ_0 is given by

$$\frac{\mu}{\mu_0} = \left(\frac{E - eV}{E}\right)^{1/2}. \tag{5.3}$$

Here E is the energy of the electron when it is outside the metal, the charge of the electron is $-e$, and V is known as the inner potential and has a constant value independent of E for a given metal.

Now in the X-ray case, the frequency of the radiation is constant and not dependent on whether the X-rays are inside or outside the material. We therefore *assume* that the same is true of the electron wavefunction, although our results are in no way dependent on the assumption. We merely wish to introduce the potential in some way other than through its effect on the kinetic energy, and hence on the wave number k, and this seems the most obvious way of doing it. If we make this assumption, then the relations between k and ω given in Eq. (4.15) must be replaced by

$$\hbar\omega = \frac{\hbar^2 k^2}{2m} + V. \tag{5.4}$$

The Schrödinger equation (5.1) is now the simplest differential equation which has plane wave solutions in regions where V is constant and in which ω and k satisfy Eq. (5.4). It is now assumed that this equation is still the correct one even though V may be a rapidly varying function of space and time.

5.4 THE TIME-INDEPENDENT SCHRÖDINGER EQUATION

The Schrödinger equation (5.1) is a very general equation applicable in all circumstances. Given an initial wavefunction it determines the time development of the wavefunction, and it is thus equivalent to the classical dynamical equations of motion. In the classical mechanics of a particle in a given field of force (which may be time-dependent) one starts with the position and momentum of the particle at time $t = 0$, say, and using Newton's laws of motion (maybe in a sophisticated form) one then calculates the position and momentum of the particle as functions of time. In quantum mechanics one cannot know simultaneously the position and momentum, and thus the initial conditions are less precise. They consist of the initial wave function, which gives us the initial probability distribution and which also contains, through its Fourier transform, the initial momentum distribution, as explained in Appendix A. Eq. (5.1) allows us to predict the wavefunction at later times, and so enables us to determine both the position and momentum probability distributions at later times.

In many problems in classical physics one is not interested explicitly in the time development from some initial conditions, but in the general nature of the path followed by the particle. One is interested to know, for example, that a particle acting under an attractive inverse square law of force moves in an ellipse with the centre of the force at one focus; one is further interested in how the shape of the ellipse depends on the energy of the particle and its angular momentum. There are many similar such problems. In these the potential is usually not a function of time, and the particle has a definite total energy. We would like to see if we can cast Eq. (5.1) in such a form that it brings out the corresponding quantum mechanical features.

Consider a particle moving in the potential of Fig. 5.1. Classically it would have a definite energy E which would be prescribed by the initial conditions, and, as a result its motion between the points A and B, where the potential is constant, would be at a constant velocity, and hence at a constant momentum, given by

$$p^2 = 2m(E - V_0)\ . \tag{5.5}$$

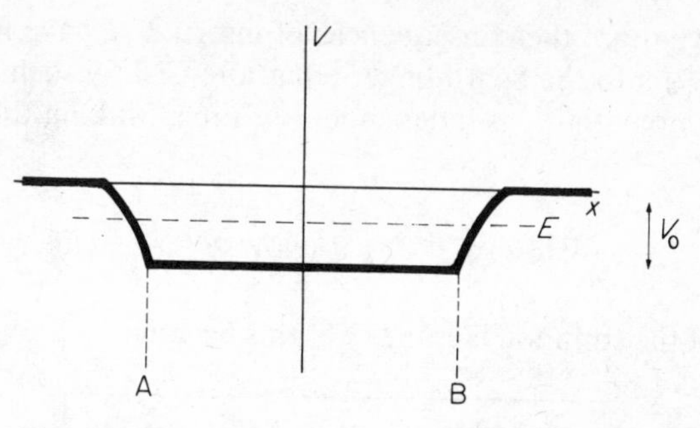

Fig. 5.1. A particle of total energy E moves in the potential shown. Classically between A and B it would have a constant velocity.

In quantum mechanics the particle need not have a definite energy. The initial conditions, for example, might prescribe a wavepacket centred closely around the origin, with a correspondingly large spread in initial momenta. Let us assume, however, that the particle does have a definite energy E. Then one would expect that the wavefunction, in the region from A to B, would have the form

$$\psi(\mathrm{A} \to \mathrm{B}) = A(\mathrm{e}^{\mathrm{i}(kx - \omega t)} + \mathrm{e}^{\mathrm{i}(-kx - \omega t)}) \tag{5.6}$$

where

$$\hbar^2 k^2 = 2m(E - V_0) \ .$$

The first term in Eq. (5.6) represents the particle moving from left to right, and the second term moving from right to left. If we insert (5.6) into Eq. (5.1) we get

$$\frac{\hbar^2 k^2}{2m} + V_0 = \hbar\omega$$

i.e.

$$\hbar\omega = E \ . \tag{5.2}$$

Thus Eq. (5.1) implies that the frequency ω is to be given by Eq. (5.2) in a region of constant potential. We now assume that this is always the case, i.e. that if we have a state of definite energy E it can be written in the form

$$\Psi(\mathbf{r}, t) = \psi(\mathbf{r})\,\mathrm{e}^{-\mathrm{i}Et/t} \tag{5.7}$$

where $\psi(\mathbf{r})$ is independent of time.

Such states are called *stationary states*, as the probability density $\Psi^*\Psi$ is independent of time. Once we have a particle in such a state it will stay in that state, unless and until it is perturbed by some outside influence.

Let us see what are the consequences of inserting a wave function of the form of Eq. (5.7) into the Schrödinger equation. Clearly such a form is only possible if the potential V is independent of time. Making this substitution we get

$$-\frac{\hbar^2}{2m}\nabla^2\psi(\mathbf{r})\,\mathrm{e}^{-\mathrm{i}Et/\hbar} + V(\mathbf{r})\psi(\mathbf{r})\,\mathrm{e}^{-\mathrm{i}Et/\hbar} = E\psi(\mathbf{r})\,\mathrm{e}^{-\mathrm{i}Et/\hbar} .$$

Cancelling out the common factor $\mathrm{e}^{-\mathrm{i}Et/\hbar}$ this becomes

$$\boxed{-\frac{\hbar^2}{2m}\nabla^2\psi(\mathbf{r}) + V(\mathbf{r})\psi(\mathbf{r}) = E\psi(\mathbf{r}) .} \tag{5.8}$$

This is the time-independent Schrödinger equation, the solutions of which are stationary states possessing a definite energy.

★ 5.5 AN ALTERNATIVE APPROACH TO THE TIME-INDEPENDENT EQUATION

In this section we attempt to solve Eq. (5.1) by looking for solutions which are separable functions of $\mathbf{r}$ and t, so that $\Psi(\mathbf{r}, t) = \psi(\mathbf{r})\phi(t)$. This will only be possible if the potential V is not a function of both $\mathbf{r}$ and t. We assume therefore that it is a function of $\mathbf{r}$ only.

Assuming therefore that $\Psi(\mathbf{r}, t) = \psi(\mathbf{r})\phi(t)$ Eq. (5.1) becomes

$$-\frac{\hbar^2}{2m}\phi(t)\nabla^2\psi(\mathbf{r}) + V(\mathbf{r})\psi(\mathbf{r})\phi(t) = \mathrm{i}\hbar\psi(\mathbf{r})\frac{\partial\phi(t)}{\partial t} . \tag{5.9}$$

Dividing through by $\psi(\mathbf{r})\phi(t)$ we get

$$-\frac{\hbar^2}{2m}\frac{\nabla^2\psi(\mathbf{r})}{\psi(\mathbf{r})} + V(\mathbf{r}) = \mathrm{i}\hbar\frac{1}{\phi(t)}\frac{\partial\phi(t)}{\partial t} . \tag{5.10}$$

The left-hand side of (5.10) is a function of $\mathbf{r}$ only, while the right-hand is a function of t only. They must therefore both be equal to a constant E which, from the equation, must have the dimensions of energy.

Thus Eq. (5.10) becomes

$$\mathrm{i}\hbar\frac{1}{\phi(t)}\frac{\partial\phi(t)}{\partial t} = E \tag{5.10a}$$

$$-\frac{\hbar^2}{2m}\frac{\nabla^2\psi(\mathbf{r})}{\psi(\mathbf{r})} + V(\mathbf{r}) = E . \tag{5.10b}$$

Integration of (5.10a) gives us

$$\phi(t) = C\,\mathrm{e}^{-\mathrm{i}Et/\hbar}$$

where C is a constant of integration, while Eq. (5.10b) is just (5.8) rearranged.

5.6 BOUNDARY CONDITIONS AND THE SOLUTIONS OF THE TIME-INDEPENDENT SCHRÖDINGER EQUATION

The possible solutions of Eq. (5.8) are determined not only by the potential $V(\mathbf{r})$ but also by the boundary conditions that are imposed by the nature of the problem. The most important result that comes from consideration of these boundary conditions occurs in situations where classically a particle would be confined to a specific volume in space, i.e. it would be bound, and not free. In these circumstances we find that, contrary to the classical case, *only certain discrete energies are allowed.*

This is a very important result, as it accounts for the experimental observations, discussed in section 3.9, that atoms possess only certain discrete energy levels.

We shall illustrate the influence of the boundary conditions on the solutions of Eq. (5.8) by taking a specific example, and we shall then generalize the results so obtained. The example we choose is that of a particle moving in the one-dimensional potential shown in Fig. 5.2. For this we require the

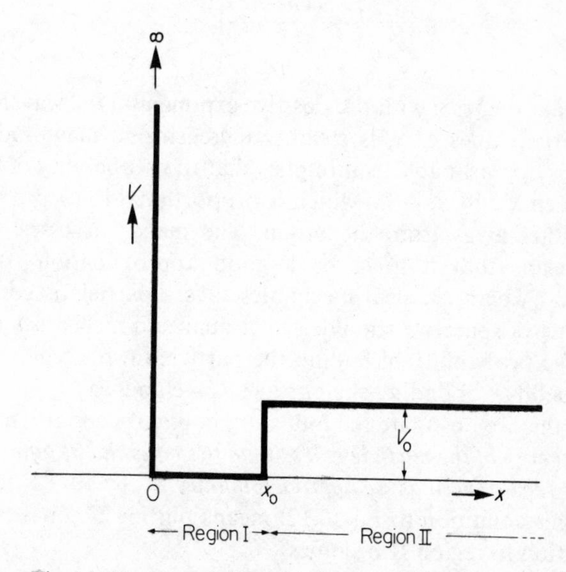

Fig. 5.2. The potential for which Schrödinger's equation is solved in the text.

one-dimensional version of Eq. (5.8) namely,

$$-\frac{\hbar^2}{2m}\frac{d^2\psi}{dx^2} + V(x)\psi = E\psi \ . \tag{5.8a}$$

Classically there are two distinct energy ranges. The first range is when the energy E is less than V_0, when the particle would be confined to the region between $x = 0$ and $x = x_0$, and the second range is for E greater than V_0 when the particle could be anywhere to the right of the origin. We start by considering the solution of Eq. (5.8a) for E less than V_0 when the particle is classically bound. Consider first the solution of Eq. (5.8a) in that part of space labelled region II in Fig. 5.2, i.e. $x \geqslant x_0$. In this region Eq. (5.8a) becomes

$$-\frac{\hbar^2}{2m}\frac{d^2\psi}{dx^2} + (V_0 - E)\psi = 0, \qquad (x \geqslant x_0) \ . \tag{5.11}$$

The general solution of this equation is

$$\psi_{II} = C\,e^{\gamma x} + D\,e^{-\gamma x} \tag{5.12}$$

where C and D are arbitrary constants and

$$\gamma = \left(\frac{2m(V_0 - E)}{\hbar^2}\right)^{1/2} .$$

Now if we retain the term with the positive exponential the wavefunction, at sufficiently large values of x, is steadily increasing in magnitude as x increases. This is unreasonable as it implies that the probability of finding the particle between x and $x + dx$, which is proportional to $|\psi_{II}|^2\,dx$, increases as we go further away from the origin. The many successes of classical mechanics means that it must be a good approximation to quantum mechanics, and where classical mechanics says a particle is confined to a specific volume of space we should expect quantum mechanics to say that there is a large probability of finding the particle in this region, and a decreasing probability of finding the particle as we go away from this region. To deal with this we postulate the following boundary condition.

When the energy of the particle is less than the potential at infinity the wave function must tend to zero as we approach infinity.

Applying this condition to Eq. (5.12) means putting C equal to zero, and the wavefunction in region II becomes

$$\psi_{II} = D\,e^{-\gamma x}, \qquad (x \geqslant x_0) \ . \tag{5.12a}$$

Now let us consider the solutions in the part of space labelled region I, i.e. $0 \leqslant x \leqslant x_0$. Here Eq. (5.8a) becomes

$$-\frac{\hbar^2}{2m}\frac{d^2\psi}{dx^2} - E\psi = 0, \qquad (0 \leqslant x \leqslant x_0) \tag{5.13}$$

whose general solution is

$$\psi_I = A \sin kx + B \cos kx \tag{5.14}$$

where A and B are arbitrary constants and $k = (2mE/\hbar^2)^{1/2}$.

We must now consider the conditions the wavefunction must satisfy as we cross the boundaries at $x = 0$ and $x = x_0$. The relative probability $|\psi|^2\,dx$ of finding the particle near to the point x clearly cannot depend on how we approach the point x. To ensure this we postulate as follows: *the wavefunction ψ must be a continuous single-valued function of position.*

To the left of the origin the potential is everywhere infinite. This means the wavefunction must be everywhere zero as can be seen by putting $V(x)$ equal to infinity in Eq. (5.8a). Hence it follows that $\psi_I(x)$ must be zero at $x = 0$, and to achieve this the constant B in Eq. (5.14) must be put equal to zero. Eq. (5.14) therefore becomes

$$\psi_I = A \sin kx, \qquad (0 \leqslant x \leqslant x_0)\ . \tag{5.14a}$$

The continuity of the wavefunction through the point $x = x_0$ means we must have

$$A \sin kx_0 = D\,e^{-\gamma x_0}\ . \tag{5.15}$$

However this condition, though necessary, is not sufficient. Not only must the wavefunction be continuous, but so must its first derivative, as can be seen as follows. Eq. (5.8a) together with the continuity of ψ implies that wherever there is a discontinuity in the potential $V(x)$ there is a corresponding discontinuity in the second derivative $d^2\psi/dx^2$. Specifically, in this case

$$\lim_{\delta x \to 0}\left[\left(\frac{d^2\psi}{dx^2}\right)_{x+\delta x} - \left(\frac{d^2\psi}{dx^2}\right)_{x-\delta x}\right] = \frac{2m}{\hbar^2}V_0\psi(x_0)\ .$$

We show such a discontinuity in Fig. 5.3(a). If we integrate such a function to obtain the first derivative $d\psi/dx$ the effect of the step at $x = x_0$ in $d^2\psi/dx^2$ will be merely to produce a kink in $d\psi/dx$, as shown in Fig. 5.3(b); so $d\psi/dx$ must also be continuous, as well as ψ. The only occasion when $d\psi/dx$ need not be continuous is when the potential step is infinite, as occurs at the origin. Thus, in addition to Eq. (5.15) at $x = x_0$ we also have

$$Ak \cos kx_0 = -D\gamma\,e^{-\gamma x_0}\ . \tag{5.16}$$

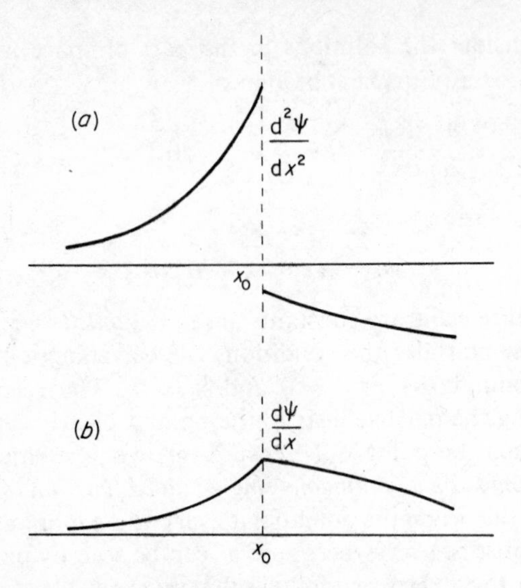

Fig. 5.3. A discontinuity in a potential at x_0 produces a discontinuity in $d^2\psi/dx^2$ at x_0 shown in (a), but only a kink in $d\psi/dx$ as in (b).

Dividing Eq. (5.15) by Eq. (5.16) we get

$$\tan kx_0 = -k/\gamma.$$

Inserting the expressions for k and γ, namely

$$k = \left(\frac{2mE}{\hbar^2}\right)^{1/2}$$

and

$$\gamma = \left(\frac{2m(V_0 - E)}{\hbar^2}\right)^{1/2},$$

we get

$$\tan\left[\left(\frac{2mE}{\hbar^2}\right)^{1/2} x_0\right] = -\left(\frac{E}{V_0 - E}\right)^{1/2} \tag{5.17a}$$

which can only be satisfied for certain values of E. The graphical solution of this equation for the case when

$$\left(\frac{2mV_0}{\hbar^2}\right)^{1/2} x_0 = 20$$

is shown in Fig. 5.4.

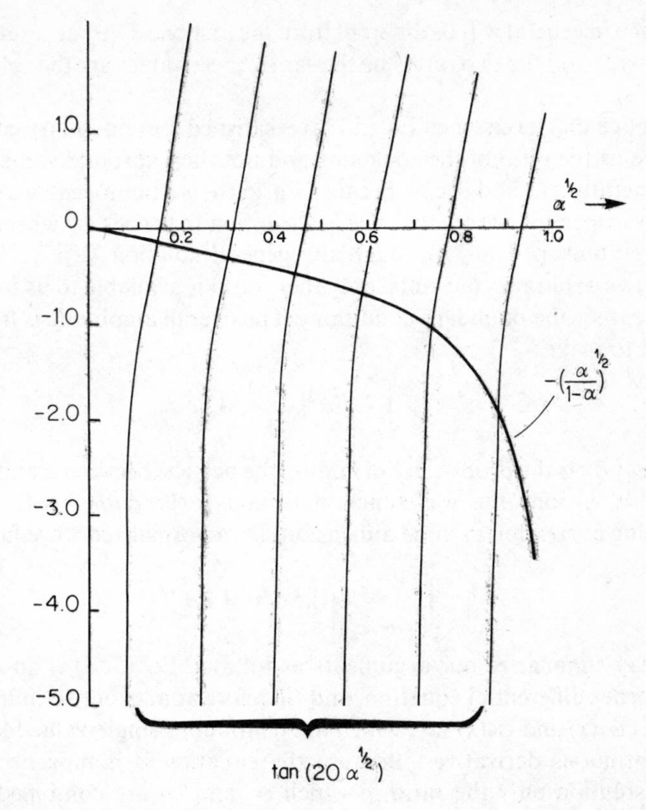

Fig. 5.4. Graphical solution of

$$\tan\left[\left(\frac{2mE}{\hbar^2}\right)^{1/2} x_0\right] = -\left(\frac{E}{V_0 - E}\right)^{1/2}$$

for

$$\left(\frac{2mV_0}{\hbar^2}\right)^{1/2} x_0 = 20 \ .$$

We put $E = \alpha V_0$ when the equation becomes

$$\tan(20\alpha^{1/2}) = -\left(\frac{\alpha}{1-\alpha}\right)^{1/2} .$$

Both sides are plotted against $\alpha^{1/2}$ and the values of $\alpha^{1/2}$ at the intersections give us the solutions for E.

Let us see how Eq. (5.17a) has arisen. We note that the matching of ψ_{I} and ψ_{II} at $x = x_0$, Eq. (5.15), gives us a value for the ratio A/D of the two arbitrary constants, this value being a function of the energy. Similarly the matching of the derivatives according to Eq. (5.16) gives us another value for the ratio

A/D which in general will be different from the first one. Only at certain values of the energy will the two ratios be the same; these values are the solutions of Eq. (5.17a).

We notice that, even when Eq. (5.17a) is satisfied, our boundary conditions only give us the ratio of the constants and not their absolute values. This is quite generally true and occurs because Eq. (5.8a) is a homogeneous equation for ψ. By inspection we see that if ψ is a solution then so is $\alpha\psi$ where α is any arbitrary number. Thus, although the general solution to Eq. (5.8a) will contain two arbitrary constants, only their ratio is available to us for adjustment to satisfy the boundary conditions. The overall amplitude is frequently adjusted to make

$$\int_{-\infty}^{+\infty} |\psi^2| \, dx = 1 \tag{5.18}$$

so that $|\psi|^2 \, dx$ is the probability of finding the particle between x and $x + dx$. When this is done the wavefunction is said to be *normalized*. The corresponding expression in three dimensions for a normalized wave function is

$$\int_{\text{all space}} |\psi(\mathbf{r})|^2 \, d^3\mathbf{r} = 1 \, . \tag{5.18a}$$

We may summarize our arguments as follows. Eq. (5.8a) is an ordinary second order differential equation, and therefore admits of two independent solutions $\phi_1(x)$ and $\phi_2(x)$ say, which are continuous single-valued functions with continuous derivatives. Because the equation is homogeneous, in a general solution only the ratio in which ϕ_1 and ϕ_2 are combined matter, that is

$$\phi(x) = A[\phi_1(x) + \alpha\phi_2(x)] \, .$$

If the energy is less than the potential at both plus infinity and minus infinity $\phi(x)$ must go to zero at plus infinity and minus infinity. With only α available for adjustment this is not normally possible except at certain values of the energy. The bound state energy spectrum is therefore discrete. These conclusions are quite general and independent of the particular form of the potential. The extension to three dimensions is fairly obvious and will be taken up where appropriate later in the text.

We now return to the case of Fig. 5.2 and briefly discuss the situation when the energy E is greater than V_0. In this case the solution for region II is

$$\psi_{\text{II}} = C \sin kx + D \cos kx, \qquad (x \geqslant x_0) \, , \tag{5.19}$$

where $k = (2m(E - V_0)/\hbar^2)^{1/2}$. Both terms in Eq. (5.19) remain finite as x tends to infinity and there is no objection to finding the particle at infinity. Both terms in Eq. (5.19) may therefore be retained and the ratio C/D is available for adjustment to satisfy the matching conditions at $x = x_0$.

A solution is always possible and the energy spectrum is continuous.

We end this section by restating the conditions that must be satisfied by acceptable solutions of Schröndinger's equation. They are

(i) The wave function must be a single-valued, continuous function of position with continuous derivatives.

(ii) If the energy is less than the potential at infinity the wavefunction must tend to zero as we approach infinity.

5.7 EIGENVALUES AND EIGENFUNCTIONS

We saw in the last section that the combination of the time-independent Schrödinger equation

$$\left(-\frac{\hbar^2}{2m}\nabla^2 + V\right)\psi = E\psi \tag{5.8}$$

together with the boundary conditions appropriate to the problem does not necessarily have solutions for all values of the energy E; that is, we had to treat E as a parameter. When such a situation occurs in mathematics the values of the parameter for which solutions occur are known as the *eigenvalues* of the problem, and the solutions are known as *eigenfunctions.* This is terminology; but it has been found very useful to have such a terminology for denoting the values of E and the functions which are allowed. Also the concept of an eigenvalue and a corresponding eigenfunction is not confined to this one case and is extremely important in other quantum mechanical situations, as we shall see.

We may regard the factor in brackets in Eq. (5.8) as an *operator* H, operating on the wave function ψ so that we have

$$H\psi = E\psi\ , \tag{5.20}$$

where

$$H = -\frac{\hbar^2}{2m}\nabla^2 + V$$

is known as the Hamiltonian operator. It receives this name because it is clearly the quantum mechanical analogue of the sum of the kinetic energy T and the potential energy V. In classical mechanics the quantity $H = T + V$ (the total energy) is called the Hamiltonian, and the same name has been taken over in quantum mechanics for

$$H = -\frac{\hbar^2}{2m}\nabla^2 + V\ .$$

Once we regard H as an operator, it is clear we can operate with it on any arbitrary function we like to choose. Normally the effect of this will be to produce a new function. Only if the function is one of the eigenfunctions of H will we get the same function multiplied by an energy E. Thus we can regard our problem as one of finding the eigenvalues and eigenfunctions of the operator H. The values of E which satisfy Eq. (5.20), including the boundary conditions, are known as the eigenvalues of the operator H for the problem, and the corresponding functions are said to be eigenfunctions of H, *belonging* to the eigenvalue E (it is possible, as we shall see, for an eigenvalue to have more than one eigenfunction).

To summarize the situation in these terms we may say that in quantum mechanics the energy is represented by an operator H, where

$$H = -\frac{\hbar^2}{2m}\nabla^2 + V\,. \tag{5.21}$$

For stationary states, which always have a definite energy, the effect of operating with H on the wave function for that state is to produce the same wave function multiplied by the energy of the state. The problem is, of course, to find the functions for which this occurs, i.e. the eigenfunctions of H, which means solving the differential equation.

5.8 ORTHOGONALITY AND COMPLETENESS OF THE EIGENFUNCTIONS OF H

The eigenfunctions of H possess two extremely important properties; they are *orthogonal*, and they form a *complete set* of functions.

Two functions are said to be orthogonal if the integral over all space of the product of one of them times the complex conjugate of the other, has the value zero; that is

$$\int_{\text{all space}} \psi_1^*(\mathbf{r})\psi_2(\mathbf{r})\,\mathrm{d}^3\mathbf{r} = 0\,. \tag{5.22}$$

(In future integrals involving $\mathrm{d}^3\mathbf{r}$ over all space will have no limits attached to the integral sign.)

It can readily be shown that this property holds for eigenfunctions of H belonging to *different* eigenvalues E_1 and E_2 say. If we have two eigenfunctions ϕ_1 and ϕ_2 which both belong to the *same* eigenvalue E, then $a_1\phi_1 + a_2\phi_2$ is also an eigenfunction of H belonging to the eigenvalue E, for *any* choice of the two constants a_1 and a_2. We may use this freedom of choice to construct orthogonal eigenfunctions if ϕ_1 and ϕ_2 are not already orthogonal. Such a procedure can be extended to any number of linearly independent eigenfunctions which all belong to the same eigenvalue. When we have more than one solution corresponding to a given eigenvalue, then we say *degeneracy* occurs, and the different eigenfunctions are said to be *degenerate*. These aspects will be dealt with further in section 5.9.

Eigenfunctions which are *orthogonal* and *normalized* are said to be *orthonormal.*

Equation (5.22) expresses the most general case of functions that are orthogonal over all space. Clearly we can have more limited functions, e.g. functions of, say, x only, which would arise in one-dimensional problems. In that case the orthogonality condition would be

$$\int_{-\infty}^{+\infty} \psi_1^*(x)\psi_2(x)\,\mathrm{d}x = 0\ . \tag{5.22a}$$

In such cases we say the functions form an orthogonal set in a certain *domain.* The domain can be a region of space, or of one or more of the coordinates, or a combination of these. The sine and cosing functions ϕ_p and ϕ_q given by

$$\left.\begin{aligned} \phi_p &= \sin px \\ \phi_q &= \cos qx \end{aligned}\right\}, \tag{5.23}$$

where p and q are integers going from nought to infinity, form an orthogonal set of functions in the domain $0 \leqslant x \leqslant 2\pi$.

A set of functions is said to be complete if it is possible to expand any arbitrary function in terms of members of the set. Suppose we label the members by $\phi_1, \phi_2, \ldots \phi_n, \ldots$ then, if they are complete, it means any arbitrary function Φ can be expressed as

$$\Phi = \sum_n a_n \phi_n, \tag{5.24}$$

where $a_1, a_2, \ldots a_n, \ldots$ are constants.

Once again the functions ϕ_n may refer to a certain domain, and Φ must then refer only to that domain. The functions of Eq. (5.23), for example, are complete in the domain $0 \leqslant x \leqslant 2\pi$, as well as being orthogonal. (They are, however, not normalized.) They are, of course, the functions used when we expand an arbitrary function, in the range from 0 to 2π, in a *Fourier series.*

If the functions which form a complete set are also orthonormal, then it is easy, in principle, to find the coefficients of the expansion expressed by Eq. (5.24). Let us assume that the functions in Eq. (5.24) have all space as their domain. We multiply both sides by ϕ_p^* and integrate over all space; we then get

$$\begin{aligned} \int \phi_p^*(\mathbf{r})\Phi(\mathbf{r})\,\mathrm{d}^3\mathbf{r} &= \int \phi_p^*(\mathbf{r}) \sum_n a_n \phi_n(\mathbf{r})\,\mathrm{d}^3\mathbf{r} \\ &= \sum_n a_n \int \phi_p^*(\mathbf{r})\phi_n(\mathbf{r})\,\mathrm{d}^3\mathbf{r} \\ &= a_p. \end{aligned} \tag{5.25}$$

The last step follows because the integrals are zero if n is not equal to p, due to the orthogonality property, and the integral with n equal to p is one because the functions are normalized.

A set of functions which are complete, orthogonal and normalized, are said to form a *complete orthonormal set.*

It is usually very difficult to demonstrate the property of completeness, and hence this property is usually just assumed. We assume it for eigenfunctions of the Hamiltonian operator H.

★ 5.9 THE ORTHOGONALITY OF EIGENFUNCTIONS OF H BELONGING TO DIFFERENT EIGENVALUES

We may demonstrate the orthogonality of the wavefunctions belonging to two different *bound* states, with energies E_1 and E_2, as follows. Let ψ_1 be an eigenfunction belonging to the value E_1, and ψ_2 an eigenfunction belonging to the value E_2. Thus the time-independent Schröndinger equations for these two functions are

$$-\frac{\hbar^2}{2m}\nabla^2\psi_1 + V\psi_1 = E_1\psi_1 \tag{5.26a}$$

and

$$-\frac{\hbar^2}{2m}\nabla^2\psi_2 + V\psi_2 = E_2\psi_2 \ . \tag{5.26b}$$

As the eigenvalues E_1 and E_2 are the energies of the states involved, they must be real numbers. This is proved following Eq. (5.29).

We multiply Eq. (5.26a) on the *left* by ψ_2^* and the complex conjugate of (5.26b) on the left by ψ_1 and subtract. Multiplication on the left means that the operator ∇^2 does not act on the function in this position. The term in V vanishes, and we get, remembering E_1 and E_2 are real,

$$-\frac{\hbar^2}{2m}(\psi_2^*\nabla^2\psi_1 - \psi_1\nabla^2\psi_2^*) = (E_1 - E_2)\psi_2^*\psi_1 \ . \tag{5.27}$$

The left hand side of Eq. (5.27) is

$$-\frac{\hbar^2}{2m}\nabla\cdot(\psi_2^*\nabla\psi_1 - \psi_1\nabla\psi_2^*) \ .$$

Now Gauss' theorem states that the integral of the normal component of any vector $\mathbf{F}$ over a closed surface S is equal to the integral of the divergence of that vector throughout the volume enclosed by the surface

$$\int_S \mathbf{F}\cdot \mathrm{d}\mathbf{s} = \int_v \nabla\cdot\mathbf{F}\,\mathrm{d}^3\mathbf{r}$$

where v is the volume enclosed by the surface S. We apply Gauss's theorem to Eq. (5.27) to obtain

$$-\frac{\hbar^2}{2m}\int_S (\psi_2^*\nabla\psi_1 - \psi_1\nabla\psi_2^*)\cdot \mathrm{d}\mathbf{S} = (E_1 - E_2)\int_v \psi_2^*\psi_1\,\mathrm{d}^3\mathbf{r}\ . \quad (5.28)$$

We now let the surface S go to infinity, when the volume v will become all space. As ψ_1 and ψ_2 refer to bound states, according to section 5.6 they go to zero at infinity. It can be shown that they go to zero sufficiently rapidly for the integral over the surface S also to go to zero as $\mathbf{r}$ tends to infinity. Hence the expression on the right hand side of Eq. (5.28) goes to zero as v expands to cover all space. By hypothesis $E_1 \neq E_2$, therefore

$$\int_{\text{all space}} \psi_2^*\psi_1\,\mathrm{d}^3\mathbf{r} = 0\ . \quad (5.29)$$

It should be noted that if we had not assumed that the energies E_1 and E_2 were real we would have had $(E_1 - E_2^*)$ on the right hand side of Eq. (5.28). If we now take ψ_2 to be the same function as ψ_1 so that E_2 equals E_1, the left hand side of Eq. (5.28) still vanishes as S expands to infinity. The integral on the right hand side of Eq. (5.28) is now certainly *not* zero, and, in fact, if ψ_1 is normalized it has the value unity, so we have proved that $(E_1 - E_1^*)$ equals zero, i.e. that E_1 is real.

The situation when we have degeneracy may be treated as follows. Suppose we have two solutions ϕ_1 and ϕ_2, belonging to the same eigenvalue E_1, which we will assume to be normalized, but not orthogonal. As ϕ_1 and ϕ_2 are degenerate, and belong to the same eigenvalue, then so does $a_1\phi_1 + a_2\phi_2$ for any choice of the numbers a_1 and a_2. This extremely wide choice makes it possible for us to find two orthonormal combinations ψ_1 and ψ_2 in an infinite number of ways. The problem is exactly analogous to that of constructing two perpendicular unit vectors from two given non-collinear unit vectors. One choice we may make is as follows

$$\psi_1 = \phi_1$$

$$\psi_2 = a_1\phi_1 + a_2\phi_2$$

and we evaluate a_1 and a_2 to satisfy the conditions

$$\int \psi_1^*\psi_2\,\mathrm{d}^3\mathbf{r} = 0$$

$$\int \psi_2^*\psi_2\,\mathrm{d}^3\mathbf{r} = 1\ .$$

Let

$$\int \phi_1^*\phi_2\,\mathrm{d}^3\mathbf{r} = p\ .$$

Then

$$\int \psi_1^* \psi_2 \, d^3\mathbf{r} = \int \phi_1^*(a_1\phi_1 + a_2\phi_2) \, d^3\mathbf{r}$$
$$= a_1 + pa_2$$

remembering that ϕ_1 and ϕ_2 are both normalized. Hence ψ_2 will be orthogonal to ψ_1 if the ratio $a_1/a_2 = -p$.

Therefore we write ψ_2 as

$$\psi_2 = a_2(\phi_2 - p\phi_1) \ .$$

The normalization condition now becomes

$$\begin{aligned} 1 &= \int a_2^*(\phi_2^* - p^*\phi_1^*)a_2(\phi_2 - p\phi_1) \, d^3\mathbf{r} \\ &= a_2^* a_2 \left[\int \phi_2^* \phi_2 \, d^3\mathbf{r} + \int p^* p \phi_1^* \phi_1 \, d^3\mathbf{r} \right. \\ &\quad \left. - p^* \int \phi_1^* \phi_2 \, d^3\mathbf{r} - p \int \phi_2^* \phi_1 \, d^3\mathbf{r} \right] \\ &= a_2^* a_2 [1 + p^* p - 2p^* p] \\ &= a_2^* a_2 [1 - p^* p] \ . \end{aligned}$$

Therefore

$$\psi_2 = \frac{1}{\sqrt{1 - p^* p}} [\phi_2 - p\phi_1]$$

is orthogonal to ψ_1 ($\equiv \phi_1$) and normalized.

5.10 OPERATORS, EXPECTATION VALUES AND EIGENVALUES

In section 5.7 we discussed the idea of treating Schrödinger's time-independent equation as an operator equation

$$H\psi = E\psi \tag{5.20}$$

in which one had to determine the eigenvalues E and the eigenfunctions ψ. Section 5.8 was devoted to the important ideas of orthogonality and completeness needed in the further development of the subject.

Suppose ψ_1 is a normalized eigenfunction of H belonging to the value E_1, and we insert this function in Eq. (5.20), multiply on the left by ψ_1^* and integrate over all space. Eq. (5.20) then becomes

$$\int \psi_1^* H \psi_1 \, d^3\mathbf{r} = E_1 \int \psi_1^* \psi_1 \, d^3\mathbf{r} = E_1 \ . \tag{5.30}$$

Thus if ψ_1 is a normalized eigenfunction of H the left-hand side of Eq. (5.30) gives us the energy of the state E_1.

Suppose, now, we do the same thing with some arbitrary function ϕ which is *not* an eigenfunction of H, that is, we evaluate the integral

$$\langle H \rangle \equiv \int \phi^* H \phi \, \mathrm{d}^3\mathbf{r} \, . \tag{5.31}$$

Can we attach any meaning to the value of $\langle H \rangle$? Let us see what happens if we take a particular case. Let ϕ be given by

$$\phi = a\psi_1 + b\psi_2 \tag{5.32}$$

where ψ_1 and ψ_2 are normalized eigenfunctions of H belonging to the eigenvalues E_1 and E_2 respectively, and let $a^*a + b^*b = 1$ so that ϕ is normalized.

Intuitively one would say that this represents a condition where the system has a probability a^*a of having energy E_1, and a probability b^*b of having an energy E_2. The *average* measured energy would be

$$a^*aE_1 + b^*bE_2 \, . \tag{5.33}$$

If we insert Eq. (5.32) into Eq. (5.31) we get

$$\begin{aligned}
\langle H \rangle &\equiv \int \phi^* H \phi \, \mathrm{d}^3\mathbf{r} \\
&\int (a^*\psi_1^* + b^*\psi_2^*) H (a\psi_1 + b\psi_2) \, \mathrm{d}^3\mathbf{r} \\
&= \int (a^*\psi_1^* + b^*\psi_2^*)(aE_1\psi_1 + bE_2\psi_2) \, \mathrm{d}^3\mathbf{r} \\
&= a^*aE_1 + b^*bE_2
\end{aligned}$$

remembering the orthogonal property of ψ_1 and ψ_2.

This strongly suggests that $\langle H \rangle$, which is known as the *expectation value* of H, is the average value of the energy for a system in the state described by the wavefunction ϕ.

Let us pursue this idea further. Consider a potential which is a function of position only. If $\phi(\mathbf{r})$ is the wavefunction for a particle, the probability of finding it in a small region near the point $\mathbf{r}$ is $\phi^*(\mathbf{r})\phi(\mathbf{r}) \, \mathrm{d}^3\mathbf{r}$, and hence it will experience the potential $V(\mathbf{r})$ with a probability $\phi^*(\mathbf{r})\phi(\mathbf{r}) \, \mathrm{d}^3\mathbf{r}$. Thus the average value of the potential experienced by the particle will be given by

$$\begin{aligned}
\int V(\mathbf{r})\phi^*(\mathbf{r})\phi(\mathbf{r}) \, \mathrm{d}^3\mathbf{r} &= \int \phi^*(\mathbf{r}) V(\mathbf{r}) \phi(\mathbf{r}) \, \mathrm{d}^3\mathbf{r} \\
&\equiv \langle V \rangle
\end{aligned} \tag{5.34}$$

where the first line follows because $V(\mathbf{r})$ is a function of position only. Thus the expectation value of the potential $\langle V \rangle$ certainly represents the average value of the potential experienced by a particle.

Let us return to Eq. (5.30) where ψ_1 is an eigenfunction of H. We have

$$\begin{aligned} E_1 &= \langle H \rangle \\ &\equiv \int \psi_1^* H \psi_1 \, \mathrm{d}^3\mathbf{r} \\ &= \int \psi_1^* \left(-\frac{\hbar^2}{2m} \nabla^2 \right) \psi_1 \, \mathrm{d}^3\mathbf{r} + \int \psi_1^* V \psi_1 \, \mathrm{d}^3\mathbf{r} \\ &\equiv \left\langle -\frac{\hbar^2}{2m} \nabla^2 \right\rangle + \langle V \rangle \, . \end{aligned} \tag{5.35}$$

As $\langle V \rangle$ represents the average value of the potential energy, it would appear that if we are to satisfy conservation of energy we must regard

$$\left\langle -\frac{\hbar^2}{2m} \nabla^2 \right\rangle$$

as representing the average value of the kinetic energy. But this implies that the kinetic energy T in quantum mechanics is to be represented by an operator, $-(\hbar^2/2m)\,\nabla^2$.

Continuining this train of thought, we have $T = p^2/2m$. As T is to be represented by an operator, it follows that p^2 is also represented by an operator, namely $-\hbar^2\,\nabla^2$, and hence presumably $\mathbf{p}$ will also be represented by an operator. Now

$$p^2 = p_x^2 + p_y^2 + p_z^2 \, .$$

If we put $p_x \to -\mathrm{i}\hbar(\partial/\partial x)$ and correspondingly for p_y and p_z we get

$$\begin{aligned} p^2 &= -\hbar^2 \frac{\partial^2}{\partial x^2} - \hbar^2 \frac{\partial^2}{\partial y^2} - \hbar^2 \frac{\partial^2}{\partial z^2} \\ &= -\hbar^2\,\nabla^2 \, . \end{aligned}$$

Thus we get no inconsistencies if we say that in quantum mechanics the momentum $\mathbf{p}$ is to be represented by the operator $-\mathrm{i}\hbar\boldsymbol{\nabla}$. We assume, in conformity with what we have discussed above, that the expectation value $\langle -\mathrm{i}\hbar\,\boldsymbol{\nabla} \rangle$ will give us the average value of the $\mathbf{p}$.

Under what circumstances will we have a definite value of the momentum $\mathbf{p}$? We know the answer to this already, of course; it is when the particle is in a plane wave state of the form $\mathrm{e}^{\mathrm{i}(\mathbf{k}\cdot\mathbf{r} - \omega t)}$, as discussed in Chapter 4. However let us think about it from the point of view of our present considerations. We have found it necessary to represent momentum as an operator. Dynamical quantities in classical mechanics are functions of position and momentum, and it would appear that these too will have to be represented by operators.

If one looks on the position vector $\mathbf{r}$ as an operator multiplying the wave function, all dynamical quantities will be represented by operators. We assume therefore that in quantum mechanics dynamical quantities (position, momentum, angular momentum, kinetic energy etc.) are represented by operators, and that the expectation value $\langle A \rangle$ of any operator A in a particular state ϕ gives us the average value of the quantity it represents when the particle is in the state given by ϕ.

Under what circumstances does such a quantity have a *definite* value? If we measure any physical quantity, say the length of a body l and repeat the measurement many times we can arrive at an average length $\bar{l}$. If the body has a definite length, all the measurements of l will yield the same value assuming we have eliminated experimental errors. As a result we find that

$$\overline{l^2} = (\bar{l})^2$$

i.e. the square of the average equals the average of the square. If l does not have a definite value, and there is a spread in the measurements, then we get

$$\overline{l^2} > (\bar{l})^2 \ .$$

We assert, therefore, that a quantity which in quantum mechanics is represented by an operator A has a definite value in a given normalized state ϕ if $\langle A^2 \rangle = (\langle A \rangle)^2$ in that state, or

$$\int \phi^* A^2 \phi \, \mathrm{d}^3\mathbf{r} = \left(\int \phi^* A \phi \, \mathrm{d}^3\mathbf{r} \right)^2 . \tag{5.36}$$

Suppose ϕ is a normalized eigenfunction of the operator A, so that we have

$$A\phi = a\phi ;$$

then

$$\begin{aligned} A^2\phi &= A(A\phi) \\ &= A(a\phi) \\ &= aA\phi \\ &= a^2\phi \ . \end{aligned}$$

Therefore

$$\int \phi^* A^2 \phi \, \mathrm{d}^3\mathbf{r} = a^2$$

and

$$\left(\int \phi^* A \phi \, \mathrm{d}^3\mathbf{r} \right)^2 = a^2$$

remembering that ϕ is normalized. Thus we have the situation that the eigenfunctions of an operator are states with a definite value of the quantity to which the operator refers, and this value is the eigenvalue to which they belong. We have not proved that these are the only states with a definite value, but shall assume that this is the case. Thus, if we wish to know what values a quantity can have we have to find the eigenvalues and eigenfunctions of its operator analogue in quantum mechanics. Since the result of any measurement must be a *real* number, as this represents a physical quantity, it follows that the eigenvalues of operators which correspond to *measurable* quantities must be real. Operators whose eigenvalues are all real, are referred to as *hermitian* operators. It is possible to show that eigenfunctions of hermitian operators belonging to different eigenvalues are orthogonal. The orthogonality of eigenfunctions of H belonging to different eigenvalues of H, referred to in section 5.8, is a particular case of this general rule.

Let us now return to the case of linear momentum. What states have a definite momentum in the x-direction, say? To answer this we have to find the eigenvalues and eigenfunctions of the operator corresponding to p_x. This, we have suggested, is $-\mathrm{i}\hbar(\partial/\partial x)$. Therefore we want to solve

$$-\mathrm{i}\hbar\frac{\partial\psi}{\partial x} = p\psi \; . \tag{5.37}$$

The solution is

$$\begin{aligned}\psi &= C\,\mathrm{e}^{\mathrm{i}px/\hbar}\\ &= C\,\mathrm{e}^{\mathrm{i}kx}\end{aligned} \tag{5.38}$$

where C is an arbitrary constant and $\hbar k = p$. This is in precise agreement with what we postulated in chapter 4 for the space portion of the plane wave representing a particle moving uniformly in the x-direction.

We have assumed above that the expectation value $\langle -\mathrm{i}\hbar\,\nabla\rangle$ gives us the mean value of the momentum. Suppose we have a normalized wave packet $\psi(x, t)$ and we express this in terms of plane waves as in Eq. (4.21):

$$\psi(x, t) = \frac{1}{\sqrt{(2\pi)}}\int_{-\infty}^{+\infty} A(k)\,\mathrm{e}^{\mathrm{i}(kx-\omega t)}\,\mathrm{d}k \; . \tag{4.21}$$

In this expression $A(k)$ is the amplitude of the plane wave component with $p = \hbar k$. As we also have

$$\int_{-\infty}^{+\infty} A(k)A^*(k)\,\mathrm{d}k = 1 \; ,$$

as $\psi(x, t)$ is assumed to be normalized, the mean value of the momentum will clearly be given by

$$\int_{-\infty}^{+\infty} A(k)A^*(k)\hbar k\,\mathrm{d}k \; . \tag{5.39}$$

We now demonstrate that this is also the expression we obtain for $\langle -i\hbar(\partial/\partial x)\rangle$:

$$\left\langle -i\hbar\frac{\partial}{\partial x}\right\rangle = \frac{1}{2\pi}\int_{-\infty}^{+\infty} dx \left\{\int_{-\infty}^{+\infty} A^*(k')\, e^{-i(k'x-\omega' t)}\, dk' \left(-i\hbar\frac{\partial}{\partial x}\right) \times \int_{-\infty}^{+\infty} A(k)\, e^{i(kx-\omega t)}\, dk\right\}$$

$$= \frac{1}{2\pi}\int_{-\infty}^{+\infty} dx \int_{-\infty}^{+\infty}\int_{-\infty}^{+\infty} A^*(k')\, e^{-i(k'x-\omega' t)}\hbar k A(k)\, e^{i(kx-\omega t)}\, dk\, dk' \ . \tag{5.40}$$

We now invert the order of integration. According to the discussion in Appendix A the integration over x gives us $2\pi\,\delta(k-k')$ and hence Eq. (5.40) becomes

$$\left\langle -i\hbar\frac{\partial}{\partial x}\right\rangle = \int_{-\infty}^{+\infty}\int_{-\infty}^{+\infty} A^*(k')A(k)\hbar k\,\delta(k-k')\, e^{i[(k-k')x-(\omega-\omega')t]}\, dk\, dk' \ .$$

The presence of the δ-function means that if we now integrate over k' what happens is that k' must be replaced by k everywhere, and correspondingly ω' by ω. Hence we get

$$\left\langle -i\hbar\frac{\partial}{\partial x}\right\rangle = \int_{-\infty}^{+\infty} A^*(k)A(k)\hbar k\, dk$$

as required.

5.11 SUMMARY

In this chapter we have established most of the principles of quantum mechanics which will be needed for later work. In this section we provide a summary of what has been said in this chapter.

(1) In quantum mechanics we cannot simultaneously give the position and momentum of a particle as a function of time, as in classical mechanics. All one can do is to give the probability distribution of the position and momentum, through the wavefunction $\psi(\mathbf{r}, t)$ and its Fourier transform.

(2) The wavefunction $\psi(\mathbf{r}, t)$ is a solution of the Schrödinger equation

$$-\frac{\hbar^2}{2m}\nabla^2\psi + V(\mathbf{r})\psi = i\hbar\frac{\partial\psi}{\partial t} \ . \tag{5.13}$$

(3) The wavefunction must be a single-valued continuous function of position with continuous derivatives, and if it refers to a bound state it must go to zero at infinity.

(4) States of a particle with a definite energy E have a time dependence $e^{-iEt/\hbar}$ so that their wavefunction $\Psi(\mathbf{r}, t)$ can be written

$$\Psi(\mathbf{r}, t) = \psi(\mathbf{r})\, e^{-iEt/\hbar} .$$

$\psi(\mathbf{r})$ is a solution of the time-independent Schrödinger equation

$$-\frac{\hbar^2}{2m}\nabla^2\psi + V(\mathbf{r})\psi = E\psi . \tag{5.8}$$

(5) Due to the boundary conditions mentioned in (3) the bound state spectrum of Eq. (5.8) is discrete.

(6) Eq. (5.8) is an example of an eigenvalue problem. It can be written in the form $H\psi = E\psi$. H is called the Hamiltonian. The values of E which give solutions satisfying the boundary conditions are called the eigenvalues of H, and the corresponding solutions are called eigenfunctions of H belonging to the eigenvalue E.

(7) In quantum mechanics dynamical variables such as position, momentum, kinetic energy, total energy, etc., are represented by operators. In particular the operator for position is $\mathbf{r}$, for momentum is $-i\hbar\boldsymbol{\nabla}$, for kinetic energy is $-(\hbar^2/2m)\,\boldsymbol{\nabla}^2$ and for total energy is

$$H = -\frac{\hbar^2}{2m}\nabla^2 + V(\mathbf{r}) .$$

(8) To find the value of any variable we operate on the wavefunction with the operator corresponding to the variable. If the variable has a definite value in this state the wave function will be an eigenfunction of the operator, and the value of the variable is the corresponding eigenvalue. If the variable does *not* have a definite value in this particular state the result of the operation will be to produce a new wave function, but the *average* value of the variable will be given by the expectation value of the operator

$$\langle A \rangle = \int \psi^* A \psi \, d^3\mathbf{r}$$

provided the wave function ψ is normalized.

(9) If the postulates in (8) are to make sense it follows that the eigenvalues of operators corresponding to measurable quantities must be real.

(10) The eigenfunctions of a hermitian operator belonging to different eigenvalues of the operator are orthogonal. If two or more eigenfunctions exist for a given eigenvalue, degeneracy is said to occur. It can always be arranged that degenerate eigenfunctions are orthogonal.

PROBLEMS 5

5.1 A metal has an inner potential of 5 eV. Calculate the wavelength of electrons inside and outside the metal if their kinetic energy outside the metal is 45 eV.

5.2 Electrons of energy 50 eV are incident on a metal with an inner potential of 6 eV. The spacing of layers of atoms parallel to the surface is 2.0 Å. At what angle would you observe the first Bragg reflection?

5.3 If a material has a refractive index $\mu = 1+\delta$ where δ is small show the Bragg condition becomes $2d \sin\theta[1+(\delta/\sin^2\theta)] = n\lambda$.

5.4 Particles of energy E move in a potential of the form $V = 0$ for $x < 0$, and $V = V_0$ for $x > 0$ where $E > V_0 > 0$. Show that a particle incident on the step at $x = 0$ from either direction has a probability of being reflected of

$$\left(\frac{E^{1/2} - (E - V_0)^{1/2}}{E^{1/2} + (E - V_0)^{1/2}}\right)^2.$$

5.5 Show that the energy levels for a particle of mass m moving in a potential given by $V = 0$ when $-a/2 < 0 < a/2$, and $V = \infty$ otherwise, are given by $E = n^2h^2/8ma^2$ where n is a positive integer and m is the mass of the particle.

5.6 Determine the normalized eigenfunctions for Problem 5.

5.7 Determine the expectation value of x^2 for the various stationary states of Problem 5 and compare your result with what you would obtain for a classical particle.

5.8 Show explicitly that the eigenfunctions in Problem 6 are orthogonal to each other.

CHAPTER 6

Angular momentum in a central field and electron spin

6.1 INTRODUCTION

In the last three chapters we have studied how the concepts of classical physics fail to account for physical phenomena at the atomic level and we have postulated that instead of using the laws of classical mechanics we must use those of quantum mechanics if we wish to predict the correct results. We were guided towards the postulates of quantum mechanics largely through a study of the phenomenon of electron diffraction which we concluded must be described by some sort of wave equation. The very introduction of this concept meant that we had already abandoned the classical ideas of determinancy, for, as was discussed in detail in section 4.10, the introduction of a wave picture results in an uncertainty principle. Thus the solutions of the Schrödinger equation do not provide us with the path of a particle as do the corresponding classical equations, but only tell us how the probability of finding the particle at various points varies as a function of time.

Our belief in the correctness of non-relativistic quantum mechanics, as was pointed out in section 5.1, lies not merely in the enormous range of physical phenomena over which it has predicted the correct results, but also in the fact that we have gone beyond non-relativistic quantum mechanics to relativistic quantum electrodynamics which has been shown to give answers to the energy levels of atomic hydrogen and the magnetic moment of the electron within the experimental errors, which in many cases are less than one part in a million.

From now on, therefore, we shall assume the Schrödinger equation is correct and proceed to see how it accounts for the phenomena of physics, and in particular, of atomic physics. We have already seen in section 5.6 that the application of quantum mechanics in the one-dimensional case leads to discrete energy levels for bound states. We now wish to proceed to more specific problems, and these will, of course, bring out specific features. In doing this we shall find that we shall have to introduce two new concepts, namely electron spin and the Pauli exclusion principle. Each of these will be first introduced from an experimental standpoint.

In this chapter we will be concerned with one of the most common situations in physics where we have a particle bound to a centre, which we conveniently place at the origin, by a force which depends only on the distance of the particle from the origin. Such a problem is known as a central force problem. We could deal with this straight away using Schrödinger's equation, but it will be much more enlightening if we first discuss the classical features of such a problem and look for their quantum mechanical analogues.

6.2 CLASSICAL FEATURES OF PLANETARY MOTION

Let us consider the classical motion of a particle in a central field of force. Such a motion is typified by the movement of the planets round the sun. If electrons obeyed the laws of classical mechanics, the motion of the electron round the proton in the hydrogen atom would be exactly the same as the motion of one of the planets round the sun, apart from the scale, as they are both acted on by an attractive inverse square law force. However, most of what we have to say is independent of the law of force acting on the particle, except we shall always assume that it is attractive and central.

The physically important features of the motion of a particle in a central field can be characterized by two quantities.

(i) *The energy*. Because the particle moves in a conservative field the energy of the system, once given, is fixed. In the case of a planet moving under an inverse square law the orbit is an ellipse with the centre of force at one focus, and the energy determines the length of the major axis.

(ii) *The angular momentum*. As the force is always directed towards a particular point there is no couple acting on the particle and its angular momentum remains constant. This means the particle moves in a plane, and the components of the angular momentum completely specify the orientation of the plane. In the case of the inverse square law the value of the angular momentum together with the energy in fact completely defines the shape of the elliptical orbit.

The situation for the case of the inverse square law is shown in Fig. 6.1. If we know the energy E and the angular momentum $\mathbf{l}$ then we know the shape of the orbit and the orientation of the plane of the orbit. The only

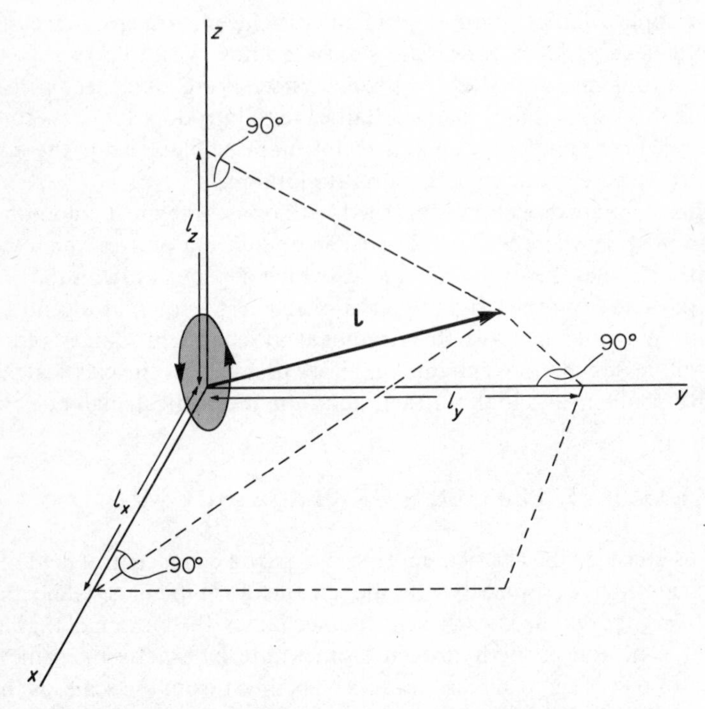

Fig. 6.1. A particle acted on by an inverse square law force moves in an elliptic orbit, with the origin of the force at one focus. The energy and angular momentum together specify the major and minor semi-axes and the components of angular momentum specify the orientation of the plane of the ellipse.

thing we do not know is the orientation of the orbit in its plane. The situation for other laws of force is essentially similar. It is only for special laws of force that the path of the particle forms a closed orbit, but the *shape* of the path, even when the orbit is not closed, is specified by the energy and angular momentum.

The classical equations of motion place no restrictions on the values of the energy a particle may have, and in particular there is no restriction on the negative values which can occur; that is for those values of the energy for which the system is bound and the particle cannot escape to infinity. In the case of atoms, though, we have direct experimental evidence that only certain energies are allowed, and we have seen in section 5.6, how this arises through the combination of the Schrödinger equation and the appropriate boundary conditions in one particular case.

In the same way all values of angular momentum are allowed in the classical case, subject only to the restrictions placed on it by the value of the energy. But if the application of boundary conditions can lead to a quantization of the energy when we go to the microscopic level, then it is perfectly possible that they could also place restrictions on the allowed values of angular momentum. We have direct experimental evidence for the quantization of energy. We would like to investigate the angular momenta of atoms and *see if there is any experimental evidence for restrictions on the values of angular momentum that are allowed.*

Fortunately this is possible through the relation that exists between angular momentum and the magnetic moment due to the motion of a charged particle.

6.3 THE RELATION BETWEEN MAGNETIC MOMENT AND ANGULAR MOMENTUM FOR CHARGED PARTICLES

Let us consider the magnetic moment due to the classical motion of an electron moving in a closed loop under the influence of a central force.

The magnetic moment of a small current loop is given by* (see Fig. 6.2)

$$\boldsymbol{\mu} = i\mathbf{A} \tag{6.1}$$

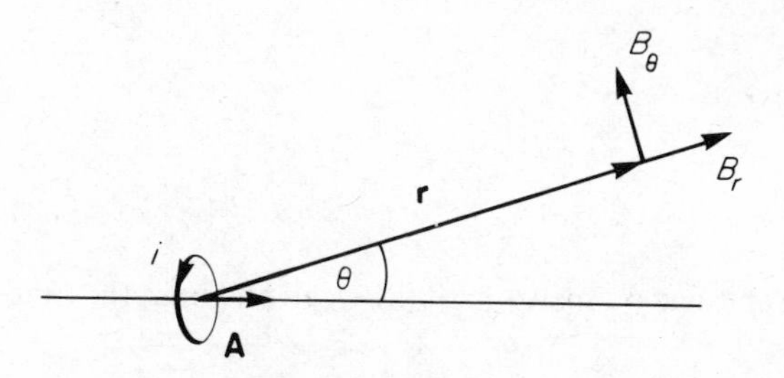

Fig. 6.2. A small loop of area A carrying a current i produces a field at a point $\mathbf{r}$ with components $B_r = (\mu_0 2iA\cos\theta)/r^3$ and $B_\theta = (\mu_0 iA\sin\theta)/r^3$ where θ is the angle between $\mathbf{r}$ and the normal to the loop. It thus behaves as if it had a magnetic moment $\boldsymbol{\mu} = i\mathbf{A}$.

* Some authors put $\boldsymbol{\mu} = \mu_0 i\mathbf{A}$ where μ_0 is the permittivity of free space. In this case Eq. (6.6) for the couple is changed to $\mathbf{C} = \boldsymbol{\mu} \wedge \mathbf{B}/\boldsymbol{\mu}_0$.

where **A** is a vector normal to the plane of the loop and equal to its area and i is the current. The positive direction of **A** is such that the current circulates round it in a clockwise direction when looking along **A**. If the current loop is due to the motion of an electron we have

$$i = -\nu e \tag{6.2}$$

where ν is the frequency of revolution and $(-e)$ is the charge on an electron.

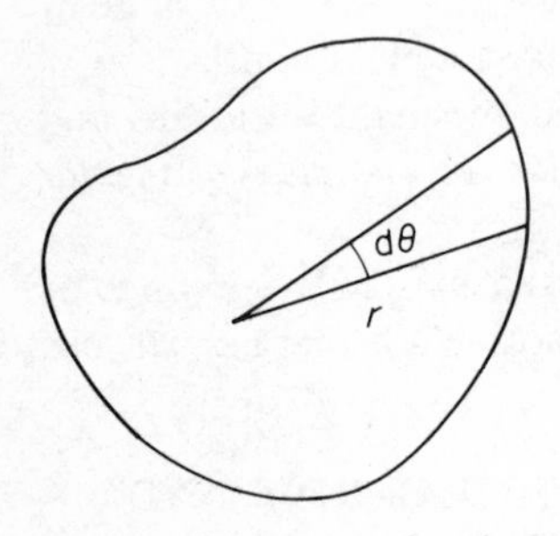

Fig. 6.3. An element of area is given by $\frac{1}{2}r^2\,d\theta$. Hence the area of a closed loop is given by

$$A = \int_0^{2\pi} \tfrac{1}{2}r^2\,d\theta\ .$$

The magnitude of the area A is given by (see Fig. 6.3)

$$A = \int_0^{2\pi} \tfrac{1}{2}r^2\,d\theta$$

$$= \int_0^{T} \tfrac{1}{2}r^2\frac{d\theta}{dt}\,dt$$

where T is the period. Now

$$\tfrac{1}{2}r^2\frac{d\theta}{dt} = \frac{l}{2m}$$

where l is the angular momentum of the electron about the centre of its orbit. Hence we have

$$A = \int_0^{T} \frac{l}{2m}\,dt = \frac{lT}{2m}\ . \tag{6.3}$$

As the convention for the direction of **A** mentioned after Eq. (6.1) is the same as the convention for the direction of the angular momentum vector of a particle moving in a closed orbit we may write Eq. (6.3) in vector form as

$$\mathbf{A} = \frac{\mathbf{l}T}{2m}\ . \tag{6.3a}$$

Putting together Eqs. (6.1), (6.2) and (6.3a), and remembering that $\nu = 1/T$ we get

$$\boldsymbol{\mu} = -\frac{e\mathbf{l}}{2m} \,. \qquad (6.4)$$

Eq. (6.4) is a relation between classical quantities. To make the transition to quantum mechanics we need to replace the classical quantities by corresponding operators which will act on an appropriate wavefunction ψ. This, in fact, is quite straightforward as the angular momentum $\mathbf{l}$ is given by $\mathbf{r} \wedge \mathbf{p}$, and we gave the prescription for the operators corresponding to $\mathbf{r}$ and $\mathbf{p}$ under heading 7 in section 5.11. However, it is often unnecessarily clumsy to go to these lengths, and it is a great convenience to introduce a notation to indicate the operator corresponding to a classical quantity. We do this by putting a circumflex over the symbol for the classical quantity. Thus, if $\mathbf{p}$ is the classical momentum of a particle we denote by $\hat{\mathbf{p}} \equiv -i\hbar\,\boldsymbol{\nabla}$ the corresponding quantum mechanical operator. In this notation the corresponding quantum mechanical equation to Eq. (6.4) is

$$\hat{\boldsymbol{\mu}} = -\frac{e\hat{\mathbf{l}}}{2m} \qquad (6.5)$$

where $\hat{\boldsymbol{\mu}}$ and $\hat{\mathbf{l}}$ are the magnetic moment and angular momentum operators respectively. To find out the actual values of $\boldsymbol{\mu}$ and $\mathbf{l}$ which we measure in any particular set of circumstances, these operators must operate on the wave function appropriate to these circumstances. However if we measure the magnetic moment of an atom we have *experimentally* found the results of acting with $\hat{\boldsymbol{\mu}}$ on the atom under these circumstances, and through Eq. (6.5) we have hence found the results of acting with $\hat{\mathbf{l}}$. *Thus a measurement of the values of* $\boldsymbol{\mu}$ *will tell us directly about the values of* $\mathbf{l}$.

Note that the relations of Eqs. (6.4) and (6.5) have depended only on the existence of a central force, and have in no way depended on the particular law of force involved. Thus we may obtain information from any system in which angular momentum is conserved and apply it to any other system in which angular momentum is conserved.

6.4 DETERMINATION OF THE MAGNETIC MOMENTS OF ATOMS

We have seen that there is a relation between the magnetic moment and the angular momentum of a system of charged particles. Thus we can acquire information about the angular momentum of an atom by studying its magnetic moment. To do this we place the atom in a non-uniform magnetic field where it will experience a translational force by virtue of any dipole moment it may possess. By measuring this force we gain information on *one* component of the magnetic moment, and hence of the angular momentum, as we now show.

The forces acting on a small current loop placed in a magnetic field are the same as those acting on a magnetic dipole formed by displacing two equal and opposite poles so as to give the same dipole strength as the loop. Consider the forces acting on a magnetic dipole which has been placed in a non-uniform magnetic field as shown in Fig. 6.4. The field is assumed to have no

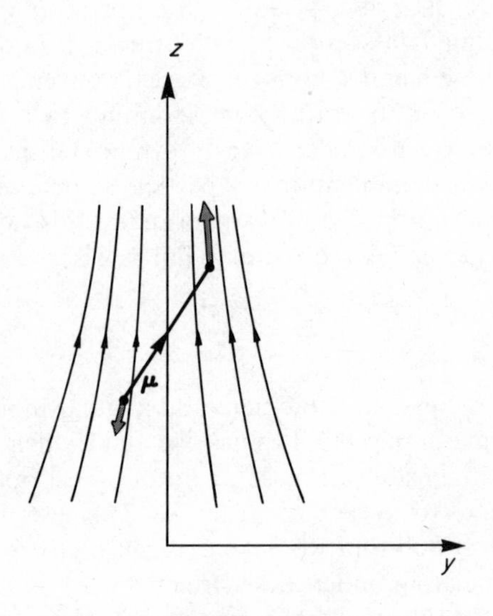

Fig. 6.4. The forces on a dipole in a non-uniform field.

component in the x-direction and to be symmetrical about the plane $y = 0$. The centre of the dipole is placed somewhere in the $y = 0$ plane and initially we assume its axis is parallel to the plane $x = 0$. Due to the two ends of the dipole being in different positions the following forces act on the dipole.

(i) A couple $\mathbf{C}$ tending to turn the dipole in the direction of the z-axis, given by

$$\mathbf{C} = \boldsymbol{\mu} \wedge \mathbf{B} . \tag{6.6}$$

(ii) A force due to the difference in $\mathbf{B}$ at the two ends of the dipole. In the conditions of Fig. 6.4 this force has both z- and y-components, as can be seen from the force vectors attached to each end of the dipole in Fig. 6.4.

Let us first consider the effect of the couple. The magnetic moment of an orbiting electron is due to an angular momentum which points in the same direction. The effect of the couple is to change the angular momentum $\mathbf{l}$ according to the relation

$$\frac{d\mathbf{l}}{dt} = \mathbf{C}$$

$$= \boldsymbol{\mu} \wedge \mathbf{B} \tag{6.7}$$

$$= -\frac{e}{2m}\mathbf{l} \wedge \mathbf{B} .$$

The change in $\mathbf{l}$ is *always* at right angles to $\mathbf{l}$ and $\mathbf{B}$. The effect of this is to cause $\mathbf{l}$ and hence $\boldsymbol{\mu}$ to precess round the direction of $\mathbf{B}$, as shown in Fig. 6.5. Due to

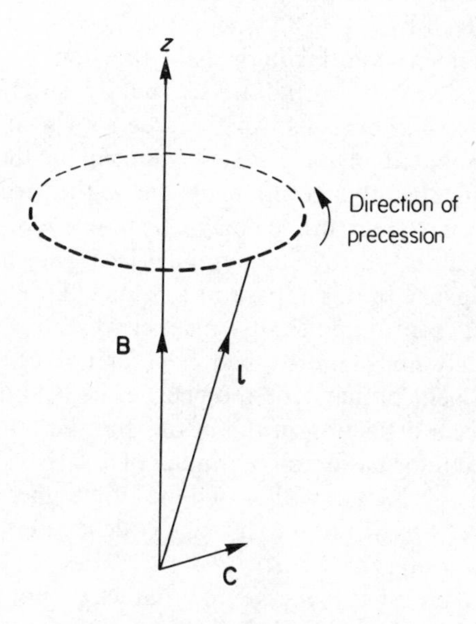

Fig. 6.5. The precession of an angular momentum vector $\mathbf{l}$ due to the couple between the field $\mathbf{B}$ and the magnetic moment associated with $\mathbf{l}$. $\mathbf{C}$ is at right angles to both $\mathbf{B}$ and $\mathbf{l}$.

this precession the force in the y-direction averages to zero and so do the components of $\boldsymbol{\mu}$ in the x- and y-directions. Thus we are left with a situation in which the time average of $\boldsymbol{\mu}$ is in the z-direction, with magnitude μ_z. The force in a non-uniform field on a magnetic moment μ_z pointing in the z-direction is

$$F_z = \mu_z \frac{\partial B_z}{\partial z} \quad . \tag{6.8}$$

In the case of an atom the magnetic moment is proportional to the resultant angular momentum of the atom, wherever it comes from. If only orbital angular momentum existed the constant of proportionality between the magnetic moment and the angular momentum would still be $-e/2m$. However, as we shall see in sections 6.11 and 6.12, electrons also have an intrinsic angular momentum. The effect of this is to modify somewhat the constant of proportionality, but otherwise to leave things the same. Consequently the arguments we have used so far in this section can be applied, equally, to a whole atom. Hence the force an atom will experience in a magnetic field is also given by Eq. (6.8), where μ_z is now the component of the atom's magnetic moment pointing in the field direction.

Eq. (6.8) is our basic result. From it we see that by finding the force acting on an atom in a non-uniform field of the type shown in Fig. 6.4 we can determine the component of its magnetic moment in the field direction. We cannot determine the other components due to the precessional motion.

The first experiment to measure the magnetic moment of an atom, namely silver, was carried out in 1921 by Stern and Gerlach using apparatus similar to that shown in Fig. 6.6. In this apparatus a beam of atoms from an oven is allowed to pass between the poles of a magnet which are so shaped as to produce an extremely non-uniform field. The force due to the interaction between the component of magnetic moment in the field direction and the field gradient causes a deflection of the beam, and the range of deflections gives us the required information on the values of μ_z.

Due to the range of velocities with which the atoms emerge from the oven, any particular value of μ_z will lead to a range of deflections, but these effects can be disentangled from each other.

When we do such an experiment we find that we do not get a continuous distribution of deflections on the deflector plate D when we allow for the velocity distribution of the atoms emerging from the oven. We find that we can account for the results if we assume that the atoms of a particular species can only have certain, definite, discrete values of μ_z and hence of the z-component of their angular momenta.

The sort of results we should obtain at the detector plate if the atoms emerging from the oven all had the same velocity is shown in Fig. 6.6(c) for various cases. At first sight this looks like direct evidence for the quantization of the z-component of orbital angular momentum, but things are a little more complicated. When we come to the analysis of these results in section

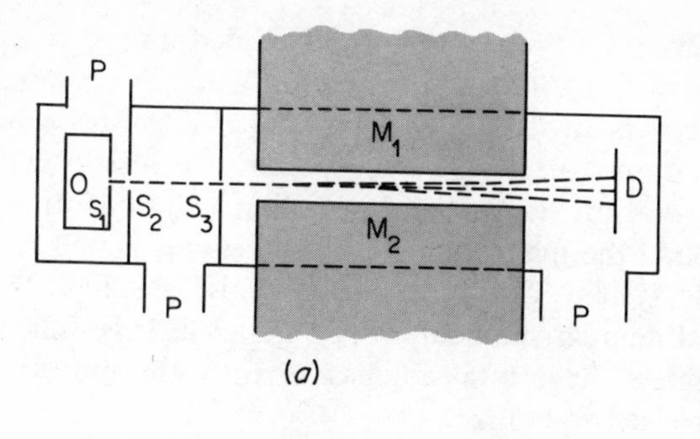

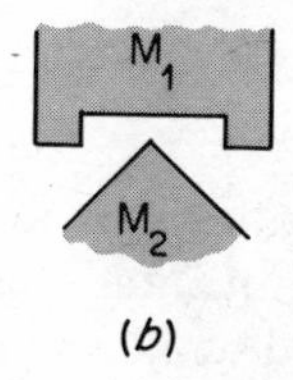

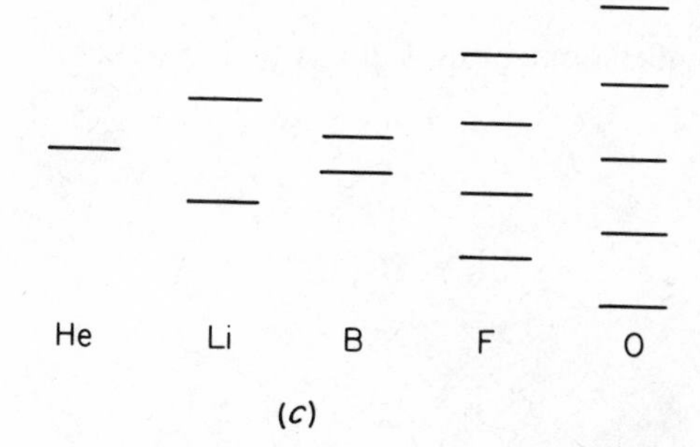

Fig. 6.6. The Stern–Gerlach experiment: (*a*) Side view of the apparatus

O—Oven
M_1, M_2—Magnet pole pieces
S_1, S_3—Beam defining slits
D—Detector plate
S_2—Isolating slit
P—Pumps

(*b*) Section through the magnet pole pieces. (*c*) The results that would be obtained for helium, lithium, boron, fluorine and oxygen if all the atoms emerging from the oven had the same velocity. The displacement in each case is proportional to the magnetic moment.

6.10 we shall find that they cannot be explained unless we assume that, in addition to orbital angular momentum, an electron possesses an intrinsic angular momentum of its own, which we call electron spin. The orbital angular momentum of an electron will combine, vectorially, with the electron spin to form a total angular momentum, and it is this quantity which is made to precess around the magnetic field by the couple acting on the magnetic moment of the system. Our experiments show, therefore, that the z-component of total angular momentum is quantized. It is difficult to see how this can be unless the orbital angular momentum and electron spin are themselves separately quantized.

6.5 THE REPRESENTATION OF ANGULAR MOMENTUM IN QUANTUM MECHANICS

The Stern–Gerlach experiment allows us to determine one component of the angular momentum of an atom, and the experimental results show that this component always has a discrete value. Conventionally this component is nearly always taken to be the z-component, and we shall follow this convention. We must now see how quantum mechanics accounts for these results.

Classically angular momentum is given by the vector

$$\mathbf{l} = \mathbf{r} \wedge \mathbf{p} \tag{6.9}$$

with components

$$l_x = yp_z - zp_y \tag{6.9a}$$

$$l_y = zp_x - xp_z \tag{6.9b}$$

$$l_z = xp_y - yp_x \ . \tag{6.9c}$$

In section 5.10 we pointed out that, in quantum mechanics, dynamical variables are represented by operators. Angular momentum is such a dynamical variable and hence must be represented by an operator in quantum mechanics whose form we must now find. We do this by replacing $\mathbf{r}$ and $\mathbf{p}$ in Eq. (6.9) by their operator equivalents $\mathbf{r}$ and $-i\hbar\nabla$ obtaining

$$\hat{\mathbf{l}} = -i\hbar\mathbf{r} \wedge \nabla \tag{6.10}$$

with components

$$\hat{l}_x = -i\hbar\left(y\frac{\partial}{\partial z} - z\frac{\partial}{\partial y}\right) \tag{6.10a}$$

$$\hat{l}_y = -i\hbar\left(z\frac{\partial}{\partial x} - x\frac{\partial}{\partial z}\right) \tag{6.10b}$$

$$\hat{l}_z = -i\hbar\left(x\frac{\partial}{\partial y} - y\frac{\partial}{\partial x}\right) . \tag{6.10c}$$

The nature of these operators can be seen more clearly if we use polar coordinates, but first let us consider the angular momentum operators from a different point of view.

In section 5.10 we saw that linear momentum is given by the operators $p_x \rightarrow -i\hbar(\partial/\partial x)$, etc. That is, the linear momentum is determined by the variation of the wavefunction along the corresponding linear coordinate direction. This suggests that the angular momentum around some axis may be determined by the variation of the wavefunction as we go round that axis.

In Fig. 6.7 we show the relation between Cartesian coordinates and polar

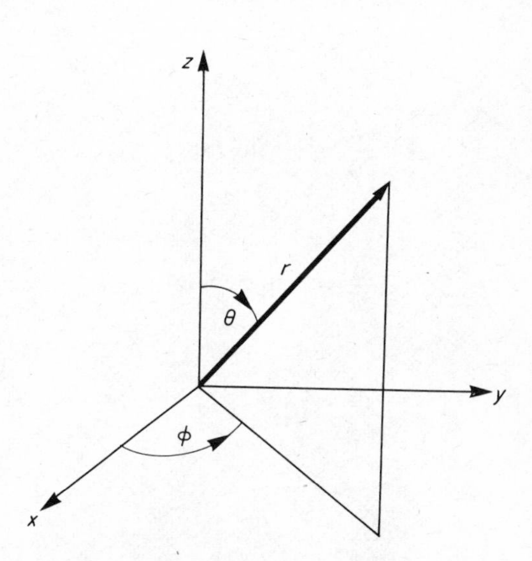

Fig. 6.7. The relation between cartesian and spherical polar coordinates

$x = r\sin\theta\cos\phi$ $\qquad r = (x^2 + y^2 + z^2)^{1/2}$

$y = r\sin\theta\sin\phi$ $\qquad \tan\theta = (x^2 + y^2)^{1/2}/z$

$z = r\cos\theta$ $\qquad \tan\phi = y/x$.

coordinates. If we have a wavefunction ψ, its variation as we go round the z-axis is $\partial\psi/\partial\phi$. If the angular momentum around the z-axis, $\hat{l}_z$, is related to the variation of the wave function as we go round the z-axis, in the same way as the x-component of linear momentum is related to the variation of the wavefunction along the axis, then we would expect the quantum mechanical operator form of l_z to be

$$\hat{l}_z = -i\hbar\frac{\partial}{\partial\phi} . \tag{6.11}$$

We may express the partial derivative $\partial\psi/\partial\phi$ in the cartesian frame as follows:

$$\frac{\partial\psi}{\partial\phi} = \frac{\partial\psi}{\partial x}\frac{\partial x}{\partial\phi} + \frac{\partial\psi}{\partial y}\frac{\partial y}{\partial\phi} + \frac{\partial\psi}{\partial z}\frac{\partial z}{\partial\phi} .$$

From the relations between cartesian and polar coordinates (see Fig. 6.7)

$$\left.\begin{aligned} x &= r\sin\theta\cos\phi \\ y &= r\sin\theta\sin\phi \\ z &= r\cos\theta . \end{aligned}\right\} \tag{6.12}$$

We get

$$\begin{aligned} \frac{\partial x}{\partial\phi} &= -r\sin\theta\sin\phi \\ &= -y \\ \frac{\partial y}{\partial\phi} &= r\sin\theta\cos\phi \\ &= x \\ \frac{\partial z}{\partial\phi} &= 0 . \end{aligned}$$

Hence

$$\begin{aligned} \frac{\partial\psi}{\partial\phi} &= -y\frac{\partial\psi}{\partial x} + x\frac{\partial\psi}{\partial y} \\ &= \left(x\frac{\partial}{\partial y} - y\frac{\partial}{\partial x}\right)\psi \\ &= \frac{-\hat{l}_z}{i\hbar}\psi . \end{aligned}$$

Therefore we have

$$\hat{l}_z = -i\hbar\frac{\partial}{\partial\phi} \tag{6.13}$$

as was proposed. A similar sort of relation holds for the component of angular momentum around any axis, as illustrated in Fig. 6.8.

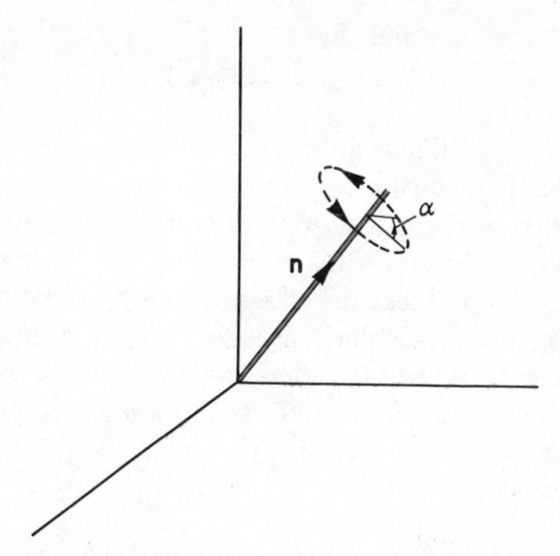

Fig. 6.8. The component of angular momentum around any arbitrary axis specified by a unit vector **n** is represented in quantum mechanics by the operator $\hat{l}_n = -i\hbar(\partial/\partial\alpha)$ where α measures the angle of rotation around **n**.

6.6 THE EIGENVALUES OF THE z-COMPONENT OF ANGULAR MOMENTUM

It was pointed out in section 5.10 and item 8 of the summary in section 5.11 that if a quantity has a definite value the wave function describing the state must be one of the eigenfunctions of the operator corresponding to the observed quantity, and the measured value will be one of its eigenvalues. To find the eigenfunctions and eigenvalues of the z-component of angular momentum we have, therefore, to solve the equation

$$-i\hbar\frac{\partial\psi}{\partial\phi} = m\hbar\psi \tag{6.14}$$

where the eigenvalue of $\hat{l}_z$ has been written as $m\hbar$.

The solution of Eq. (6.14) can be written down at once as

$$\psi = A\,e^{im\phi}\ . \tag{6.15}$$

However, the values of m in Eq. (6.15) are constrained by the condition of section 5.6 that the wavefunction must be single valued everywhere. If we start with given values of r, θ and ϕ, and let ϕ increase, when we get to the value $\phi' = \phi + 2\pi$ we are back at the same point in space from which we started, and this happens for every 2π increase in ϕ. For the wavefunction to be single-valued, then, we must have

$$e^{im\phi} \equiv e^{im(\phi + 2\pi)} \tag{6.16}$$

and this will only be true if m is an integer which may be positive, negative or zero. The eigenvalue spectrum of $\hat{l}_z$ is therefore *discrete*, and must have one of the values $m\hbar$. From the discussions of sections 6.3 and 6.4 one would expect the measured values for the magnetic moments of atoms to be discrete, and equally spaced, as the spectrum of $\hat{l}_z$ is equally spaced. Referring to the results shown in Fig. 6.6, we see that this is so, except that in some cases there is an undeflected beam, but not in others. As m is integral we should always have the first case where the undeflected beam corresponds to $m = 0$. We return to this important point later.

6.7 THE MAGNITUDE AND THE OTHER COMPONENTS OF ANGULAR MOMENTUM

We have found an experimental technique for determining the value of the z-component of the angular momentum of an atom by using the Stern–Gerlach apparatus. If we now select one of the beams emerging from such an apparatus, we have a sample of atoms all with the same, known, z-component of angular momentum. The question arises, is this all we can know about their angular momentum, and if not, how much more can we find out?

Let us first consider free atoms. These are not subject to any external couple, so the angular momentum, once given, must remain fixed in magnitude, and it is therefore not unreasonable to expect that we can at least know the magnitude as well as one component. Classically we could know all three components. Quantum mechanically it can be shown, using the relations (6.24) below, that the most we can know is the square magnitude and the component in one direction. Conventionally this direction is chosen as the z-direction. There is one exception to the above statement, which is when the square magnitude is zero, and hence each of the three components must be zero too.

The determination of the allowed values of the square magnitude (i.e. the eigenvalues of the operator $\hat{l}^2$) is beyond the scope of this book. We therefore content ourselves with the assertion that the eigenvalues of $\hat{l}^2$ are $l(l + 1)\hbar^2$ where l is a positive integer or zero. That is, if ψ is an eigenfunction of $\hat{l}^2$, then we have

$$\hat{l}^2\psi = l(l + 1)\hbar^2\psi \tag{6.17}$$

where

$$\hat{l}^2 = \hat{l}_x^2 + \hat{l}_y^2 + \hat{l}_z^2 \ .$$

For any particular value of l Eq. (6.17) has several different solutions corresponding to different values of the z-component of the angular momentum, $m\hbar$. Clearly m^2 cannot be greater than $l(l + 1)$ as we would then have one component greater than the total magnitude. As m, too, must be integral it follows that its values must fall within the range from $-l$ to $+l$. It can be shown that for a particular value of l there are solutions to Eq. (6.17) for every value of m from $-l$ to $+l$, a total of $(2l + 1)$. Thus for each eigenvalue of $\hat{l}^2$, specified by an integer l, there are $(2l + 1)$ different eigenfunctions each of which is specified by a value of m, where $-l \leqslant m \leqslant l$.

We notice that the magnitude of the angular momentum $\sqrt{(l(l + 1))}\hbar$ is always greater than the maximum value of the z-component $l\hbar$. This is a necessary consequence of the fact that we can observe, at most, the value of one component. Clearly the magnitude of the angular momentum cannot be less than the value of the chosen component, and if it were equal to it the other two components would each have to be zero, and thus well defined.

The situation is summarized in Fig. 6.9 which shows the following facts:

(i) We can only know the magnitude and one component of angular momentum. This component is conventionally taken as the z-component.

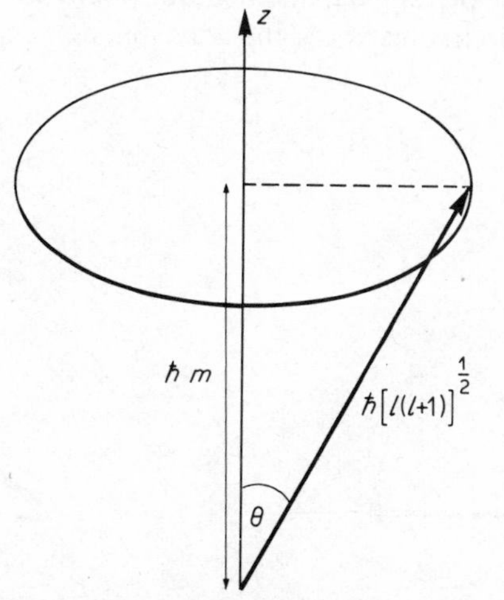

Fig. 6.9. Angular momentum in quantum mechanics. We may know the magnitude $\hbar[l(l + 1)]^{1/2}$ and one component only, conventionally called the z-component. As a result the angle θ is well defined but the orientation of the magnitude around the z-axis is completely undefined.

(ii) The magnitude of the angular momentum must be greater than the magnitude of the component we know. (Except when $l = 0$).

(iii) As we know the magnitude and the z-component, the angle the magnitude makes with the z-axis is well defined, but, as the other components are not known, the position of the magnitude around the z-axis is completely unknown.

The statements we have made and the arguments we have used in the last three paragraphs refer, strictly, to the orbital angular momentum of a single particle. It can be shown that they also apply to the total orbital angular momentum of a collection of particles, and with minor modifications to the total angular momentum of an atom, including the contribution from the intrinsic spin of the electrons.

★ 6.8 A PROOF THAT WE CANNOT KNOW ALL THREE COMPONENTS OF ANGULAR MOMENTUM

We have stated that we cannot know all three components of angular momentum, except in the case when all three are zero. We now prove this.

Let us suppose that we can know all three components. This implies the angular momentum points in a specified direction. We now choose a new frame $x'y'z'$ so that the angular momentum points in the z'-direction (see Fig. 6.10). In these circumstances the wavefunction representing the state

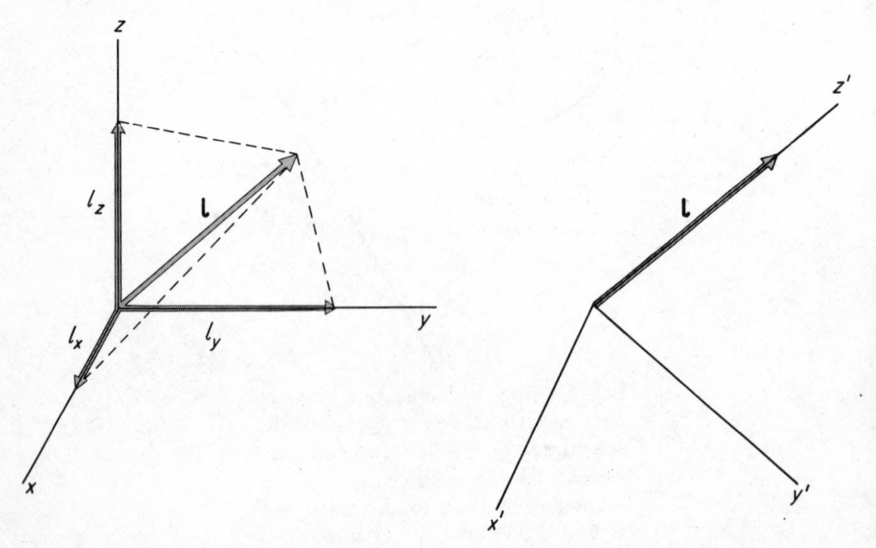

Fig. 6.10. If we know all three components of $\mathbf{l}$, l_x, l_y and l_z we may choose new coordinate x', y' and z' such that $l'_x = l'_y = 0$ and $l'_z = |\mathbf{l}|$.

will be an eigenfunction of $\hat{l}_{x'}$, $\hat{l}_{y'}$, and $\hat{l}_{z'}$ as follows

$$\hat{l}_{x'}\psi = 0 \tag{6.18a}$$

$$\hat{l}_{y'}\psi = 0 \tag{6.18b}$$

$$\hat{l}_{z'}\psi = A\psi \tag{6.18c}$$

where A is the magnitude of the angular momentum. If we put in the explicit expressions for $\hat{l}_{x'}$ and $\hat{l}_{y'}$ from Eqs. (6.10), Eqs. (6.18a) and (6.18b) become

$$-i\hbar\left(y'\frac{\partial\psi}{\partial z'} - z'\frac{\partial\psi}{\partial y'}\right) = 0 \tag{6.19a}$$

$$-i\hbar\left(z'\frac{\partial\psi}{\partial x'} - x'\frac{\partial\psi}{\partial z'}\right) = 0\ . \tag{6.19b}$$

Eliminating $\partial\psi/\partial z'$ between these two equations we obtain

$$-i\hbar z'\left(y'\frac{\partial\psi}{\partial x'} - x'\frac{\partial\psi}{\partial y'}\right) = 0$$

from which it follows that

$$\begin{aligned}\hat{l}_{z'}\psi &\equiv -i\hbar\left(y'\frac{\partial\psi}{\partial x'} - x'\frac{\partial\psi}{\partial y'}\right)\\ &= 0\end{aligned}$$

contrary to Eq. (6.18c) unless $A = 0$. Thus the assumption that we can know all three components is not self-consistent unless the angular momentum is zero.

6.9 THE CONVENTIONAL DESCRIPTION OF STATES WITH A DEFINITE ANGULAR MOMENTUM

In the last two sections we have concluded that a particle in a central field can have a definite value for the magnitude of its angular momentum, and also a definite value for one component of its angular momentum. Conventionally this component is chosen to be the z-component.

These two quantities cannot however take on any value. The z-component can have only integral values when measured in units of $\hbar$ and the magnitude is given by $\sqrt{[l(l+1)]}$, where l is an integer, when measured in the same units.

The quantity $\sqrt{[l(l+1)]}$ is an irrational number for integral values of l and is therefore somewhat inconvenient as a label for describing the angular momentum. It is therefore conventional to label the magnitude of the angular momentum by the integral value l. Thus when we say that a particle in a certain state has angular momentum l we mean that the magnitude of its angular momentum is $\sqrt{[l(l+1)]}\hbar$. If we say that a particle is in a state with l equal to three, we mean that the magnitude of its angular momentum is $\sqrt{(12)}\hbar$. The quantity l describing the angular momentum of the state is referred to as *the angular momentum quantum number*. Similarly the quantity m, denoting the value of the component along the z-axis, is called the *magnetic quantum number* for reasons which will appear later.

For historical reasons certain letters (s, p, d etc.) are associated with definite values of the angular momentum. Thus if we have a particle in a state with $l = 0$, i.e. no angular momentum, it is said to be in a s-state, and the wave-function describing it is called an s-wave. Similarly a state with $l = 1$ is called a p-state, and the corresponding wave function is called a p-wave. As indicated in Table 6.1 the first four values of l are labelled by the letters s, p, d and f, and thereafter progress alphabetically.

Table 6.1

Angular momentum quantum number l	Associated descriptive letter
0	s
1	p
2	d
3	f
4	g
5	h

6.10 ANGULAR MOMENTUM AND THE RESULTS OF THE STERN–GERLACH EXPERIMENT

Now that we have studied the behaviour of orbital angular momentum in quantum mechanics we return to the Stern–Gerlach experiment and see to what extent we can explain the results we quoted at the end of section 6.4. When we do this we shall find we are only partially successful, and that in order to explain all the observed results we shall have to postulate that the electron has an intrinsic angular momentum of its own, and that the total angular momentum of an atom is made up of a combination of the angular momentum due to the spatial movement of the electrons and the angular momentum arising from the intrinsic angular momentum of the electrons. We start by discussing the consequences of assuming that only *orbital* angular momentum exists.

Let us suppose we have a collection of free identical atoms, all in their lowest state. We will assume that the arguments we have applied to the angular momentum due to the motion of a single particle can equally be applied to the angular momentum of a system of particles in a central field; this is in fact the case. Classically a system of particles in a central field has a total orbital angular momentum which is a vector. Quantum mechanically the total orbital angular momenum has the same properties as the angular momentum of a single particle, i.e. it has a square magnitude $L(L+1)\hbar^2$ where L is a non-negative integer, and one can define one component $M_L\hbar$ where M is also integral, and the subscript L is added to make clear it refers to the orbital part of the angular momentum.

We will also assume that the gound state of the atom has a definite angular momentum, that is a definite value for L. This does not specify the state uniquely for, because the atoms are free, the ground state will be degenerate with respect to the magnetic quantum number M_L, as we now show.

As can be seen from Fig. 6.11 different values of M_L correspond to different

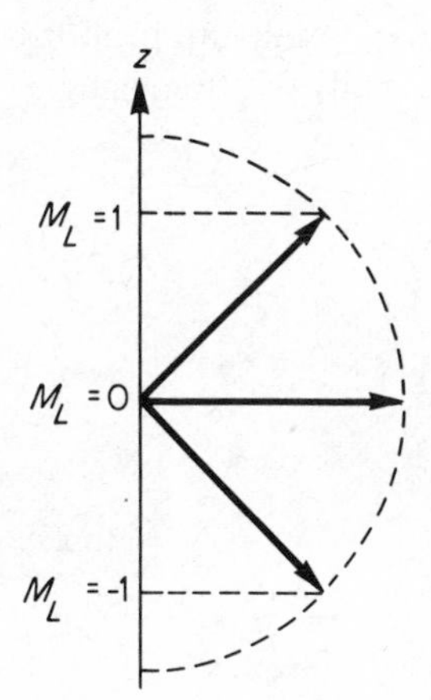

Fig. 6.11. An angular momentum with $L = 1$ can have $M_L = 1$, 0 or -1 corresponding to different possible distributions of direction in which the angular momentum can point.

distributions of direction in which the angular momentum can point. Because the atoms are free, the energy of the atoms cannot possibly depend on the direction in which the angular momentum is pointing, and all the states, corresponding to different M_L values, must have the same energy.

Another way of seeing this is to realize, that for free atoms, any one direction for choosing the z-axis is as good as any other direction, and hence the answer one gets for the energy for a specified value of M_L cannot depend on the direction chosen for the z-axis. This must mean the energy is independent of the value of M_L.

Thus, if we now investigate our collection of atoms we expect to find different atoms possessing different values of M_L, and as there is no reason for preferring one value of M_L to another, for a sufficiently large collection of atoms we shall expect to find equal numbers of atoms with each of the possible values of M_L. Because specification of the value of M_L is a finer distinction than merely specifying L, states with the same value of L but different values of M_L are referred to as the magnetic sub-states of a state with a given angular momentum. The word 'magnetic' is used because these states can be separated and identified by applying a magnetic field in the z-direction. In this language, then, we are saying that our collection of atoms will have equal numbers of atoms in each of the magnetic sub-states.

In section 6.7 we pointed out that the values of m, for a given value of l, must range in integral steps from $m = -l$ to $m = +l$, that is

$$-l \leqslant m \leqslant l \tag{6.20}$$

where l and m are both integers. There are thus exactly $2l + 1$ different values of m allowed for each value of l.

Each magnetic sub-state will have a different magnetic moment, proportional to the value of m. In section 6.3 we showed classically that for a single particle we have*

$$\boldsymbol{\mu} = -\frac{e\mathbf{l}}{2m} \tag{6.4}$$

and hence for a collection of similar particles we shall get

$$\begin{aligned} \boldsymbol{\mu} &= -\frac{e}{2m}\Sigma\mathbf{l} \\ &= -\frac{e}{2m}\mathbf{L} \end{aligned} \tag{6.21}$$

where $\mathbf{L}$ is the total angular momentum, and hence

$$\mu_z = -\frac{e}{2m}L_z\ . \tag{6.21a}$$

To make the transition to quantum mechanics we replace L_z by its operator form and let it act on the wavefunction of the atom. Since we have assumed that the atoms are in eigenstates of the operator $\hat{L}_z$ the observed values of μ_z will be

$$\mu_z = -\frac{e\hbar}{2m}M_L\ . \tag{6.22}$$

* Eq. (6.4) contains the electronic mass m in the denominator, and similarly in later equations. This should not lead to confusion with the magnetic quantum number m.

Thus the values of magnetic moment we expect to observe are equal to the magnetic quantum number M_L multiplied by the quantity $-e\hbar/2m$. As a result magnetic moments on the atomic scale are measured in units of $e\hbar/2m$ which is known as the *Bohr magneton*. Its relation to other units of magnetic dipole strength is as follows

$$\begin{aligned} e\hbar/2m &= 1 \text{ Bohr magneton} \\ &= 9.3 \times 10^{-24}\,\mathrm{J\,T^{-1}} . \end{aligned}$$

We see that if we do a Stern–Gerlach experiment with our collection of atoms, as described in section 6.4, we shall get a separate beam for each M_L-value, and hence we shall get $(2L + 1)$ beams, all equally spaced. *Thus by counting the number of beams emerging from a Stern–Gerlach apparatus we can, it would appear, determine the angular momentum of the ground state of an atomic species.* We note that for integral L, $(2L + 1)$ is always odd.

If we do such an experiment we find we can group our results into two classes, (i) cases where we observe an odd number of emergent beams, including an undeviated beam ($M_L = 0$) as required by the ideas we have developed up to now, and (ii) cases where we observe an even number of beams, and no undeviated beam. Typical among these are atomic hydrogen and the alkali metals, which each give two beams, symmetrically disposed about the centre line.

Thus our attempts to determine the orbital angular momentum of the ground states of atoms has led to an anomoly, which we must now investigate.

6.11 INTRINSIC ANGULAR MOMENTUM AND HALF INTEGRAL VALUES OF ANGULAR MOMENTUM

How are we to interpret the results where we have an even number of beams? In the case of the alkalis and atomic hydrogen, where we observe only two lines, it might be tempting to assume that we have $L = 1$, but the line corresponding to $M_L = 0$ is missing for some reason. The *positions* of the two lines are consistent with this supposition. However such an explanation will not hold water when we look at cases with four or more lines, even in number. In these cases we find that the lines are equally spaced and symmetrically displaced around the centre line, or that they can be analysed into groups of this nature. There is no sign of any missing component. Secondly, the magnitude of the spacing between adjacent lines does not necessarily correspond to adjacent beams being separated in the value of their magnetic moment by one Bohr magneton, as results from an application of Eq. (6.22).

Finally, as we shall show, the ground state of the hydrogen atom is an s-state, with $L = 0$, yet atomic hydrogen shows two beams. Clearly the situation is more complex than we have assumed hitherto.

If we apply our rule that the number of magnetic substates, for a given L, is $(2L + 1)$ to the alkali metals and hydrogen, where we get two lines, we get the result that $L = \frac{1}{2}$. According to our discussion of sections 6.6 and 6.7 such a result is not allowed. In that discussion, however, we were referring specifically to angular momentum produced by the motion of a particle in space. However one could envisage that a particle might have an internal angular momentum of its own, or intrinsic angular momentum as it is called. By this we mean an angular momentum not associated with the motion of the particle relative to some frame of reference, but which occurs even in its own centre-of-mass frame, as for example with a spinning top. Clearly such an angular momentum cannot be described by a wave function as we have so far understood it, for it is not a function of position. Can we then say anything about the values that could be taken on by an intrinsic angular momentum?

At least we expect such an intrinsic angular momentum to be quantized, as there is no evidence in any Stern–Gerlach experiment of a continuous distribution of magnetic moment values.

To proceed further, let us consider the classical case where we have both an orbital and intrinsic angular momentum. Such a case is provided by the motion of the earth round the sun. The earth has orbital angular momentum due to its orbital motion round the sun, and intrinsic angular momentum due to its rotations about its own axis. Both these angular momenta are vector quantities, with components in three directions, as is the total angular momentum produced from the sum. (See Fig. 6.12.) Further there is no distinction between these two forms of angular momenta in their dynamical properties. All forms of angular momentum obey the same equation of motion

$$\frac{\mathrm{d}\mathbf{J}}{\mathrm{d}t} = \mathbf{C}$$

where $\mathbf{J}$ is the angular momentum and $\mathbf{C}$ an applied couple. Thus we would expect the quantum mechanical angular momentum *operators* to exhibit the same, or similar properties, independently of the type of angular momentum to which they refer.

We now have to decide what these properties are. We have to look for some property of the orbital angular momentum operators which can be shared by all angular momentum operators, and does not depend on the presence of a spatial wave function representing orbital angular momentum. A study of the form of the operators $\hat{l}_x$, $\hat{l}_y$ and $\hat{l}_z$, as given in Eq. (6.10), shows

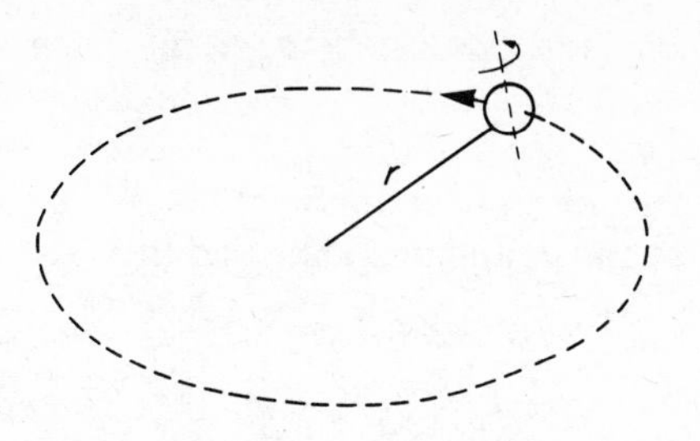

Fig. 6.12. The total angular momentum of the earth around the sun consists of two components. An orbital angular momentum **L** due to its movement round the sun and a spin angular momentum **S** due to its rotation on its own axis. These combine vectorially to form the total angular momentum **J**. (The ratio of S to L has been greatly exaggerated in the figure, in practice it is less than 10^{-6}.)

that when applied to any wavefunction we always get the result

$$\left.\begin{aligned} \hat{l}_x\hat{l}_y - \hat{l}_y\hat{l}_x &= i\hbar\hat{l}_z \\ \hat{l}_y\hat{l}_z - \hat{l}_z\hat{l}_y &= i\hbar\hat{l}_x \\ \hat{l}_z\hat{l}_x - \hat{l}_x\hat{l}_z &= i\hbar\hat{l}_y \end{aligned}\right\}. \tag{6.23}$$

This, then, is an algebraic relation between the components of the orbital angular momentum operator which always holds. We *assume*, therefore, that a relation of this sort also holds between the components of the operators representing any form of angular momentum. We denote these operator components by $\hat{J}_x$, $\hat{J}_y$ and $\hat{J}_z$ and postulate that they obey the relations:

$$\left.\begin{aligned}\hat{J}_x\hat{J}_y - \hat{J}_y\hat{J}_x &= i\hbar\hat{J}_z\\ \hat{J}_y\hat{J}_z - \hat{J}_z\hat{J}_y &= i\hbar\hat{J}_x\\ \hat{J}_z\hat{J}_x - \hat{J}_x\hat{J}_z &= i\hbar\hat{J}_y\end{aligned}\right\}. \tag{6.24}$$

It turns out that this set of relations is sufficient to determine the eigenvalues of $\hat{J}_z$ and $J^2 = (\hat{J}_x^2 + \hat{J}_y^2 + \hat{J}_z^2)$ by algebraic manipulation, although we do not prove this in this book. The result of an analysis of this sort is that the eigenvalues of $\hat{J}^2$ are given by $J(J+1)\hbar^2$ where J is integral (i.e. 0, 1, 2, . . .) or *half-integral* (i.e. $\frac{1}{2}, \frac{3}{2}, \frac{5}{2}, \ldots$), and the eigenvalues of $\hat{J}_z$, for a given value of J, range in integral steps from $-J$ to J in units of $\hbar$. For example, for $J = \frac{3}{2} J_z$ can take on the four values $-\frac{3}{2}\hbar, -\frac{1}{2}\hbar, \frac{1}{2}\hbar, \frac{3}{2}\hbar$.

Thus the algebraic properties of the angular momentum operators allow us to have either integral or half-integral values of the angular momentum, and it is the *additional* requirement, in the case of *orbital* angular momentum, that the wavefunction be single valued that limits the quantum numbers to the integral values only. The necessity for this, as far as the z-component of the orbital angular momentum is concerned, was shown in section 6.6, Eq. (6.16). For intrinsic angular momentum the quantum numbers can be either integral or half integral, and the same is true of a total angular momentum compounded from a sum of intrinsic and orbital angular momenta.

It must be emphasized that we have introduced an additional assumption in defining a general angular momentum through Eq. (6.24) and our belief in the correctness of this assumption is based on the large number of phenomena this allows us to explain and the fact that no case has ever been discovered which contradicts this assumption.

6.12 ELECTRON SPIN

Let us now return to the results of the Stern–Gerlach experiment, and in particular to the results for atomic hydrogen. Atomic hydrogen consists of a single electron moving in the Coulomb field of a proton. As we shall show in the next chapter, the ground state of atomic hydrogen is an s-state which has no orbital angular momentum. Yet in a Stern–Gerlach experiment atomic hydrogen shows two beams equally spaced about the centre line of the beam. We can account for this by assuming the electron has an intrinsic angular momentum $s = \frac{1}{2}$ measured in units of $\hbar$. We refer to this property as the electron *spin*. Spin is a more compact phrase to use than 'intrinsic angular momentum'. As angular momentum in quantum mechanics is always measured in units of $\hbar$ we say that the electron has a spin of one half; or 'the electron is a spin one half particle'.

What we mean by this statement is that the electron has an intrinsic angular momentum with the following properties.

(*a*) Its square magnitude is

$$s(s+1)\hbar^2 = \tfrac{1}{2}\cdot\tfrac{3}{2}\hbar^2 = \tfrac{3}{4}\hbar^2.$$

(*b*) The allowed values of the z-component are given by $s_z = \pm\frac{1}{2}\hbar$. The two possible eigenstates are illustrated in Fig. 6.13.

Of course to account for the splitting of atomic hydrogen into two beams in a non-uniform magnetic field we must associate a magnetic moment with the electron spin. The magnitude of the observed splitting tells us the magnetic moment of the electron is one Bohr magneton. What we mean by this is that

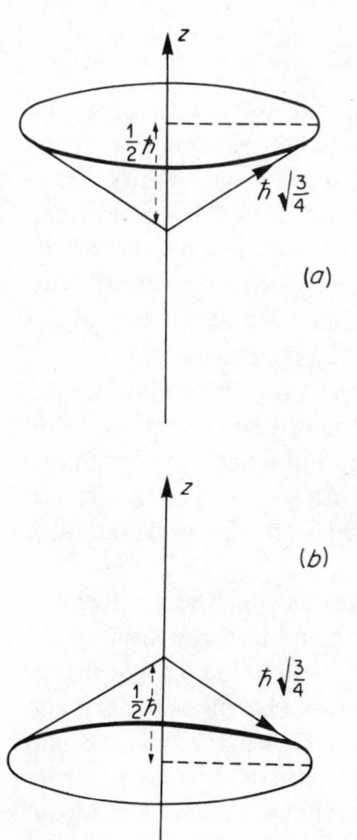

Fig. 6.13. The intrinsic spin states of the electron. The magnitude of the spin angular momentum is $(\frac{3}{4})^{1/2}\hbar$. Its component along any specified direction z is either $+\frac{1}{2}\hbar$ as in (*a*) or $-\frac{1}{2}\hbar$ as in (*b*). The other two components satisfy the relation $s_x^2 + s_y^2 = \frac{1}{2}\hbar^2$ but are otherwise unspecified. Hence the spin direction points anywhere along either of the two cones indicated, according to the value of its z-component.

when the electron is in such a state that its magnetic quantum number $s_z = \frac{1}{2}$ then it has a magnetic moment in the z-direction of $-e\hbar/2m$ and when $s_y = -\frac{1}{2}$ the electron has a magnetic moment of $+e\hbar/2m$.

If we refer back to Eq. (6.22) we see that the z-component of the magnetic moment in Bohr magnetons for orbital motion is equal to the z-component of angular momentum in units of $\hbar$ whereas the results we have just stated for the electron spin give us a magnetic moment equal to twice the z-component of the spin. The ratio

$$g = \frac{\text{observed } z\text{-component of the magnetic moment in Bohr magnetons}}{z\text{-component of angular momentum in units of } \hbar}$$

is called the gyromagnetic ratio. *For the orbital motion of electrons the gyromagnetic ratio is 1, while for the electron's intrinsic spin the gyromagnetic ratio is 2* (but see section 6.15).

It is perhaps worth mentioning that the idea of electron spin was first introduced in 1925 by Uhlenbeck and Gouldsmit to account for the doublet structure of spectral lines and the effect on spectral lines of having the source in a strong magnetic field (the Zeeman effect).

★ 6.13 THE ANALYSIS OF ELECTRON SPIN

Suppose we set up a Stern–Gerlach experiment and pass a beam of neutral hydrogen atoms through the apparatus. We know experimentally that the initial beam will be split into two emergent beams, one in which the electron in each hydrogen atom has 'spin up' and the other with 'spin down'. Further, these two beams are equal in intensity. These results do not depend on the orientation of the magnetic field around the original beam axis, but the two states 'spin up' and 'spin down' do depend on the orientation, as they are always parallel and antiparallel respectively to the magnetic field.

Suppose now we take one of these beams and pass it into another, similar apparatus, but with the field at right angles to the original direction. What happens now? The situation is illustrated in Fig. 6.14 where one of the two beams produced by the first Stern–Gerlach apparatus is stopped by a screen S, and the other beam allowed to pass into a second Stern–Gerlach apparatus with its magnetic field in the x-direction.

The second magnet defines a new z-direction for us which we shall call z', which is parallel to the original x-direction. As the electron spin in the hydrogen atom can only be 'up' or 'down' in the z'-direction (i.e. parallel or antiparallel to the x-axis), we again expect to find two beams at the exit end of the magnet, and as the original z-direction is at right angles to z' we might expect to get equal intensities in the two beams. This is, indeed, what we find. But are the two beams in every way similar to the two beams coming out of the first magnet? To help us think about this let us consider a similar experiment using polarized light.

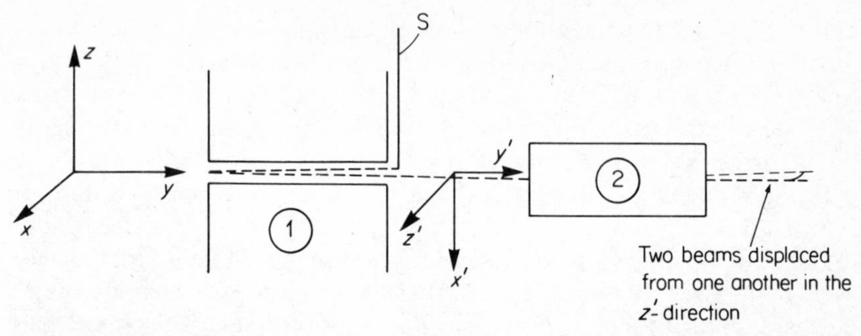

Fig. 6.14. A Stern–Gerlach apparatus 1 splits a beam of neutral hydrogen atoms into two components. One beam is stopped and the other passes into a second Stern–Gerlach apparatus 2 with its field at right angles to the field in 1. This beam is split into two components, but are these two beams similar in every way to the two beams emerging from 1?

Suppose we take an unpolarized light source and split it into two beams T_1 and R_1 of equal intensity, polarized at right angles to each other, as shown in the first part of Fig. 6.15. (This can be done in a variety of ways, one of which consists of making a mirror from a pile of thin glass plates and placing them at the Brewster angle $\tan i = \mu$). The transmitted beam T_1 in Fig. 6.15 is polarized in the plane defined by T_1 and the reflected beam R_1, and R_1 is polarized at right angles to this plane. We could now pass one of these beams into a second analyser which again splits the beam into two,

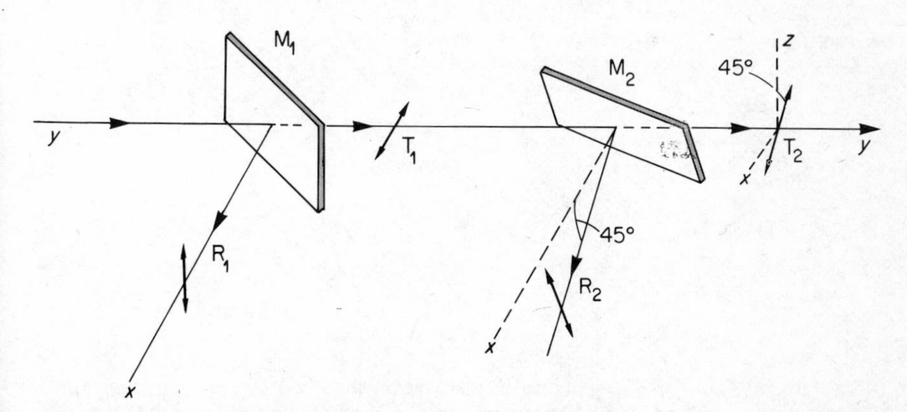

Fig. 6.15. Incident unpolarized light travelling in the y-direction is split into two beams T_1 and R_1 by the mirror M_1. The transmitted beam T_1 is polarized in the plane defined by T_1 and R_1 and the reflected beam is polarized at right angles to this plane. The two beams, although polarized, have no phase relation between them. The transmitted beam falls in a second mirror M_2 which is rotated about y through 45° relative to M_1. The transmitted beam T_2 is polarized at 45° to the z-axis and the reflected beam R_2 is polarized at right angles to this. There is now, however, a phase relation between the two beams.

but with different polarization directions. In Fig. 6.15 the second analyser has been oriented so that the transmitted beam T_2 has its polarization direction at 45° to that of T_1, and similarly the reflected beam R_2 is polarized at 45° to R_1. In this case the two beams R_2 and T_2 are again of equal intensity, but nevertheless they are different from R_1 and T_1.

In the original case, because the incident beam is unpolarized, the two beams into which it is split, although both polarized, have no phase relation to each other, and so if recombined we get unpolarized light again as indicated in the Fig. 6.16(a). In the second case, however the two components are in phase, and, if they are recombined after travelling equal optical path

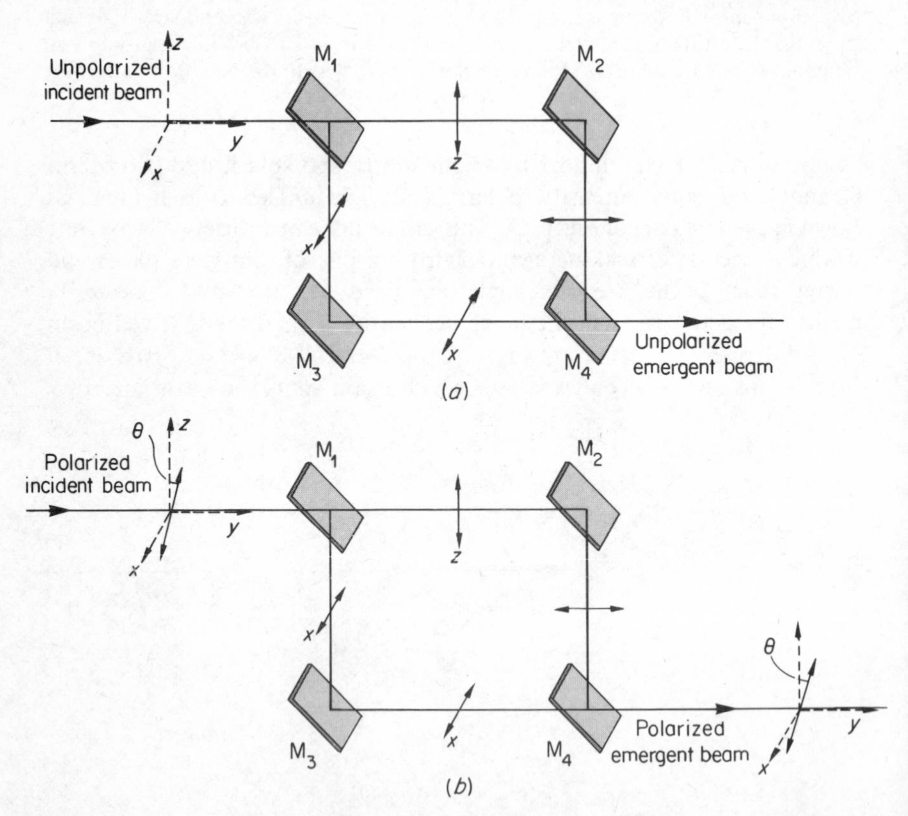

Fig. 6.16. The mirror M_1 splits an incident beam into two components, the transmitted beam polarized along the z-direction and the reflected beam polarized along the x-direction. M_2 and M_3 reflect the two beams onto the half-silvered mirror M_4 which combines them. If the incident beam is unpolarized, as in (a), the two beams produced at M_1 have no phase relation to each other, and on recombination at M_4 produce unpolarized light. If the incident light is polarized, as in (b), there is a phase relation between the two beams produced at M_1, and on recombination at M_4 (provided the two path lengths are identical) we recover light in the original state of polarization.

lengths, we recover the original light in the original state of polarization as illustrated in Fig. 6.16(*b*).

This suggests that if we could do similar experiments with our hydrogen atoms we should get similar results. We can in principle do such experiments using the modified Stern–Gerlach apparatus shown in Fig. 6.17. In this

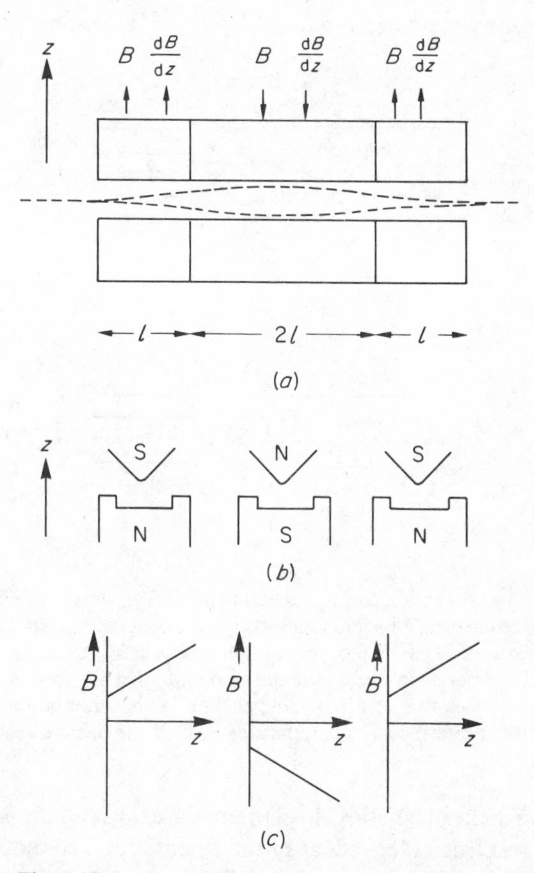

Fig. 6.17 (*a*) A modified Stern–Gerlach apparatus which splits a beam into two components and recombines them. (*b*) Arrangement of the pole pieces. (*c*) The field across the gap in the three sections.

apparatus two sections, each of length l in which the field and gradient have the values B and $\mathrm{d}B/\mathrm{d}z$ respectively are separated by a length $2l$ in which both the field and the gradient are reversed. An initial beam is thus split into two and then recombined. (It is necessary to reverse the field as well as the gradient for reasons which will be explained in the next section. For the moment we merely remark that by reversing the field as well, we ensure that the atomic electrons in both beams spend equal times aligned parallel and antiparallel to the magnetic field.) If the incident beam of atoms is unpolarized we find we get an unpolarized beam out at the far end of the apparatus. That is, if we put a conventional Stern–Gerlach apparatus at the far end, as in Fig. 6.18, we get two equal beams with spin 'up' and spin 'down' whatever

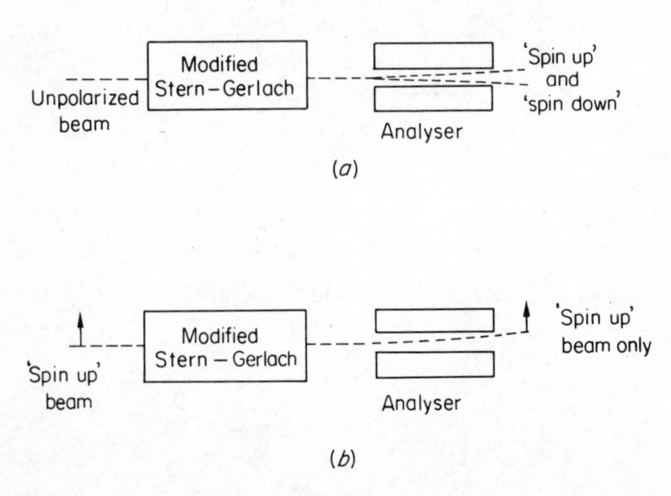

Fig. 6.18. As in the optics experiment of Fig. 6.16, unpolarized hydrogen atoms, after passing through the modified Stern–Gerlach apparatus of Fig. 6.17, emerge unpolarized as in (*a*). Atoms polarized initially emerge in their original state of polarization as in (*b*). In both cases this result is independent of the orientation of the modified Stern–Gerlach apparatus around the beam direction.

the orientation of the magnetic field in the analyser around the beam direction. Further, this result is independent of the orientation around the beam axis of the field in the modified Stern–Gerlach apparatus.

Now let our input beam be polarized by a conventional Stern–Gerlach apparatus, as shown in Fig. 6.18(*b*). In this case we find that the output beam is polarized in exactly the same way as the input beam; again independently of the orientation of the apparatus around the beam axis.

These results are obviously very closely analogous to what we would observe with light and the apparatus of Fig. 6.16. However in the latter case we know that although the light out has the same polarization as the light in, the relative intensities of the light in the two beams *inside* the apparatus depends on the relative orientation of the plane of polarization of the incident light and the plane of polarization of the light transmitted by the polarizer. We can calculate these relative intensities because we know that if we have plane polarized light we may resolve its amplitude into components in any two planes at right angles to each other and passing through the beam axis, as illustrated in Fig. 6.19.

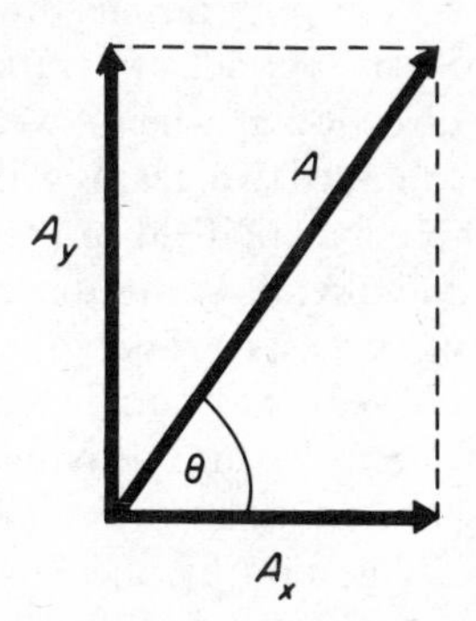

Fig. 6.19. A plane-polarized beam of amplitude A, polarized at an angle θ to an arbitrarily chosen x-axis may be analysed into two components at right angles, $A_x = A\cos\theta$ and $A_y = A\sin\theta$. As the intensities I_0, I_x and I_y are proportioned to the squares of the amplitudes we have $I_x = I_0\cos^2\theta$, $I_y = I_0\sin^2\theta$, and $I_x + I_y = I_0$.

The analogy with optics and the results we have obtained so far suggests that a similar resolution may be possible for electron spin. We must however be careful about trying to press the analogy too far. In the optical case we perform our resolution into two directions at right angles to each other, as components in two such directions do not interfere with each other; components in two directions at right angles are therefore orthogonal to each other in the same sense as we have talked about orthogonal functions in quantum mechanics. But for electron spin the orthogonal eigenfunctions represent states with the measurable components of the spin in *opposite* directions. Secondly the possible polarization directions in the optical case are confined to a plane at right angles to the direction of propagation. If, as we are suggesting, a coherent superposition of the spin up and spin down states represents a spin pointing in some other direction, then there is *a priori* no obvious reason for confining the possible directions in which the spin can point. (By spin 'pointing in some direction' we mean that if we measure the component of spin in that direction we get with certainty the result $\frac{1}{2}\hbar$.)

Let us now return to the experiments with the modified Stern–Gerlach apparatus. We have discovered that if we put in a beam of neutral atomic hydrogen with the electron spin pointing in some specified direction, then we get out of the modified Stern–Gerlach apparatus a beam with the electron spin pointing in the same direction as when it went in. Inside the apparatus we expect the atoms to be split into two beams with spins pointing along and against the direction of the magnetic field in the first part of the modified Stern–Gerlach apparatus and obviously this direction need not be along the original spin direction. This suggests strongly that inside the apparatus the original spin has been 'resolved' into components along and against the field in the apparatus, but when they are recombined at the exit we recover the original spin direction, as with polarization in the optical case.

We can investigate this suggestion by injecting into the modified Stern–Gerlach apparatus a polarized beam of hydrogen atoms and analysing the polarization of the emergent beam, and its intensity, when we block off the first one, and then the other, of the two beams into which we expect the incident beam to be split inside the apparatus. (See Fig. 6.20). Remembering

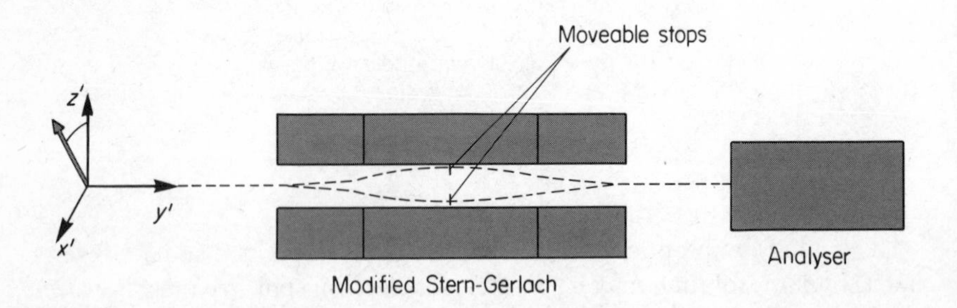

Fig. 6.20. By putting moveable beam stops in the modified Stern–Gerlach apparatus we may find the polarization directions of the two beams inside the apparatus, and their intensities, as a function of the angle θ between the initial spin polarization axis z and the z'-direction. We find the upper beam is polarized in the $-z'$-direction and the lower beam in the $+z'$-direction and that

$$I_{(+z')} = I_0 \cos^2 \theta/2$$
$$I_{(-z')} = I_0 \sin^2 \theta/2\ .$$

These should be compared with the optical case quoted in Fig. 6.19.

that the electron spin and magnetic moment point in opposite directions because of the sign of the electron charge, we expect the upper beam to have spin down and *vice versa*. This is indeed what we find. If we allow both beams through we get the original spin direction but if we let only the lower beam through we find the emergent beam is polarized along the z'-direction in the apparatus, and if we let only the upper beam through the polarization is in the direction of $-z'$.

We also investigate the intensities of the two beams as a function of the angle θ between the direction of the spin polarization of the initial beam and z'. Denoting the two intensities by $I(+z')$ and $I(-z')$ and the intensity with no shutter in either beam as I'_0 the intensities are

$$\left.\begin{aligned} I(+z') &= I_0 \cos^2 \theta/2 \\ I(-z') &= I_0 \sin^2 \theta/2 \end{aligned}\right\} \qquad (6.25)$$

Following the discussion in section 6.11 we expect the components of electron spin to be represented by three operators $\hat{s}_x, \hat{s}_y, \hat{s}_z$ which obey the algebraic relations Eq. (6.24). Once a z-direction has been chosen we find an electron has a component of spin along this direction of either $\frac{1}{2}\hbar$ or $-\frac{1}{2}\hbar$, that is, the operator has two eigenvalues, namely $\frac{1}{2}\hbar$ and $-\frac{1}{2}\hbar$. The eigenstates of $\hat{s}_z$ cannot be represented by conventional wavefunctions as they are not functions of position. Let us denote the eigenstate belonging to the eigenvalue $\frac{1}{2}\hbar$ by the symbol α and the eigenstate belonging to the eigenvalue $-\frac{1}{2}\hbar$ by the symbol β. The results we have obtained demonstrate quite clearly that we can have electrons polarized in directions other than the chosen z-direction, but these can always be represented by a superposition, with suitable amplitudes and phases, of the two eigenstates α and β of the operator $\hat{s}_z$ for the component of the spin along the chosen z-direction. Alternatively we may say that any actual state of polarization can be expressed as a superposition of just two basic states α and β referred to any arbitrarily chosen direction. These two statements are entirely equivalent, and are of course analogous to the situation in optics, where, once one has chosen two directions at right angles, one can express any state of polarization in terms of the amplitudes and phases along these directions. Alternatively a specific state of polarization in optics can be expressed as a superposition of two plane polarized waves with the direction of polarization arbitrarily chosen to be along any two directions at right angles in the plane normal to the beam.

Thus any spin state of an electron can be described in terms of the two basic states α and β referred to some arbitrarily chosen z-axis. A change in the direction of the z-axis of course changes the states α and β that we use in our description, and a given *physical* situation will have different coefficients in front of the two terms when we choose different z-directions.

In the last experiment we described we did not specify whether the variation of the angle θ between z and z' was to be achieved by changing z, i.e. putting in atoms with the electron spin pointing in different directions and thus finding out how different spin states are to be analysed along a *given* direction, or by changing z' and thus determining how a *given physical* situation is to be analysed when referred to different directions for the basic states. Not very surprisingly these two are very closely related as the rotations required in the two cases are equal but of opposite sign (see Fig. 6.21).

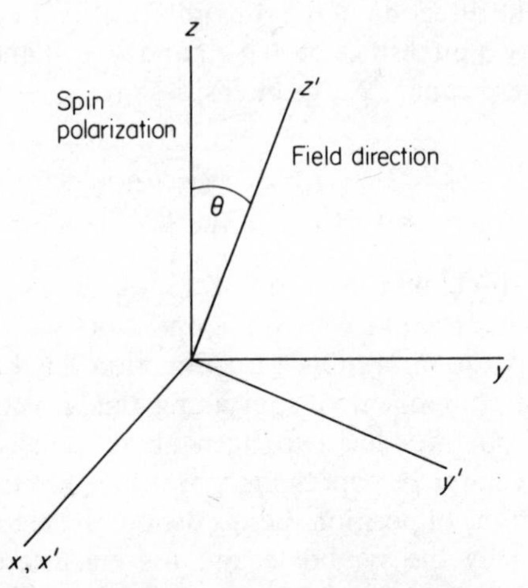

Fig. 6.21. If z represents the polarization direction of an atom and z' the field direction inside the modified Stern–Gerlach apparatus we may bring them into coincidence either by rotating z *anti-clockwise* around the x-axis through the angle θ or by rotating z *clockwise* around x' (the same as x) through θ.

It is important to emphasize that an unpolarized collection of electrons will consist of half the electrons with spin 'up', and half with spin 'down' whatever direction we choose for our z-axis. Further, there is no phase relation between the 'spin up' and 'spin down' states. Consequently we cannot describe an unpolarized beam by a wavefunction! It is, in fact, an incoherent superposition of wavefunctions. Mathematical techniques exist for handling such situations, though they are well beyond the scope of this book.

We may summarize the last few paragraphs by saying that a complete description of the state of an electron requires a description of its spin wavefunction as well as of its space wavefunction. For a given space wavefunction there are always two possible spin states, one with spin 'up' and one with spin 'down' with respect to some chosen direction. The choice of this direction is quite arbitrary. There may exist a preferred direction in space due to some external constraint (e.g. the imposition of a magnetic field) and in this case it is usually *convenient* to take this direction for one's chosen axis, but it is not necessary and any direction will do, though the mathematics may be more complicated if an unwise choice is made.

★ 6.14 THE PRACTICAL FEASIBILITY OF EXPERIMENTS USING THE MODIFIED STERN–GERLACH APPARATUS

In the last section we described a series of experiments using what we called a modified Stern–Gerlach apparatus in which the field and gradient in the central region was reversed compared with the same quantities in the outer regions. There is no objection *in principle* to these experiments, but it is interesting to enquire whether they are feasible in practice. First we pointed out in section 6.13 that it is necessary to reverse the field as well as the gradient in the middle section. The reason for this is as follows. Classically a particle with a magnetic moment pointing along the direction of its angular momentum would, on entering the magnetic field, precess round the field direction. Referring to Fig. 6.22, the field exerts a couple given by Eq. (6.6)

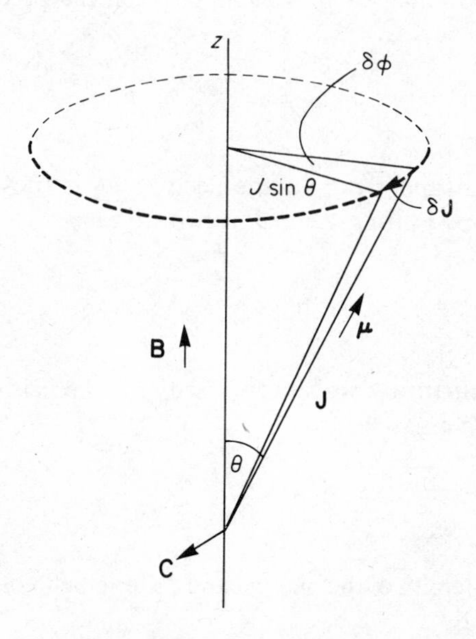

Fig. 6.22. The precession of an angular momentum **J** due to a torque **C** trying to align it with the *z*-axis.

which in this case can be written as

$$C = \mu B \sin \theta \tag{6.26}$$

where C points *out* of the page. Hence in a time δt we get a change in angular momentum

$$\delta J = \mu B \sin \theta \, \delta t$$

where δJ is at right angles to J. Thus, under the action of the couple the magnitude of J does not change but the plane containing B and J turns through an angle $\delta \phi$ given by

$$\delta \phi = \frac{\delta J}{J \sin \theta} = \frac{\mu B \delta t}{J} .$$

Hence the angular frequency of precession $\omega = \mathrm{d}\phi/\mathrm{d}t$ is

$$\omega = \frac{\mu B}{J} . \tag{6.27}$$

A quantum mechanical analysis of a particle with spin $\frac{1}{2}$ reveals precisely the same situation, i.e. there is a direction, which precesses around the field direction, along which one observes with certainty a component of angular momentum equal to $\frac{1}{2}\hbar$. For an electron the precession frequency is found to be

$$\omega_L = g \frac{eB}{2m}$$

where g is the gyromagnetic ratio of the electron. ω_L is known as the Larmor precession frequency. Putting g equal to two we get

$$\omega_L = \frac{eB}{m} . \tag{6.28}$$

Thus if, in our apparatus, we had reversed only the field gradient the spin would turn through an angle

$$\theta = \frac{eBL}{mv} \tag{6.29}$$

where L is the total length of the magnet and v the velocity of the atom passing through the magnet.

Let us assume B is 0·1 T and L is 1 m. For hydrogen emerging from an oven at 2,500 °C (required to produce atomic hydrogen) v is of the order of 5×10^3 m s^{-1}.

Hence

$$\theta = \frac{1\cdot6 \times 10^{-19} \times 0\cdot1 \times 1}{9\cdot1 \times 10^{-31} \times 5 \times 10^{3}} \simeq 3 \times 10^{6} \text{ radians} . \tag{6.30}$$

As the atoms emerging from the oven have a wide range of velocities they will precess through a wide range of angles and all sense of polarization will be lost. By reversing the field as well as the gradient in the centre section we reverse the direction of precession in this region. If the lengths of the magnets are sufficiently closely matched (to better than one part in ten million) the electron spin will emerge pointing in the same direction as when it entered. To achieve this degree of matching over all the paths followed by the atoms is clearly very difficult.

There is, however, a further and even more serious objection from an experimental point of view. Consider an atom with its electron spin pointing up and let us consider its passage from a region with the field pointing along the spin direction (call this region A) to a region B, where the field is pointing against the original spin direction. The field distribution in the transition region is shown in Fig. 6.23. We have assumed that in passing from region A to region B the electron spin continues in the 'up' direction. Will it do this or will it turn over by following the direction of the field lines? Again it is

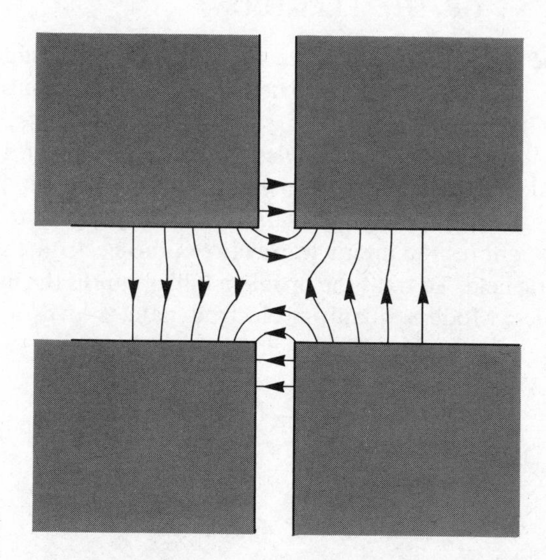

Fig. 6.23. The field in the transition region between two magnets which have fields in opposite directions.

useful to think classically. Classically the electron would precess around the field lines, and hence its precession axis would follow the field lines, providing the rate at which they turn as the atom moves through them is slow compared with the Larmor precession rate. Alternatively the precession axis will be preserved if the change of field direction takes place in a time that is short compared with the Larmor period. Similarly for an electron in quantum mechanics we may say the spin direction will be preserved if the field changes direction in a time short compared with the Larmor period ω_L^{-1} where ω_L is given by Eq. (6.29).

For a field of 0.1 T we have $\omega_L^{-1} \approx 6 \times 10^{-11}$ secs. Hence for an atom with a velocity of 5×10^3 m s^{-1} the transition from field up to field down must be made in a distance of less than 3×10^{-7} m which is clearly not a practical proposition. Thus in going from region A to region B the spin will turn over and the atom will not be refocused.

There are, therefore, severe practical difficulties in manufacturing the modified Stern–Gerlach apparatus. Matching the lengths is difficult but not impossible, but reversing a field in 10^{-7} m is beyond our present day experimental abilities. There is, however, no objection in principle to the apparatus, and the results deduced are confirmed by conceptually more complex experiments which are feasible.

★ 6.15 AN ACCURATE MEASUREMENT OF THE MAGNETIC MOMENT OF THE ELECTRON

Towards the end of section 6.12 we stated that the gyromagnetic ratio for the electron's spin is 2. In non-relativistic quantum mechanics electron spin is a postulate and there is no reason to suppose the gyromagnetic ratio should have any particular value. In relativistic quantum mechanics as developed by Dirac the idea of electron spin arises quite naturally and in Dirac's theory the electron has a gyromagnetic ratio, g, exactly equal to 2. However, Dirac's theory ignores the interaction between the electron and a quantized electromagnetic field. To treat the problem fully requires the use of quantum electrodynamics. Modern quantum electrodynamics gives a value for the gyromagnetic ratio

$$g = 2\left(1 + \frac{\alpha}{2\pi} - 0.328\frac{\alpha^2}{\pi} + \cdots\right) \tag{6.31}$$

where

$$\alpha = e^2/4\pi\varepsilon_0\hbar c \ .$$

If we insert the latest and most accurate value for α we get a value

$$g = 2(1.001159617 \pm 3 \times 10^{-9}) \ .$$

It is thus clearly important to measure g with the highest possible accuracy. The techniques developed for doing this are due to H. R. Crane and his coworkers.

The principle is as follows. Consider the situation in Fig. 6.24 where we

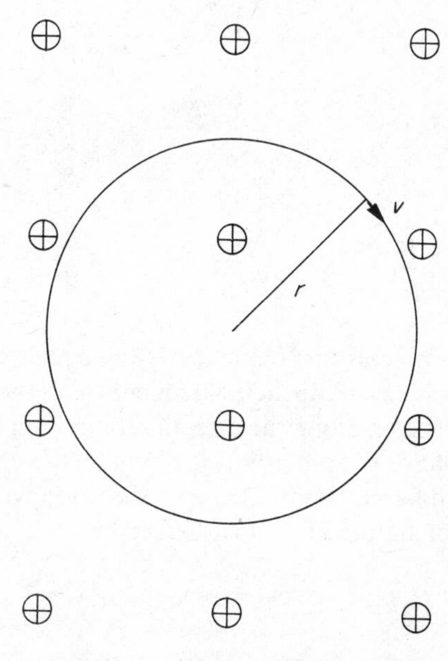

Fig. 6.24. An electron of velocity v in a uniform magnetic field B moves in a circle with an angular frequency $\omega_C = eB/m$. The Larmor precession frequency

$$\omega_L = \frac{eB}{2m}g \ .$$

If g is exactly equal to 2, then $\omega_L = \omega_C$, and an electron with spin initially pointing, say, along its direction of motion will stay in this condition.

have an electron moving in a plane at right angles to a magnetic field. Under the influence of the field the electron will move in a circle of radius r given by

$$\frac{mv^2}{r} = Bev \ .$$

Hence

$$r = \frac{mv}{Be} \, .$$

The angular revolution frequency ω_C is given by

$$\begin{aligned}\omega_C &= \frac{2\pi}{T} \\ &= \frac{2\pi}{(2\pi r/v)} \\ &= \frac{v}{r} \\ &= \frac{eB}{m} \, . \end{aligned} \tag{6.32}$$

Now the Larmor precession frequency ω_L is equal to $g(eB/2m)$. If g is exactly equal to 2 these two frequencies are identical. If therefore the electron was polarized with its spin along the direction of motion at one point in the circle, it will always have its spin pointing along this direction. On the other hand if g is slightly different from 2 the spin direction will slowly be turned out of the direction of motion at a rate ω_D given by

$$\omega_D = a\omega_C \tag{6.33}$$

where $g = 2(1 + a)$. A measurement of the frequency ω_D in a given field B will therefore give us a value for a.

To perform this experiment we require

(i) A source of electrons polarized along the direction of motion.

(ii) A method of trapping these electrons in a magnetic field for a specified (and variable) length of time.

(iii) A method of measuring the polarization of the electrons after ejection from the trap.

A completely polarized group of electrons is difficult to obtain, but also unnecessary. If we define the polarization P by the equation

$$P = \frac{N_+ - N_-}{N_+ + N_-} \tag{6.34}$$

where N_+ is the number of electrons with their spins pointing along their direction of motion and N_- is the number of electrons with their spins pointing against their direction of motion, then after a time t in the magnetic field we shall have

$$P(t) = P(0) \cos \omega_{D} t \ . \tag{6.35}$$

To both produce and detect the polarization of the electrons advantage was taken of the fact that a beam of electrons scattered through an angle in the region of 90° from a gold foil emerges partially polarized along the emergent direction of motion. Similarly the *fraction* of a polarized beam of electrons which are scattered through 90° depends on this state of polarization (see Fig. 6.25).

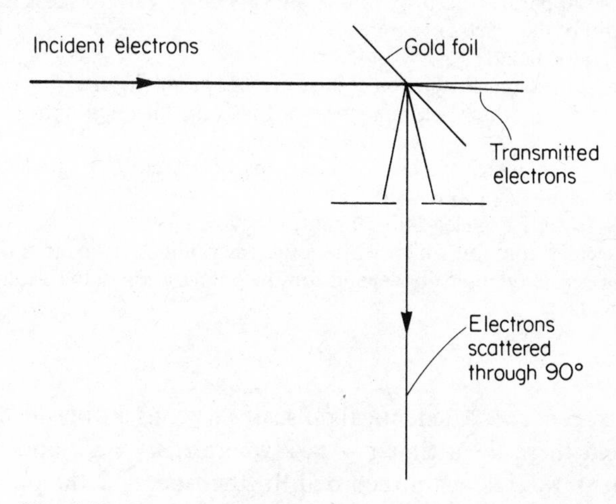

Fig. 6.25. Unpolarized electrons scattered by a gold foil through an angle in the region of 90° are partially polarized along their emergent direction of motion. If the incident beam is polarized along its direction of motion the intensity of electrons scattered through angles close to 90° depends on the degree and direction of polarization.

The experimental arrangement used is shown in Fig. 6.26. A pulse of electrons from an electron gun is scattered by a gold foil F_1 and a collimator C picks out electrons scattered through 89°. The energy of the electrons was in the range from 50 to 100 keV.

The scattered electrons spiral in the longitudinal field with a pitch of 1° and enter the trap formed by the two cylinders E and I within the field. The operation of the trap will be described shortly. After being held in the trap for a time t the electrons are ejected from the trap and spiral out to the right where they are scattered by a second gold foil F_2 through approximately 90° into the detector D. The number of electrons $N(t)$ detected in the

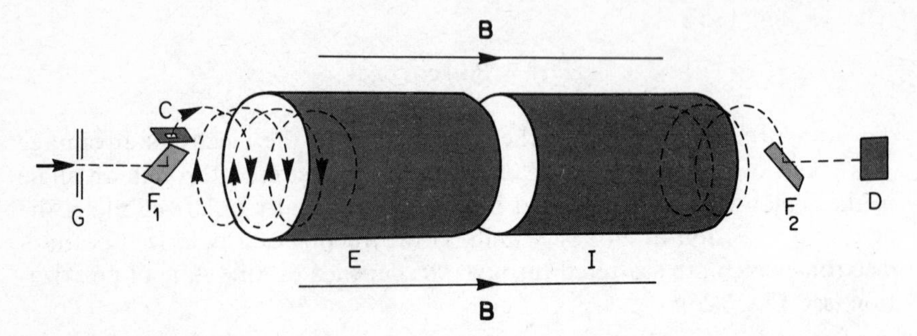

Fig. 6.26. The apparatus used by Crane and his coworkers to measure the gyromagnetic ratio of the electron.

B—Magnetic field.

G—an electron gun sending pulses of electrons along the direction of the field B.

F_1—a gold foil which scatters some of the beam through approximately 90° to produce partially polarized electrons.

C—collimator which picks out electrons scattered through 89° and thus producing a spiral with a 1° pitch.

E and I—cylindrical electrodes for trapping and ejection.

F_2—a second gold foil which scatters the emerging electrons into the detector. The scattered intensity depends on the polarization of the electrons.

D—detector.

detector D is recorded. An identical pulse of electrons is now injected into the trap and held there for a time $t + t_D/2$ where $t_D\omega_D = 2\pi$ and the number detected $N(t + t_D/2)$ is again recorded. In the time $t_D/2$ the polarization of the electrons will reverse, hence the quantity

$$Q(t) = \frac{N(t) - N(t + t_D/2)}{N(t) + N(t + t_D/2)} \tag{6.36}$$

is directly proportional to the polarization of the electrons after a time t in the trap. Thus by plotting $Q(t)$ against t Crane *et al.* obtain a sinusoidal function whose period equals $2\pi/\omega_D$. It is not necessary to know t_D with any accuracy in Eq. (6.36) as only the amplitude of $Q(t)$ and not its period is affected by the value chosen. Provided the same value is always used t_D need only be known to $\pm 50\%$.

We must now describe the operation of the trap. Consider first the motion of electrons spiralling in a slightly non-uniform field, as indicated in Fig. 6.27, which has its weakest point in the middle. The lines of force are therefore concave towards the axis. Electrons spiralling round the z-axis in a clockwise sense experience a force at right angles to the field, and this force, therefore, always has a component towards the weakest point of the field as well as the inwards force which keeps them moving in a circle round the z-axis. If the

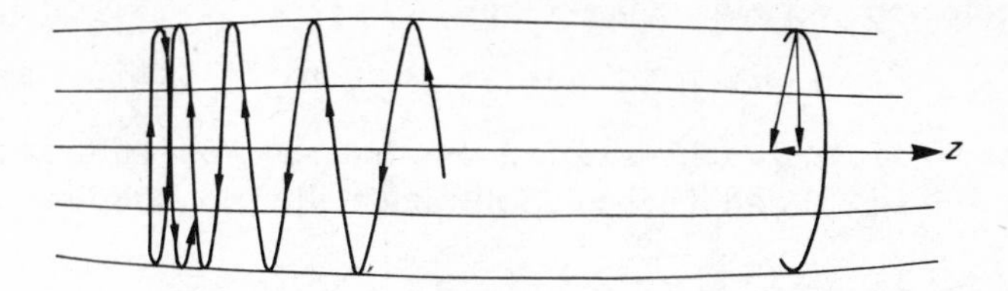

Fig. 6.27. The trapping effect of a field concave to the axis: On the right we show the force on a charged particle moving normal to the field. Due to the inclination of the field the force acting on the particle has a component which pushes it towards the central region. On the left we show the resultant motion of the particle projected onto a plane through the z-axis. The pitch becomes steeper as the particle moves towards the stronger field region until it is eventually reflected back to the right.

electron spiral has a certain pitch at the centre of the field, this pitch is reduced as they move away, and, under the action of the axial component of the force the motion along the z-axis is eventually reversed. The same thing occurs as they move in the opposite direction. Provided, therefore that the pitch of the spiral is sufficiently small, a locally non-uniform field of the type described will act as a trap for the electrons. The electrons are injected into the trap as follows. After being scattered by F_1 (see Fig. 6.26) the electrons spiral to the right with a pitch that increases slightly as they move towards the central portion of the field. The cylinder I has a negative potential of about 70 volts applied to it, relative to cylinder E. As the electrons cross from cylinder E to cylinder I their longitudinal velocity v_z is reduced by the field between the cylinders to a value such that they will be turned back inside the cylinder I. When they are well inside the cylinder the potential on cylinder I is removed so both E and I are at the same potential. This change of potential of I does not affect the motion of the electrons as they are well inside the cylinder. The timing of the removal of this potential is critical. When the electrons spiral back to the left there is no field to accelerate them back to the pitch they originally had within cylinder E; rather they now have the same pitch as when they were in the corresponding position in I. The electrons are therefore trapped.

The electrons are ejected from the trap to the right by applying a negative potential of about 20 to 30 volts to cylinder E.

A full analysis of the results has to include the influence of relativistic effects, and possible stray radial electric fields. After considering all these effects Crane and Wilkinson, in 1963, arrived at a value for a of

$$a = 0.001\,159\,622 \pm 27 \times 10^{-9}$$

to be compared with a theoretical value of

$$a = 0.001\,159\,617 \pm 50 \times 10^{-9} \ .$$

Thus the gyromagnetic ratio is known to within nearly one part in a hundred million, and agrees with the theoretical value well within this limit of error!

PROBLEMS

6.1 A particle of mass m moves round the origin under the action of a force $F = -\alpha/r^2$. Its maximum and minimum distances from the origin are r_1 and r_2 respectively. Show the total energy of the particle is $-\alpha/(r_1 + r_2)$ and its angular momentum around the origin is given by

$$J^2 = 2m\alpha\left(\frac{r_1 r_2}{r_1 + r_2}\right) .$$

6.2 A particle of mass m and charge $-e$ rotates in a circular orbit around a massive particle of charge $+e$. Show that the magnitude of the magnetic moment is given by

$$\mu = \frac{e^2}{2}\left(\frac{r}{4\pi\varepsilon_0 m}\right)^{1/2} ,$$

where r is the radius of the orbit.

6.3 An atom whose component of magnetic moment in the z-direction is 1 Bohr magneton is in a magnetic field whose gradient is 10 T m^{-1}. Calculate the force on the atom. Assuming that the atom has an atomic weight of 109, compare this force with that due to gravity.

6.4 Show that

$$x\frac{\mathrm{d}\psi}{\mathrm{d}x} - \frac{\mathrm{d}}{\mathrm{d}x}(x\psi) = -\psi$$

where ψ is any function of x. Hence show that

$$[x\hat{p} - \hat{p}x]\psi(x) = \mathrm{i}\hbar\psi(x)$$

where $\hat{p}$ is the quantum mechanical momentum operator $-\mathrm{i}\hbar(\mathrm{d}/\mathrm{d}x)$.

6.5 Denote $(x\hat{p}_x - \hat{p}_x x)$ by $[x, \hat{p}_x]$, etc. Then, as in Problem 6.4, we find

$$[x, \hat{p}_x] = [y, \hat{p}_y] = [z, \hat{p}_z] = \mathrm{i}\hbar$$

and

$$[x, \hat{p}_y] = [x, \hat{p}_z] = [y, \hat{p}_x] = [y, \hat{p}_z] = [z, \hat{p}_x] = [z, \hat{p}_y] = 0 \ .$$

Use these relations and the definition of angular momentum to obtain the relations of Eq. (6.23).

6.6 When a beam of neutral atoms passes through a Stern–Gerlach apparatus five equally-spaced lines are observed. What is the total angular momentum of the atom? Calculate the angle between the total angular momentum vector and the z-axis for each of the magnetic substates.

6.7 Silver atoms (atomic weight 109) emerge from an oven at 1300 K. Calculate the most probable velocity with which the atoms emerge. The atoms then pass into a magnet 50 cm long with a field gradient of $10\,\mathrm{T\,m^{-1}}$ and strike a screen 25 cm from the end of the magnet. Calculate the position on the screen at which silver atoms arrive which emerge with the most probable velocity and travel initially along the axis of the apparatus. The magnetic moment of neutral silver atoms is 1 Bohr magneton.

6.8 A beam of electrons is polarized with its spins perpendicular to the direction of motion. A magnetic field bends the beam through 90°. By treating the motion of the spin in the field as a classical precession, show that the spins are still perpendicular to the direction of motion after emerging from the field. (Assume $g = 2$ for electron spin.)

6.9 A ball has a diameter of 6 cm and a mass of 200 g. Calculate its angular momentum in units of $\hbar$ if it is rotating at 1 cycle per second. Repeat the calculation for a particle of the same density with a diameter of 6 microns rotating at 1 cycle per year.

CHAPTER 7

The energy levels of the hydrogen atom and single-electron ions

7.1 INTRODUCTION

In the last chapter we discussed the general features of the motion of a particle in a central field, and in particular its angular momentum properties. We discovered that the angular momentum properties are independent of the radial variation of the field, and that, whatever the nature of this variation, the angular momentum is quantized (we neglect intrinsic spin for the moment) both in magnitude and its component along some chosen direction, these two quantities being the most we can know.

We have also pointed out previously that there is experimental evidence for the quantization of energy and discussed qualitatively how this could arise. Clearly it is desirable to give a quantitative solution to at least one central field problem as these form such an important class of problems. We choose, for this purpose, the hydrogen atom as it is the only neutral atom for which an exact solution can be given (exact, of course, within the limits of non-relativistic quantum mechanics) and the wavefunctions which arise from its solution form the basis for a large number of qualitative and semi-quantitative discussions of more complex atoms.

Equally simple are the cases of atoms which have been ionized to such a degree that they have only one electron left. The solutions will clearly be of the same form, and differ only in the scale of lengths and energies involved. Thus the problem we shall tackle is that of a single electron moving in the field of a point nucleus with charge Ze. We shall, for the moment, ignore the influence of electron spin and hence of the magnetic moment of the electron.

7.2 THE CLASSICAL AND QUANTUM MECHANICAL HAMILTONIAN OF THE PROBLEM

The classical Hamiltonian is given by $H_c = T + V$, which, for this problem, is

$$H_c = \tfrac{1}{2}mv^2 - \frac{Ze^2}{4\pi\varepsilon_0 r} \tag{7.1}$$

where e is the magnitude of the electron charge, and the minus sign in front of the second term indicates that the potential is attractive. The quantum-mechanical problem is to be solved by finding the solution of the Schrödinger equation

$$H\Psi = E\Psi$$

where H is the quantum-mechanical Hamiltonian

$$H = -\frac{\hbar^2}{2m}\nabla^2 - \frac{Ze^2}{4\pi\varepsilon_0 r}\,. \tag{7.2}$$

In both cases we have the term $-[Ze^2/(4\pi\varepsilon_0 r)]$ which depends only on the radial distance r, and it is clear that it will be easier to find solutions in both cases if we express the first term in each case in radial and angular components, rather than in cartesian components.

The classical problem is particularly simple as we know that under the influence of a central force a particle moves in a plane orbit around the centre. We therefore resolve the velocity v into a radial component v_r and an angular component v_α, as indicated in Fig. 7.1. (We use α to indicate the angular position of the particle rather than θ to avoid confusion with the angle in spherical polar coordinates.) When this is done, we obtain

$$H_c = \tfrac{1}{2}m(v_r^2 + v_\alpha^2) - \frac{Ze^2}{4\pi\varepsilon_0 r}\,. \tag{7.1a}$$

Now

$$v_\alpha = r\frac{\mathrm{d}\alpha}{\mathrm{d}t}$$

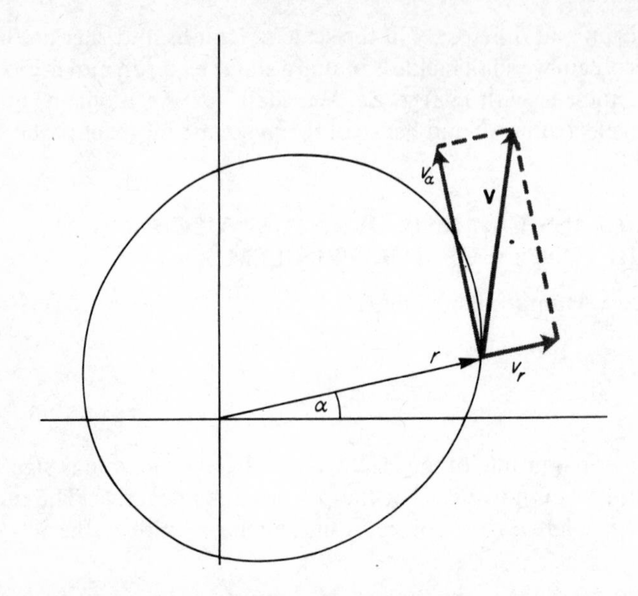

Fig. 7.1. Resolution of the orbital velocity of a particle moving in a plane orbit into a radial component v_r and an angular component v_α.

and

$$mr^2\frac{\mathrm{d}\alpha}{\mathrm{d}t} = l\ ,$$

where l is the angular momentum of the particle about the centre, so

$$v_\alpha^2 = \frac{l^2}{m^2r^2}\ . \tag{7.3}$$

Similarly

$$v_r^2 = \frac{p_r^2}{m^2} \tag{7.4}$$

where p_r is the radial component of the linear momentum. Thus Eq. (7.1) may be written in the form

$$H_c = \frac{p_r^2}{2m} + \frac{l^2}{2mr^2} - \frac{Ze^2}{4\pi\varepsilon_0 r}\ . \tag{7.5}$$

In the quantum-mechanical case, no such thing as an orbit in the classical sense exists, but it would appear sensible to express the operator ∇^2 in Eq. (7.2) in spherical polar coordinates rather than in cartesian coordinates as we have done so far.

The derivation of the expression for ∇^2 in polar coordinates is somewhat tedious and lengthy and is given in many mathematical texts (e.g. M. Margenau and G. M. Murphy, *The Mathematics of Physics and Chemistry*, D. Van Nostrand, New York, 1943, Chapter 5). The student is advised to accept the result as a fact. The formula is

$$\nabla^2 = \frac{1}{r^2}\frac{\partial}{\partial r}\left(r^2\frac{\partial}{\partial r}\right) + \frac{1}{r^2 \sin\theta}\frac{\partial}{\partial\theta}\left(\sin\theta\frac{\partial}{\partial\theta}\right) + \frac{1}{r^2\sin^2\theta}\frac{\partial^2}{\partial\phi^2}\,. \tag{7.6}$$

If we insert Eq. (7.6) into Eq. (7.2) we obtain for the quantum-mechanical Hamiltonian

$$H = -\frac{\hbar^2}{2m}\frac{1}{r^2}\frac{\partial}{\partial r}\left(r^2\frac{\partial}{\partial r}\right) - \frac{\hbar^2}{2mr^2}\left[\frac{1}{\sin\theta}\frac{\partial}{\partial\theta}\left(\sin\theta\frac{\partial}{\partial\theta}\right) + \frac{1}{\sin^2\phi}\frac{\partial^2}{\partial\phi^2}\right] - \frac{Ze^2}{4\pi\varepsilon_0 r}\,. \tag{7.7}$$

We see that there is a considerable similarity between Eq. (7.7) and Eq. (7.5). Thus in Eq. (7.7) the first term involves only *radial* derivatives and the first term of Eq. (7.5) involves only the radial momentum. Likewise the second term of Eq. (7.7) involves only *angular* derivatives and the second term in Eq. (7.5) involves only the angular momentum. Finally the third term, being the potential, is the same in both cases. This suggests that the term

$$-\frac{\hbar^2}{2m}\frac{1}{r^2}\frac{\partial}{\partial r}\left(r^2\frac{\partial}{\partial r}\right)$$

corresponds to $p_r^2/2m$, and that

$$-\frac{\hbar^2}{2mr^2}\left[\frac{1}{\sin\theta}\frac{\partial}{\partial\theta}\left(\sin\theta\frac{\partial}{\partial\theta}\right) + \frac{1}{\sin^2\theta}\frac{\partial^2}{\partial\phi^2}\right]$$

corresponds to $l^2/2mr^2$. By a procedure similar to that used in deriving ∇^2 in polar coordinate form one can show that the operator $\hat{l}^2 = \hat{l}_x^2 + \hat{l}_y^2 + \hat{l}_z^2$, where $\hat{l}_x$, $\hat{l}_y$ and $\hat{l}_z$ are given by Eqs. (6.10a, b, c), is indeed given in polar coordinate form by

$$\hat{l}^2 = -\hbar^2\left[\frac{1}{\sin\theta}\frac{\partial}{\partial\theta}\left(\sin\theta\frac{\partial}{\partial\theta}\right) + \frac{1}{\sin^2\theta}\frac{\partial^2}{\partial\phi^2}\right]\,. \tag{7.8}$$

Thus we can write Eq. (7.7) in the form

$$H = -\frac{\hbar^2}{2m}\frac{1}{r^2}\frac{\partial}{\partial r}\left(r^2\frac{\partial}{\partial r}\right) + \frac{\hat{l}^2}{2mr^2} - \frac{Ze^2}{4\pi\varepsilon_0 r} \tag{7.7a}$$

where $\hat{l}^2$ is the operator (7.8).

Returning to Eq. (7.5) for the classical Hamiltonian, we see that this is a one-dimensional equation for the motion of a particle moving in a potential

$$V(r) = -\frac{Ze^2}{4\pi\varepsilon_0 r} + \frac{l^2}{2mr^2} \tag{7.9}$$

as the angular momentum l is a constant of the motion. The first term of Eq. (7.9) is the actual central potential; the second term is the equivalent radial potential which arises from the centrifugal force due to the angular motion of the particle about the origin. This may be verified as follows. The force due to this term in Eq. (7.9) is

$$F = -\frac{\partial}{\partial r}\left(\frac{l^2}{2mr^2}\right) = \frac{l^2}{mr^3} .$$

Substituting for l^2 from Eq. (7.3) we obtain

$$F = mv_\alpha^2/r$$

which is the centrifugal force.

The same sort of thing happens in the quantum-mechanical case also. We know that atoms possess definite values for their angular momentum, as we discussed in Chapter 6, and that the eigenvalues of $\hat{l}^2$ are $l(l+1)\hbar^2$. Thus if $\Psi(r, \theta, \phi)$ is an eigenfunction of H, as expressed by Eq. (7.7a), belonging to some eigenvalue E, we know it will also be an eigenfunction of $\hat{l}^2$ belonging to some eigenvalue $l(l+1)\hbar^2$ i.e.

$$\left[-\frac{\hbar^2}{2m}\frac{1}{r^2}\frac{\partial}{\partial r}\left(r^2\frac{\partial}{\partial r}\right) + \frac{\hat{l}^2}{2mr^2} - \frac{Ze^2}{4\pi\varepsilon_0 r}\right]\Psi(r, \theta, \phi) = E\Psi(r, \theta, \phi) \tag{7.10}$$

and

$$\hat{l}^2\Psi(r, \theta, \phi) = l(l+1)\hbar^2\Psi(r, \theta, \phi) . \tag{7.11}$$

If we insert Eq. (7.11) into Eq. (7.10) we get

$$-\frac{\hbar^2}{2m}\frac{1}{r^2}\frac{\partial}{\partial r}\left(r^2\frac{\partial}{\partial r}\right)\Psi(r, \theta, \phi) + \left(\frac{l(l+1)\hbar^2}{2mr^2} - \frac{Ze^2}{4\pi\varepsilon_0 r}\right)\Psi(r, \theta, \phi) = E\Psi(r, \theta, \phi) . \tag{7.12}$$

This is almost, but not quite, like a one-dimensional Schrödinger equation for a particle moving in a potential

$$V(r) = -\frac{Ze^2}{4\pi\varepsilon_0 r} + \frac{l(l+1)\hbar^2}{2mr^2} \ . \tag{7.9a}$$

We shall see how to make the similarity more precise in a moment. It is, however, a differential equation for a function of r only. Hence its solution is a function of r only, say $\psi(r)$, which can be multiplied by a function of θ and ϕ, say $\chi(\theta, \phi)$ to give a solution of Eq. (7.12). So we have

$$\Psi(r, \theta, \phi) = \psi(r)\chi(\theta, \phi) \ . \tag{7.13}$$

If we insert Eq. (7.13) into Eq. (7.11), as $\hat{l}^2$ consists of differential operators which operate only on angular functions, we can cancel out the function $\psi(r)$ on either side and obtain

$$\hat{l}^2\chi(\theta, \phi) = l(l+1)\hbar^2\chi(\theta, \phi) \tag{7.14}$$

and the solutions of this equation determine the function $\chi(\theta, \phi)$ for us. We shall not pursue the nature of the functions $\chi(\theta, \phi)$ any further at the moment, nor indeed in any great detail later, but we would like to consider one special case, as it will enable us to see how to convert Eq. (7.12) into the exact analogue of a one-dimensional Schrödinger equation.

One possible solution of Eq. (7.14) is

$$\chi(\theta, \phi) = \text{constant} \ . \tag{7.15}$$

This solution belongs to the eigenvalue $l = 0$ of Eq. (7.14). If we put the constant in Eq. (7.15) equal to one and insert it into Eq. (7.13) we get

$$\Psi(r, \theta, \phi) \equiv \psi(r) \tag{7.16}$$

i.e. the solution is spherically symmetric. We shall have more to say about these solutions in the next section. The point we wish to make here is that the probability of finding a particle, described by a spherically symmetric wavefunction $\psi(r)$, between r and $r + \mathrm{d}r$ is (see Fig. 7.2)

$$\psi^*(r)\psi(r)4\pi r^2 \, \mathrm{d}r$$

as $4\pi r^2 \, \mathrm{d}r$ is the volume of space between r and $r + \mathrm{d}r$. In a true one-dimensional problem with a solution $\phi(x)$, the probability of finding the particle between x and $x + \mathrm{d}x$ is

$$\phi^*(x)\phi(x) \, \mathrm{d}x \ .$$

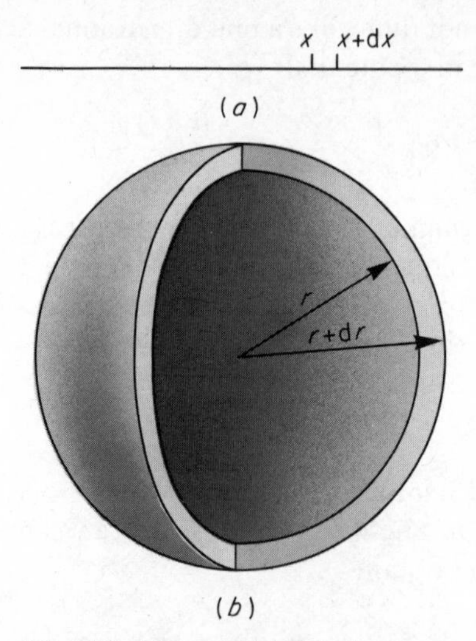

Fig. 7.2. The probability in a one-dimensional problem of finding a particle between x and $x + \mathrm{d}x$, as in (a), is given by $\psi^*(x)\psi(x)\,\mathrm{d}x$. In a three-dimensional problem the probability of finding a particle between r and $r + \mathrm{d}r$, as in (b), for a spherically symmetric wavefunction, is

$$4\pi r^2\psi^*(r)\psi(r)\,\mathrm{d}r \ .$$

In both cases we assume the wavefunctions are normalized.

Thus, the analogue of the one-dimensional wavefunction is not the function $\psi(r)$, but the function

$$u(r) = r\psi(r) \ . \tag{7.17}$$

This can be shown to be so even when $\chi(\theta, \phi)$ is not constant. If we substitute $u(r)/r$ for $\psi(r)$ in Eq. (7.13), insert this in Eq. (7.12) and cancel out the common factor $\chi(\theta, \phi)$, the first term becomes

$$\begin{aligned} -\frac{\hbar^2}{2m}\frac{1}{r^2}\frac{\partial}{\partial r}\left(r^2\frac{\partial}{\partial r}\frac{u}{r}\right) &= -\frac{\hbar^2}{2m}\frac{1}{r^2}\frac{\partial}{\partial r}r^2\left(\frac{1}{r}\frac{\partial u}{\partial r} - \frac{u}{r^2}\right) \\ &= -\frac{\hbar^2}{2m}\frac{1}{r^2}\frac{\partial}{\partial r}\left(r\frac{\partial u}{\partial r} - u\right) \\ &= -\frac{\hbar^2}{2m}\frac{1}{r}\frac{\partial^2 u}{\partial r^2} \ . \end{aligned}$$

Eq. (7.12) therefore becomes

$$-\frac{\hbar^2}{2m}\frac{1}{r}\frac{\partial^2 u(r)}{\partial r^2}+\left(\frac{l(l+1)\hbar^2}{2mr^2}-\frac{Ze^2}{4\pi\varepsilon_0 r}\right)\frac{u(r)}{r}=E\frac{u(r)}{r}$$

or multiplying through by r

$$-\frac{\hbar^2}{2m}\frac{\partial^2 u(r)}{\partial r^2}+\left(\frac{l(l+1)\hbar^2}{2mr^2}-\frac{Ze^2}{4\pi\varepsilon_0 r}\right)u(r)=Eu(r)\ . \tag{7.12a}$$

Eq. (7.12a) is *exactly* the same in form as the Schrödinger equation for a particle in one dimension moving in a potential

$$V(r)=-\frac{Ze^2}{4\pi\varepsilon_0 r}+\frac{l(l+1)\hbar^2}{2mr^2}\ . \tag{7.9a}$$

Thus the problem of finding the energy levels of the hydrogen atom, or of a hydrogen-like ion, or indeed for any central potential, reduces to solving the Schrödinger equation for a particle in one dimension moving in a potential made up of the given radial potential plus the centrifugal potential.

7.3 THE GROUND STATE SOLUTION

In Fig. 7.3 we plot the effective potential of Eq. (7.9a) for $l = 0, 1, 2$. By the uncertainty principle, the energy of the particle must be somewhat higher than the minimum of the potential energy curve, and it is immediately obvious that the ground state solution must be one with $l = 0$, i.e. an s-state solution.

We have already pointed out in the previous section that the solutions with $l = 0$ are spherically symmetric. This should be contrasted with the classical situation in which a particle with zero angular momentum moves in a straight line through the centre of force. The assertion that spherically symmetric states have zero angular momentum was based on the angular form of the operator $\hat{l}^2$ as given by Eq. (7.8). We have not derived this expression, and it is therefore worthwhile demonstrating that spherically symmetric wavefunctions have zero angular momentum.

Consider the effect of the operator $\hat{l}_z$ acting on some arbitrary spherically symmetric function $f(r)$. In Chapter 6, Eqs. (6.11) and (6.13) we showed that the operator $\hat{l}_z$ can be written as

$$\hat{l}_z=-i\hbar\frac{\partial}{\partial\phi} \tag{6.13}$$

from which is follows that $\hat{l}_z f(r) = 0$ and hence $\hat{l}_z^2 f(r) = 0$. Following Eq. (6.13) and in Fig. 6.8 we pointed out that the angular momentum operator

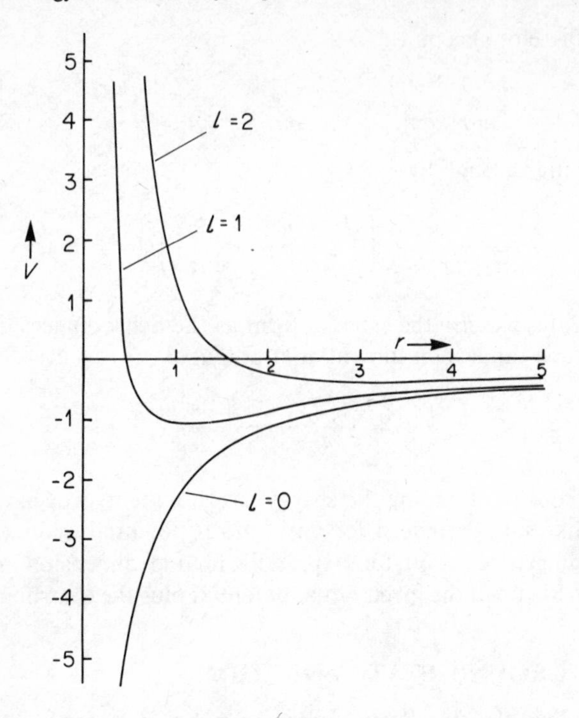

Fig. 7.3. The potential

$$V = -\frac{e^2}{4\pi\varepsilon_0 r} + \frac{l(l+1)\hbar^2}{2mr^2},$$

in units of 10^{-18} J, against r, in units of 10^{-10} m, for $l = 0$, 1 and 2.

for the component of angular momentum along any axis is given by $-i\hbar(\partial/\partial\alpha)$ where α measures an azimuthal rotation about that axis. As $f(r)$ is spherically symmetric $\partial f(r)/\partial\alpha = 0$ for all choices of the axis. In particular this means $\hat{l}_x f(r) = \hat{l}_y f(r) = 0$ and hence $\hat{l}_x^2 f(r) = \hat{l}_y^2 f(r) = 0$, and therefore

$$\hat{l}^2 f(r) = (\hat{l}_x^2 + \hat{l}_y^2 + \hat{l}_z^2) f(r) = 0 .$$

Thus $f(r)$ is an eigenfunction of the operator $\hat{l}^2$ belonging to the eigenvalue zero.

Eq. (7.12a) with $l = 0$ can be written in the form

$$\frac{\partial^2 u(r)}{\partial r^2} + \frac{2m}{\hbar^2}\left(\frac{Ze^2}{4\pi\varepsilon_0 r} + E\right)u(r) = 0 . \tag{7.18}$$

As a guide to finding a solution to Eq. (7.18) let us consider first the form of the solution of this equation at large values of r. For large r we can neglect the term $Ze^2/4\pi\varepsilon_0 r$ compared with E, and arrive at an approximate equation

$$\frac{\partial^2 u(r)}{\partial r^2} + \frac{2mE}{\hbar^2} u(r) = 0 \ . \tag{7.18a}$$

In writing the potential as $-Ze^2/4\pi\varepsilon_0 r$ we have implicitly taken our zero of energy to be when the electron is infinitely far from the origin. Thus the bound state solutions we are seeking will be ones in which the energy E is negative. The exact solutions to Eq. (7.18a) are

$$u(r) = \mathrm{e}^{\pm r/a} \tag{7.19}$$

where

$$a = \left(-\frac{\hbar^2}{2mE}\right)^{1/2} . \tag{7.20}$$

Thus we expect that the exact solutions to Eq. (7.18) will behave like Eq. (7.19) for sufficiently large values of r, and in particular will contain the exponential term of Eq. (7.19). The positive exponential tends to infinity as r tends to infinity, which offends against the boundary conditions laid down in section 5.6. Hence we must exclude this solution and only retain the negative exponential.

Let us now look at the situation at the origin. The origin is the point of lowest potential energy for the electron, and it is therefore reasonable to suppose that there is a high probability of finding the electron close to it. Hence for the ground state we would expect the electron wavefunction $\psi(r)$ at the origin to be non-vanishing. Remembering that according to section 5.6 all wavefunctions must be finite everywhere it follows that close to the origin $\psi(r)$ should be approximately constant, and hence $u(r) = r\psi(r)$ should be proportional to r.

Putting these considerations together with those for large r we guess that a likely form for the ground state function $u(r)$ might be

$$u(r) = r\,\mathrm{e}^{-r/a} \tag{7.21}$$

on the grounds that it is the simplest function that satisfies both sets of conditions.

Before inserting Eq. (7.21) into Eq. (7.18) to test whether it is a solution or not, we note that the parameter a acts as a range parameter for $u(r)$ determining its overall extent, and that a is related to the energy of the state E through Eq. (7.20). This relation is true for any potential that goes to zero as r tends to infinity, as for sufficiently large r we have, to a very close approximation, a particle moving in a uniform potential, namely zero.

We now insert Eq. (7.21) into the differential equation (7.18). Differentiating Eq. (7.21) we have

$$\frac{\partial u(r)}{\partial r} = -\frac{r}{a}\mathrm{e}^{-r/a} + \mathrm{e}^{-r/a}$$

and

$$\frac{\partial^2 u(r)}{\partial r^2} = \frac{r}{a^2}\mathrm{e}^{-r/a} - \frac{2}{a}\mathrm{e}^{-r/a}$$

$$= \frac{u(r)}{a^2} - \frac{2u(r)}{ar} .$$

Inserting this in Eq. (7.18) we obtain

$$\frac{u}{a^2} - \frac{2u}{ar} + \frac{2m}{\hbar^2}\left(\frac{Ze^2}{4\pi\varepsilon_0 r} + E\right)u = 0 .$$

This equation must be true for all r, so we must have, equating coefficients of terms in r^0 and r^{-1},

$$\frac{1}{a^2} = -\frac{2mE}{\hbar^2} \tag{7.20a}$$

$$a = \frac{\hbar^2 4\pi\varepsilon_0}{Zme^2} . \tag{7.22}$$

Eq. (7.20a) is just a restatement of the relation between the range parameter and the energy contained in Eq. (7.20) and is a consequence of the boundary condition at infinity. Eq. (7.22) is the result of satisfying the boundary condition at the origin, with the particular type of function given by Eq. (7.21), and gives a relation between a and the constants in the differential equation, and hence, through Eq. (7.20a), an expression for E. Combining Eqs. (7.20a) and (7.22) we obtain

$$E = -\frac{Z^2 e^4 m}{(4\pi\varepsilon_0)^2 \cdot 2\hbar^2} \tag{7.23}$$

as an expression for the lowest energy state for a single electron moving in the field of a point nucleus of charge Ze.

If we put in the values for e, m and $\hbar$

$$e = 1.6 \times 10^{-19}\ \mathrm{C}$$

$$m = 9.1 \times 10^{-31}\ \mathrm{Kg}$$

$$\hbar = 1.05 \times 10^{-34}\ \mathrm{J\,s}$$

we obtain

$$E = -21.9 \times 10^{-19} \times Z^2 \text{ joules}$$
$$= -13.6Z^2 \text{ eV} .$$

The ionization potential for atomic hydrogen is 13.6 V and the second ionization potential for helium (the energy required to remove the second electron, the first having been already removed) is 54.4 V in excellent agreement with the values of Eq. (7.23) with $Z = 1$ and 2 respectively.

The expression (7.22) for the range parameter for the case when Z equals one is always denoted by the *special symbol* a_0

$$\boxed{a_0 = \frac{4\pi\varepsilon_0\hbar^2}{me^2}} \tag{7.24}$$

and is known as the *Bohr radius.*

We note that Eq. (7.24) is *exactly* the length we obtained in Eq. (3.4) when we were looking for a unit of length to characterize the sizes of atoms. Its value is

$$\boxed{a_0 = 0.529 \times 10^{-10} \text{ m} .}$$

We shall show in the next section that the electron is most likely to be found in the region of $r = a_0/Z$. An electron at a distance a_0/Z from a nucleus of charge Ze would have a potential energy $-Z^2e^2/4\pi\varepsilon_0 a_0$ and this would seem a natural unit of energy in which to express E. If we multiply Eqs. (7.20a) and (7.22) together, we obtain

$$-\frac{2E \cdot 4\pi\varepsilon_0}{Ze^2} = \frac{1}{a}$$
$$= \frac{Z}{a_0} .$$

Therefore

$$\boxed{E = -\frac{1}{2}\frac{Z^2e^2}{4\pi\varepsilon_0 a_0} .} \tag{7.23a}$$

7.4 NORMALIZATION OF THE GROUND STATE WAVEFUNCTION AND OTHER FEATURES OF THE SOLUTION

The full solution to the hydrogen atom problem has the form given in Eq. (7.13)

$$\Psi(r, \theta, \phi) = \psi(r)\chi(\theta, \phi) \ . \tag{7.13}$$

We have considered the special case where $\chi(\theta, \phi)$ is a constant and found that there exists a solution where $\psi(r) \equiv u(r)/r$ has the form

$$\psi(r) = e^{-Zr/a_0} \ . \tag{7.21a}$$

As Schröndinger's equation is homogeneous, we can multiply any solution by an arbitrary constant and still have a solution. We would like to choose this constant so that the wavefunction is normalized, i.e. so that

$$\int |\Psi(r, \theta, \phi)|^2 \, d^3\mathbf{r} = 1 \ . \tag{7.25}$$

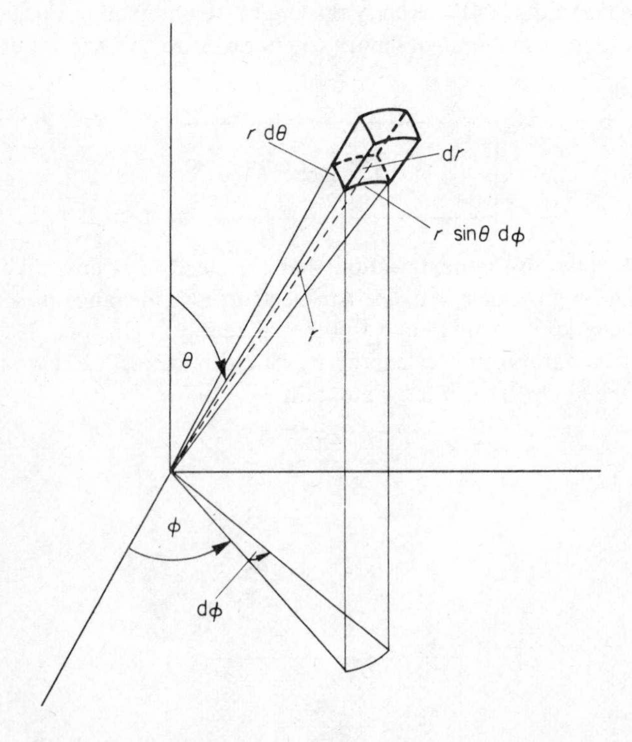

Fig. 7.4. The volume element in spherical polar coordinates
$d^3\mathbf{r} = (r \sin\theta \, d\phi)(r \, d\theta) \, dr = r^2 \, dr \sin\theta \, d\theta \, d\phi$.

From Fig. 7.4 we see that the volume element $d^3\mathbf{r}$ in a spherical polar coordinate system is given by

$$d^3\mathbf{r} = r^2\, dr \sin\theta\, d\theta\, d\phi$$

so that the left-hand side of Eq. (7.25) becomes

$$\int_0^\infty \int_0^\pi \int_0^{2\pi} |\Psi(r, \theta, \phi)|^2 r^2\, dr \sin\theta\, d\theta\, d\phi$$

$$= \int_0^\infty r^2 |\psi(r)|^2\, dr \int_0^\pi \int_0^{2\pi} |\chi(\theta, \phi)|^2 \sin\theta\, d\theta\, d\phi$$

where we have used Eq. (7.13).

It is conventional to choose the normalization of $\chi(\theta, \phi)$ so that the integration over the angular variables comes to unity. If $\chi(\theta, \phi)$ is independent of angle, as in the case we are considering, $\chi(\theta, \phi)$ must be put equal to $1/\sqrt{(4\pi)}$ to achieve this result.

With this convention the normalization condition (7.25) becomes

$$\int_0^\infty r^2 |\psi(r)|^2\, dr = \int_0^\infty |u(r)|^2\, dr = 1 \,. \tag{7.25a}$$

We shall achieve this condition if we write

$$u(r) = A^{-1/2} r\, e^{-Zr/a_0}$$

where

$$A = \int_0^\infty (r\, e^{-Zr/a_0})^2\, dr = \int_0^\infty r^2\, e^{-2Zr/a_0}\, dr \,. \tag{7.26}$$

Integrals of the form of Eq. (7.26) occur quite frequently in atomic physics, and the general result is worth stating, namely

$$\int_0^\infty x^n\, e^{-\alpha x}\, dx = \frac{n!}{\alpha^{n+1}} \qquad (\alpha > 0, n \geqslant 0) \,. \tag{7.27}$$

Using this result we have, from Eq. (7.26),

$$A = \frac{2}{(2Z/a_0)^3} = \frac{1}{4}\left(\frac{a_0}{Z}\right)^3$$

and hence the normalized version of Eq. (7.21) is

$$u(r) = 2\left(\frac{Z}{a_0}\right)^{3/2} r\,\mathrm{e}^{-Zr/a_0} \,. \tag{7.21b}$$

In Fig. 7.5 we plot the wavefunction $R(r) \equiv u(r)/r$, measured in units of $(Z/a_0)^{1/2}$, against $\rho = Zr/a_0$, and in Fig. 7.6 the radial probability density

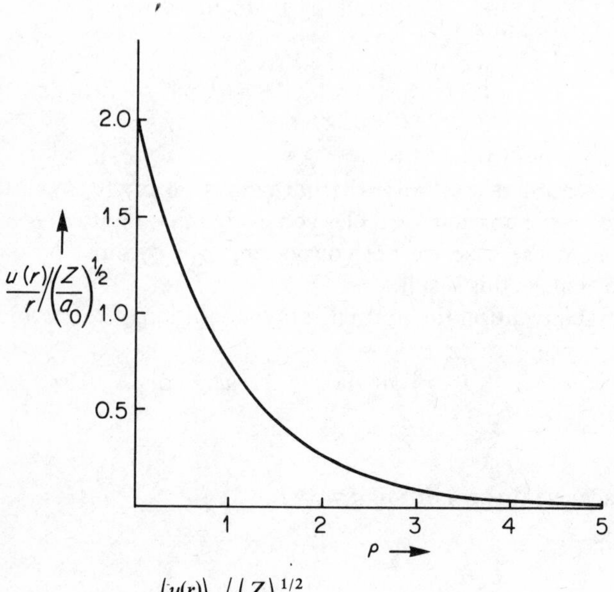

Fig. 7.5. $\left(\frac{u(r)}{r}\right)\Big/\left(\frac{Z}{a_0}\right)^{1/2}$ against $\rho = Zr/a_0$ for

$$u(r) = 2\left(\frac{Z}{a_0}\right)^{3/2} r\,\mathrm{e}^{-Zr/a_0} \,.$$

distribution $|u(r)|^2$, measured in units of (Z/a_0), against ρ. The area under the latter curve is unity, and the probability of the electron being found between ρ and $\rho + \mathrm{d}\rho$ is $\mathrm{d}\rho$ times the ordinate of the curve at ρ.

We see from Fig. 7.6 that there is a most probable value of ρ at which the electron is likely to be found. This is given by

$$\frac{\mathrm{d}}{\mathrm{d}r}(u^* u) = 0$$

i.e.

$$\frac{\mathrm{d}}{\mathrm{d}r}(r^2\,\mathrm{e}^{-2Zr/a_0}) = 0$$

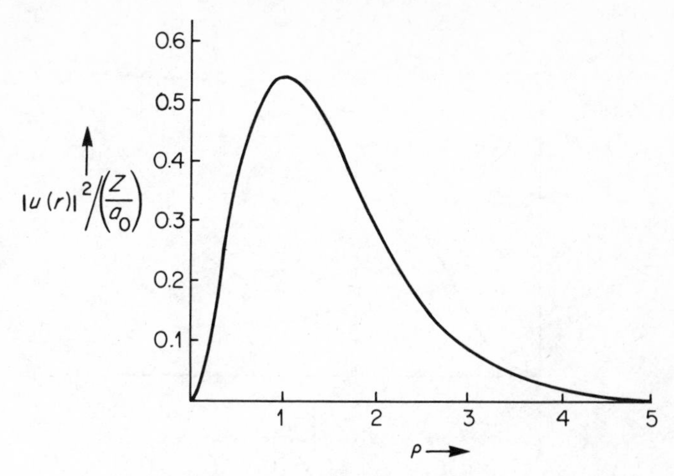

Fig. 7.6. $|u(r)|^2/(Z/a_0)$ against $\rho = Zr/a_0$ for $u(r) = 2(Z/a_0)^{3/2} r\,e^{-Zr/a_0}$. The area under the curve is unity, and the probability of finding the particle between ρ and $\rho + d\rho$ is $d\rho$ times the ordinate of the curve at ρ.

or

$$\left\{2r\,e^{-2Zr/a_0} - \frac{2r^2Z}{a_0}e^{-2Zr/a_0}\right\} = 0$$

which has the solution

$$r = a_0/Z\ . \tag{7.28}$$

Thus according to quantum mechanics the electron can be found anywhere in space, but is most likely to be found near $r = a_0/Z$. In Fig. 7.7 we plot the probability that the electron will be found within a radius r, where r is measured in units of a_0/Z. We see that there is a 94% probability that the electron will be found within a distance of $3a_0/Z$ from the origin.

Fig. 7.7 also shows clearly the usual quantum mechanical phenomenon that there is a finite probability of finding the electron in classically forbidden regions. Classically, the electron could not be further out than r_0, where

$$-\frac{Ze^2}{4\pi\varepsilon_0 r_0} = E$$

$$= -\frac{Z^2e^2}{4\pi\varepsilon_0\,.\,2a_0}$$

from Eq. (7.23a). Hence

$$r_0 = 2a_0/Z\ .$$

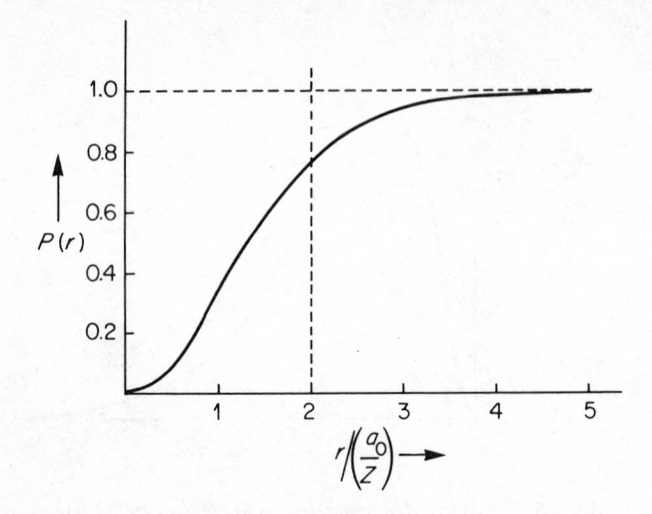

Fig. 7.7. The probability $P(r)$ that a particle will be found within a radius r, against r, measured in units of a_0/Z for the ground state hydrogen-like wavefunction. The vertical dotted line indicates the maximum excursion a particle with the ground state energy could have classically.

From Fig. 7.7 there is a 24% probability that the electron will be found further out than this.

7.5 OTHER SPHERICALLY SYMMETRIC STATES OF THE HYDROGEN ATOM

Eq. (7.18) represents a general equation for finding the allowed energies of hydrogen-like atoms in s-states, i.e. those that are spherically symmetrical. We have found one solution, but we expect that there will be other solutions corresponding to less strongly bound states which satisfy the boundary conditions at the origin and at infinity. Also according to the discussion of section 5.8, these solutions must be orthogonal to the ground state solution. These considerations supply us with enough information to guess the next simplest solution to the ground state solution. The arguments of section 7.3 leading to an exponential fall at large r still hold with equal force. Thus, we expect a factor of the form e^{-Zr/na_0}, where we expect n to be a number greater than one because of the relation between energy and range parameter given by Eq. (7.20). At small r the potential $-Ze^2/(4\pi\varepsilon_0 r)$ is much larger in magnitude than the energy E, and so the behaviour of any allowed solution of Eq. (7.18) near the origin should be the same, so we expect a factor r to appear. Finally, the ground state solution Eq. (7.21) is positive for all r. A solution that is orthogonal to Eq. (7.21) must therefore be positive in some regions and negative in others, i.e. it must cross the axis at least once, as illustrated in Fig. 7.8. Clearly the simplest solution is one which crosses the axis just once, at some point $r = b$. We therefore include a term of the form $(1 - r/b)$.

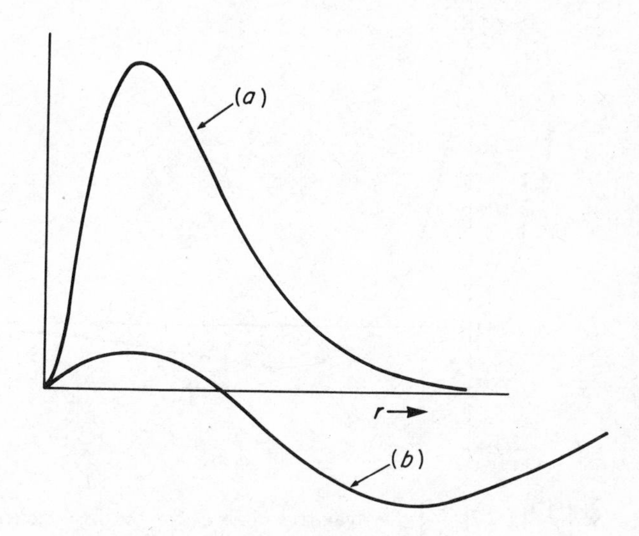

Fig. 7.8. The curve (*a*) is everywhere positive. A curve (*b*) which is orthogonal to curve (*a*) must cross the axis at least once, so that the product has both positive and negative regions.

Putting these factors together, we arrive at a suggested solution

$$u(r) = r(1 - r/b)\,e^{-Zr/na_0} \;. \tag{7.29}$$

If we insert Eq. (7.29) into Eq. (7.18) and equate the coefficients of the various powers of r separately to zero, we find the following conditions

$$\left.\begin{aligned} n &= 2 \\ b &= 2a_0/Z \\ E &= -\frac{1}{4}\frac{Z^2 e^2}{4\pi\varepsilon_0 2a_0} \;. \end{aligned}\right\} \tag{7.30}$$

After inserting these values for n and b into Eq. (7.29) we may normalize the solution in the same way as in section 7.4 to obtain

$$u(r) = 2\left(\frac{Z}{2a_0}\right)^{3/2} r\left(1 - \frac{Zr}{2a_0}\right) e^{-Zr/2a_0} \;. \tag{7.29a}$$

In Fig. 7.9 we plot $R(r) = u(r)/r$ for this state and in Fig. 7.10 we plot the radial probability function $|u(r)|^2$.

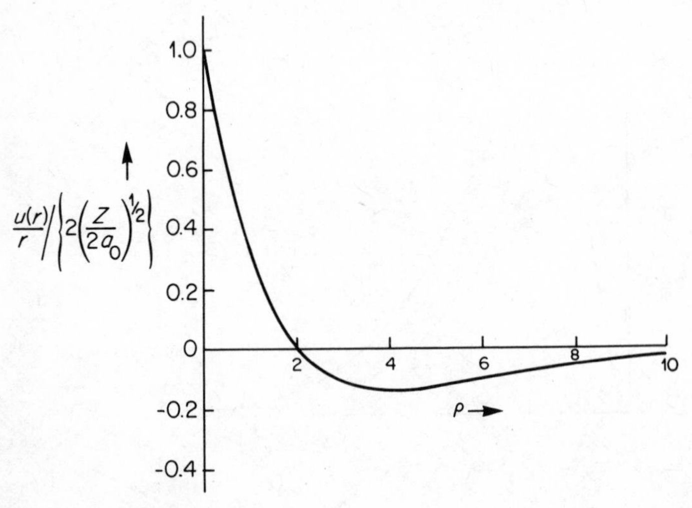

Fig. 7.9. $\left(\dfrac{u(r)}{r}\right)\bigg/\left\{2\left(\dfrac{Z}{2a_0}\right)^{1/2}\right\}$ against $\rho = Zr/a_0$ for the first excited spherically symmetric state of a hydrogen-like atom, Eq. (7.29a).

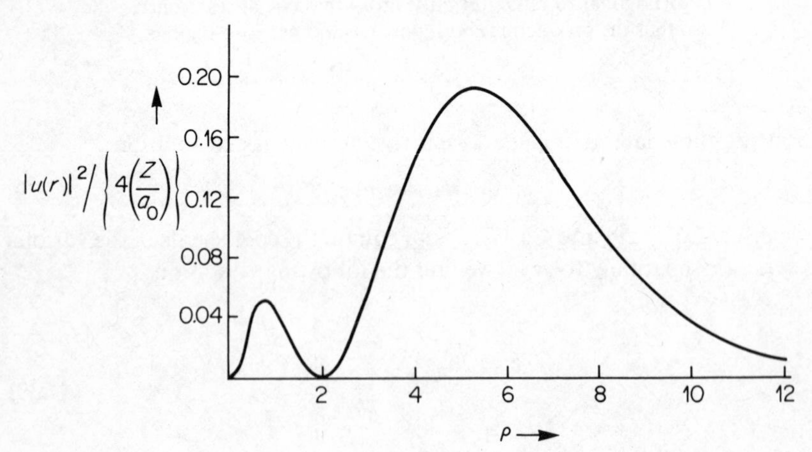

Fig. 7.10. $|u(r)|^2\bigg/\left\{4\left(\dfrac{Z}{a_0}\right)\right\}$ against $\rho = Zr/a_0$ for

$$u(r) = 2(Z/2a_0)^{3/2}r(1 - Zr/2a_0)\,e^{-Zr/2a_0}\,.$$

The area under the curve is unity, and the probability of finding the particle between ρ and $\rho + d\rho$ is $d\rho$ times the ordinate at ρ.

The success of our method of finding a second solution to Eq. (7.18) suggests a general procedure we could use. We shall not carry this procedure through, but merely indicate how it works. We have found a solution which does not cross the axis and one which crosses the axis once. This suggests that there are solutions which cross the axis twice, thrice, four times, etc. To satisfy the boundary conditions these would have to be of the form

$$u(r) = r(1 - r/b_1)(1 - r/b_2)\ldots(1 - r/b_{n-1})\,\mathrm{e}^{-Zr/pa_0},$$

where $b_1, b_2, \ldots$ and p are constants. We can write this in the form

$$u(r) = \mathrm{e}^{-Zr/(pa_0)}f(r) \tag{7.31}$$

where $f(r)$ is a polynomial in r with a highest power of r^n, and a lowest power, r, to satisfy boundary conditions at the origin, and $Z^2e^2/(4\pi\varepsilon_0 \cdot 2p^2a_0) = -E$ to satisfy the asymptotic equation for large r. When we insert Eq. (7.31) into Eq. (7.18) we find it gives us a recurrence relation between the coefficients of successive powers of r in the polynomial $f(r)$. If we take an arbitrary negative value for the energy E which corresponds to an arbitrary positive value for the number p we find that the series expression for $f(r)$ does not terminate and the ratio of successive powers makes it behave like e^{2Zr/pa_0} for large r. Thus if p is not an integer $f(r)$, as determined by the differential equation, is *not* a polynomial as we have required, and the solution as a whole does *not* satisfy the boundary condition that the wave function vanishes at infinity. If p is an integer n, the series does terminate with a highest power r^n. Thus the allowed energy levels are quantized and have energies given by

$$\boxed{E_n = -\frac{1}{n^2}\frac{Z^2e^2}{4\pi\varepsilon_0 2a_0}} \tag{7.32}$$

where n is an integer, $n = 1, 2, 3\ldots$ The function $u_n(r)$ for this state is

$$u_n(r) = \mathrm{e}^{-Zr/na_0}f(r) \tag{7.31a}$$

where $f(r)$ is a polynomial in r with a highest power of r^n, where n is the same integer as occurs in Eqs. (7.31a) and (7.32).

It should be emphasized that the quantization comes about through having to satisfy *two* boundary conditions, one at infinity and one at the origin. What we have done is to choose a form, an expansion in ascending integral powers of r, which guarantees the correct behaviour at the origin, and we then discover that only for certain energies can we also satisfy the condition at infinity. It is perfectly possible to work the other way round and choose a form that guarantees the behaviour at infinity. We then find that for arbitrary E the wave function becomes infinite at the origin, and only for the values given by Eq. (7.32) do we obtain the correct behaviour at the origin.

To illustrate these points, in Fig. 7.11 we show the solution to Eq. (7.18) in which we put

$$E = -\frac{Z^2e^2}{4\pi\varepsilon_0 2a_0(0.8)^2}$$

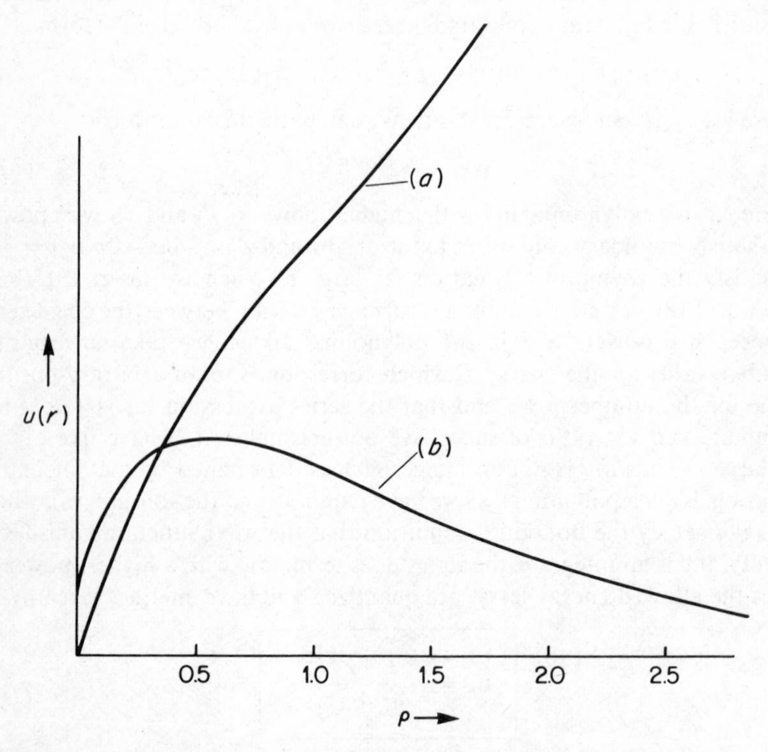

Fig. 7.11. The radial wavefunction $u(r)$ plotted against $\rho = Zr/a_0$ for the spherically symmetric case in a hydrogen-like atom with $p = 0.8$: (*a*) the case which behaves properly at the origin but diverges at infinity; (*b*) the case which behaves properly at infinity but remains finite at the origin instead of going to zero.

that is, we assume $p = 0.8$. Fig. 7.11(*a*) shows the solution which behaves properly at the origin, but which diverges at infinity, and Fig. 7.11(*b*) shows the solution with the proper behaviour at infinity, but in which $u(r)$ remains finite at the origin, with the result that $\psi(r)$ becomes infinite there.

The values for the energy levels of atomic hydrogen and hydrogen-like ions given by Eq. (7.32) are found to agree very accurately with the values deduced experimentally. Further, Eq. (7.32) accounts for the energies of *all* the levels that are found experimentally, and yet we have so far considered only the s-wave levels which have zero angular momentum, and there must surely exist levels with other values of the angular momentum. The only conclusion we can come to is that these other levels must be degenerate in energy with some of the s-wave levels, and we shall see that this is indeed the case, a situation peculiar to this problem.

7.6 NON-SPHERICALLY SYMMETRIC STATES OF THE HYDROGEN ATOM AND HYDROGEN-LIKE IONS—THE p-STATES

So far we have confined our discussion of energy levels to the spherically symmetric s-states. States that are not spherically symmetric possess angular momentum, and the effective radial potential is modified as is shown in Fig. 7.3. It would be very surprising if this modification were to result in there being no bound states. In Fig. 7.12 we redraw the effective radial

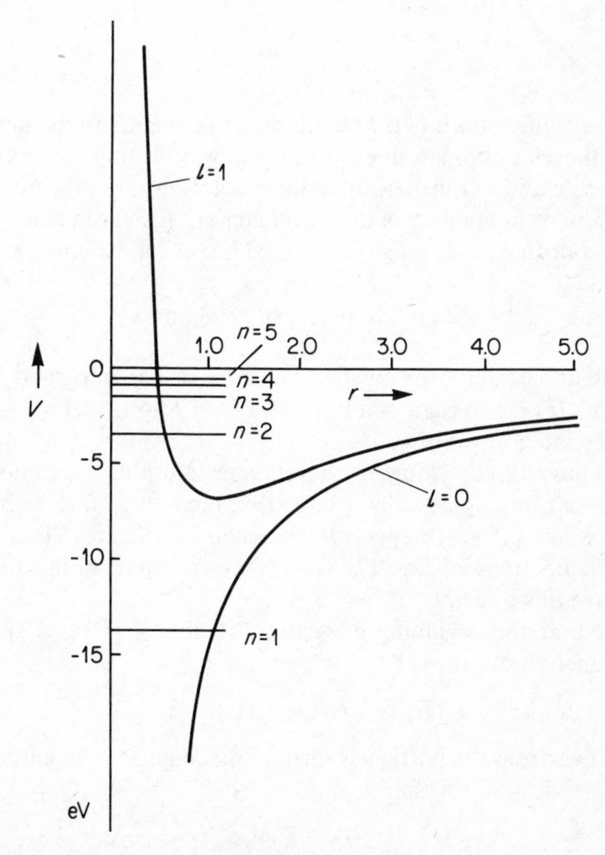

Fig. 7.12. The potential energy curves for $l = 0$ and $l = 1$ for the hydrogen atom, in electron volts, against the radius r in units of 10^{-10} m. Also shown on the left hand side are the positions of the first five energy levels for the $l = 0$ case.

potentials for hydrogen for the cases of $l = 0$ and $l = 1$, and add in the positions of the energy levels for the $l = 0$ case.

We see that only the lowest level lies below the minimum of the $l = 1$ potential, and if levels of higher energy can exist in the $l = 0$ potential there seems no *a priori* reason why we should not get similar levels in the other potentials.

The simplest non-spherically symmetric wavefunction we can write down is of the form

$$\Psi_0 = zf(r) \ . \tag{7.33a}$$

If we can find a function $f(r)$ which makes Eq. (7.33a) an eigenfunction of the Hamiltonian of Eq. (7.2), belonging to some energy E, then

$$\Psi_x = xf(r) \tag{7.33b}$$

and

$$\Psi_y = yf(r) \tag{7.33c}$$

must also be eigenfunctions of the Hamiltonian belonging to the same eigenvalue E, as the Hamiltonian does not in any way distinguish between the coordinates x, y and z. Further, these three solutions are all orthogonal to each other. The orthogonality of the wavefunctions (7.33) can readily be seen in cartesian coordinates. Taking (7.33a) and (7.33b), for example, we have

$$\iiint zx[f(x^2 + y^2 + z^2)^{1/2}]^2 \,\mathrm{d}x\,\mathrm{d}y\,\mathrm{d}z \ .$$

Consider the integration over x with y and z fixed. The integrand is an odd function of x ($F(x, y, z)$ is an odd function of x if $F(-x, y, z) = -F(x, y, z)$) and hence the integral vanishes.

Hence we have three orthogonal degenerate solutions. In section 6.7 we pointed out that in a spherically symmetric potential a state with angular momentum l has $(2l + 1)$ degenerate magnetic sub-states. This suggests that the wavefunctions of Eqs. (7.33a,b,c) have angular momentum $l = 1$. We may prove this directly.

We showed, at the beginning of section 7.3, that if $f(r)$ is a spherically symmetric function of r then

$$\hat{l}_x f(r) = \hat{l}_y f(r) = \hat{l}_z f(r) = 0 \ .$$

Hence it follows from the particular form of the angular momentum operators, as given by Eqs. (6.10), that

$$\hat{l}_z z f(r) = f(r)\hat{l}_z z$$

and correspondingly for the other angular momentum operators and the other functions.

Using explicit forms of the operators $\hat{l}_x$, $\hat{l}_y$ and $\hat{l}_z$ as given by Eqs. (6.10a,b,c) namely

$$\hat{l}_x = -i\hbar\left(y\frac{\partial}{\partial z} - z\frac{\partial}{\partial y}\right) \qquad \text{etc.}$$

we find

$$\left.\begin{array}{lll} \hat{l}_x x = 0 & \hat{l}_x y = i\hbar z & \hat{l}_x z = -i\hbar y \\ \hat{l}_y x = -i\hbar z & \hat{l}_y y = 0 & \hat{l}_y z = i\hbar x \\ \hat{l}_z x = i\hbar y & \hat{l}_z y = -i\hbar x & \hat{l}_z z = 0\ . \end{array}\right\} \qquad (7.34)$$

From Eqs. (7.34) we obtain

$$\hat{l}_x^2 zf(r) = \hat{l}_x(-i\hbar y f(r)) = \hbar^2 zf(r)$$

$$\hat{l}_y^2 zf(r) = \hbar^2 zf(r)$$

$$\hat{l}_z^2 zf(r) = 0$$

and therefore

$$\hat{l}^2 zf(r) = 2\hbar^2 zf(r)\ .$$

Thus we conclude that $\psi = zf(r)$ is an eigenfunction of $\hat{l}^2$ belonging to the eigenvalue $2\hbar^2$. But the eigenvalues of $\hat{l}^2$ are $l(l+1)\hbar^2$ and so $l = 1$ in this case.

By symmetry it follows that $yf(r)$ and $xf(r)$ are also eigenfunctions of $\hat{l}^2$ belonging to the value $l = 1$.

We would like to insert these functions into Eq. (7.12) and solve for the energy E. To do this we must express the non-spherically symmetric parts x, y, z in polar coordinates and separate out the radial and angular portions of the wave functions. Thus from

$$x = r \sin\theta \cos\phi$$

$$y = r \sin\theta \sin\phi$$

$$z = r \cos\theta$$

we can replace Eq. (7.33a,b,c) respectively by

$$\Psi_0 = \psi(r)\cos\theta$$

$$\Psi_x = \psi(r)\sin\theta\cos\phi$$

$$\Psi_y = \psi(r)\sin\theta\sin\phi$$

where

$$\psi(r) = rf(r)\ .$$

If we express the operator $\hat{l}_z$ in the polar coordinate form given by Eq. (6.13), namely $\hat{l}_z \equiv -i\hbar(\partial/\partial\phi)$, we see straight away that, as Ψ_0 is independent of ϕ,

$$-i\hbar\frac{\partial\Psi_0}{\partial\phi} = 0$$

and so Ψ_0 is an eigenfunction of $\hat{l}_z$ belonging to the eigenvalue zero.

Applying $\hat{l}_z$ to the functions Ψ_x and Ψ_y we have

$$\begin{aligned}\hat{l}_z\Psi_x &= -i\hbar\frac{\partial}{\partial\phi}\psi(r)\sin\theta\cos\phi \\ &= i\hbar\psi(r)\sin\theta\sin\phi \\ &= i\hbar\Psi_y\end{aligned}$$

and similarly

$$\hat{l}_z\Psi_y = -i\hbar\Psi_x\ .$$

Thus Ψ_x and Ψ_y are not eigenfunctions of $\hat{l}_z$. This should not surprise us, as we showed in section 6.6, Eq. (6.15), that the eigenfunctions of $\hat{l}_z$ are of the form $A\,e^{im\phi}$ where m is an integer, and Ψ_x and Ψ_y are not of this form. However the linear combinations

$$\begin{aligned}\Psi_{+1} &= \Psi_x + i\Psi_y \\ &= \psi(r)\sin\theta(\cos\phi + i\sin\phi) \\ &= \psi(r)\sin\theta\,e^{i\phi}\end{aligned}$$

and

$$\begin{aligned}\Psi_{-1} &= \Psi_x - i\Psi_y \\ &= \psi(r)\sin\theta\,e^{-i\phi}\end{aligned}$$

are of the required form and are eigenvalues of $\hat{l}_z$ belonging to the values $+\hbar$ and $-\hbar$ respectively. Thus

$$\begin{aligned}-i\hbar\frac{\partial}{\partial\phi}\Psi_{+1} &= -i\hbar\frac{\partial}{\partial\phi}\psi(r)\sin\theta\,e^{i\phi} \\ &= \hbar\psi(r)\sin\theta\,e^{i\phi} \\ &= \hbar\Psi_{+1}\end{aligned}$$

and

$$-i\hbar\frac{\partial}{\partial\phi}\Psi_{-1} = -\hbar\Psi_{-1}\ .$$

Further Ψ_{+1} and Ψ_{-1} are orthogonal to each other and to Ψ_0 as may readily be seen by integrating the products $\Psi^*_{+1}\Psi_{-1}$ etc. over ϕ from 0 to 2π.

Thus we replace the original set of functions Eq. (7.33a,b,c) by

$$\Psi_0 = \psi(r)\cos\theta \tag{7.35a}$$

$$\Psi_{+1} = \psi(r)\sin\theta\, \mathrm{e}^{\mathrm{i}\phi} \tag{7.35b}$$

$$\Psi_{-1} = \psi(r)\sin\theta\, \mathrm{e}^{-\mathrm{i}\phi}\ . \tag{7.35c}$$

We may now insert these functions into Eq. (7.12), cancel out in each case the angularly dependent factors, and, putting $\psi(r) = u(r)/r$, convert to the form of Eq. (7.12a)

$$-\frac{\hbar^2}{2m}\frac{\partial^2 u(r)}{\partial r^2} + \left(\frac{l(l+1)\hbar^2}{2mr^2} - \frac{Ze^2}{4\pi\varepsilon_0 r}\right)u(r) = Eu(r) \tag{7.12a}$$

with $l = 1$.

In section 7.3 we successfully guessed the correct form, Eq. (7.21), for the ground state radial function $u(r)$ by considering the form of $u(r)$ at large and small values of r, and putting the results together as a product. The success of that procedure suggests we should try the same method for the lowest state in the present case.

Eq. (7.12a) only differs from Eq. (7.18) in having an additional term that varies as $1/r^2$, and hence it has the same approximate form, Eq. (7.18a) at large r. As a result we expect that $u(r)$ will contain an exponential term $\mathrm{e}^{-r/a}$ (as before, we discard the positive sign) with a related to the energy through Eq. (7.20). It is convenient to express a in terms of a_0/Z, so our exponential term becomes e^{-Zr/pa_0} where $pa_0/Z = (-\hbar^2/2mE)^{1/2}$ and p is a number. As before, E will be negative, but presumably less so than for the ground state, so we expect p to be greater than one. At small r the centrifugal term dominates everything and Eq. (7.12a) becomes

$$\frac{\mathrm{d}^2 u(r)}{\mathrm{d}r^2} = \frac{l(l+1)}{r^2}u(r)\ .$$

If we try $u = r^n$ as a solution to this equation we obtain

$$n(n-1)r^{n-2} = l(l+1)r^{n-2}$$

which is satisfied if $n = l + 1$, or $n = -l$. The latter solution gives $u(r) \sim r^{-l}$ as r goes to zero and offends the boundary condition $u(r) = 0$ at $r = 0$ and must be discarded. We obtain, therefore, the result that the *radial wave function $u(r)$ for a state with angular momentum l must behave near the origin as*

$$u(r) \propto r^{l+1}\ . \tag{7.36}$$

This result is independent of the form of the radial potential provided it can be neglected compared with the centrifugal potential at small r. This will be true of all cases discussed in this book. The result Eq. (7.36) reflects the fact that states of high angular momentum are states in which the centrifugal force is large. This force tends to keep the particle away from the origin, and hence the probability of finding the particle close to the origin is smaller the higher the angular momentum. This is made particularly clear if we consider the probability per unit volume of finding the particle close to the origin. The probability per unit volume is proportional to $|\psi(r)|^2$ where $\psi(r) = (1/r)u(r)$. Thus we obtain the results that the probability per unit volume of finding the particle close to the origin is proportional to

a constant for s-waves

r^2 for p-waves

r^4 for d-waves, etc.

States with zero orbital angular momentum are distinguished by being the only states for which there is a finite probability per unit volume of finding the particle at the origin. This means the interactions between atomic and nuclear phenomena are much more significant when the electron is in an s-state.

Specializing to the case of $l = 1$, Eq. (7.12a) becomes

$$\frac{\partial^2 u(r)}{\partial r^2} - \frac{2u(r)}{r^2} + \frac{2m}{\hbar^2}\left(\frac{Ze^2}{4\pi\varepsilon_0 r} + E\right)u(r) = 0 . \tag{7.37}$$

Putting together the forms at large and small r we try as a solution

$$u(r) = r^2\,\mathrm{e}^{-Zr/pa_0} . \tag{7.38}$$

Hence

$$\frac{\partial^2 u(r)}{\partial r^2} = \left[2 - \frac{4Z}{pa_0}r + \left(\frac{Z}{pa_0}\right)^2 r^2\right]\mathrm{e}^{-Zr/pa_0} .$$

Inserting this in Eq. (7.37) and cancelling out the exponential factor we obtain

$$\left(\frac{2m}{\hbar^2}\frac{Ze^2}{4\pi\varepsilon_0} - \frac{4Z}{pa_0}\right)r + \left(\frac{2mE}{\hbar^2} + \left(\frac{Z}{pa_0}\right)^2\right)r^2 = 0 .$$

As this equation has to hold for all r the coefficients of different powers of r must vanish, i.e.

$$\frac{2me^2}{4\pi\varepsilon_0\hbar^2} = \frac{4}{pa_0} \tag{7.39a}$$

$$\frac{2mE}{\hbar^2} = -\left(\frac{Z}{pa_0}\right)^2 . \tag{7.39b}$$

If we substitute for $m/\hbar^2$ from $a_0 = 4\pi\varepsilon_0\hbar^2/(me^2)$ we obtain

$$\frac{2}{a_0} = \frac{4}{pa_0}$$

or

$$p = 2$$

from Eq. (7.39a), and

$$2 \times \frac{4\pi\varepsilon_0 E}{e^2 a_0} = -\left(\frac{Z}{pa_0}\right)^2$$

from Eq. (7.39b). Inserting the value $p = 2$ we get

$$\boxed{E = -\frac{1}{4}\frac{Z^2e^2}{4\pi\varepsilon_0 2a_0}\,.} \tag{7.40}$$

We note straight away that this is the same expression for the energy as we obtained in Eq. (7.30) for the energy of the first excited s-state. This degeneracy is peculiar to the Coulomb potential, and arises because of an additional, subtle, symmetry that exists in addition to the spherical symmetry of the problem.

Eq. (7.40) is the energy of the lowest p-state (i.e., with orbital angular momentum equal to one). Its radial function $u(r)$ (Eq. (7.38) with $p = 2$) is plotted in Fig. 7.13. We note that like the lowest s-state it is all of one sign.

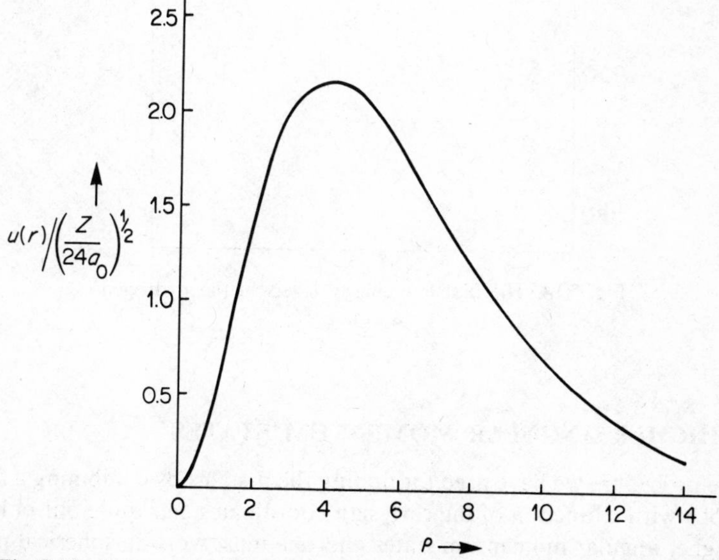

Fig. 7.13. $u(r)$ versus r, where r is measured in units of a_0/Z, for the hydrogen-like case with $n = 2$, $l = 1$, Eq. (7.38).

We can generate higher p-states in exactly the same way as we generated higher s-states in section 7.5. Once again we find an exponential factor e^{-Zr/na_0} multiplied by a function of r which must be a finite polynomial if the boundary conditions at infinity are to be satisfied. This only occurs if n is an integer, when the polynomial has a highest power of r^n. We get the same result for any value of the orbital angular momentum. As the lowest power of r occurring is r^{l+1} it follows that a specific value of the orbital angular momentum, l, first occurs for $n = l + 1$. The energies of the whole family of states for the hydrogen atom can be expressed as

$$E_{nl} = -\frac{1}{n^2}\frac{Z^2e^2}{4\pi\varepsilon_0 2a_0}, \qquad n \geqslant l + 1, \quad l = 0, 1, 2, \ldots \tag{7.41}$$

where the subscript l limits the value of n that can occur. These states are illustrated in Fig. 7.14.

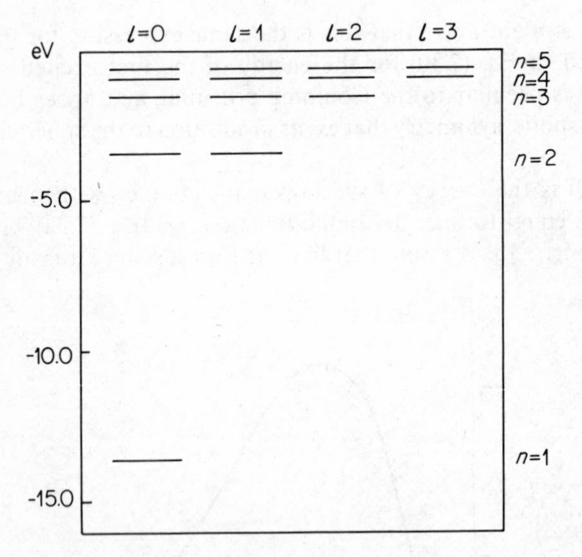

Fig. 7.14. The first few energy levels of the hydrogen atom.

7.7 HIGHER ANGULAR MOMENTUM STATES

The procedure we have used for finding the p-states by combining a function of r with a function of the cartesian coordinates gets quite out of hand for higher angular momentum states and one must work in spherical polar coordinates. Essentially this means finding the solutions of Eq. (7.11) where

$\hat{l}^2$ is given by Eq. (7.8). As these equations are independent of r they give solutions which depend on θ and ϕ only. These solutions are known as spherical harmonics and are expressed by the symbol

$$Y_{l,m}(\theta, \phi)$$

where l and m denote respectively, the eigenvalues of orbital angular momentum and component along the z-axis to which the function belongs.

The explicit expressions for the first few spherical harmonics are as follows:

$$\begin{aligned}
Y_{0,0}(\theta, \phi) &= \sqrt{(1/4\pi)} \\
Y_{1,1}(\theta, \phi) &= -\sqrt{(3/8\pi)} \sin\theta\, e^{i\phi} \\
Y_{1,0}(\theta, \phi) &= \sqrt{(3/4\pi)} \cos\theta \\
Y_{1,-1}(\theta, \phi) &= \sqrt{(3/8\pi)} \sin\theta\, e^{-i\phi} \\
Y_{2,2}(\theta, \phi) &= \sqrt{(15/32\pi)} \sin^2\theta\, e^{i2\phi} \\
Y_{2,1}(\theta, \phi) &= -\sqrt{(15/8\pi)} \sin\theta \cos\theta\, e^{i\phi} \\
Y_{2,0}(\theta, \phi) &= \sqrt{(5/4\pi)}((3\cos^2\theta - 1)/2) \\
Y_{2,-1}(\theta, \phi) &= \sqrt{(15/8\pi)} \sin\theta \cos\theta\, e^{-i\phi} \\
Y_{2,-2}(\theta, \phi) &= \sqrt{(15/32\pi)} \sin^2\theta\, e^{-i2\phi}) .
\end{aligned}$$

The spherical harmonics have the property

$$\int_0^{\pi}\int_0^{2\pi} Y^*_{l',m'}(\theta, \phi) Y_{l,m}(\theta, \phi) \sin\theta \, d\theta \, d\phi = \delta_{ll'}\, \delta_{mm'} \tag{7.42}$$

where $\delta_{ll'} = 1$ if $l = l'$ and is zero otherwise, and similarly for $\delta_{mm'}$.

The reader should note that the angular portions of $Y_{1,0}$, $Y_{1,1}$ and $Y_{1,-1}$ are the same as the angular portions of Eqs. (7.35a,b,c) except for the normalization factors. We did not include these factors in Eqs. (7.35) as they did nothing to the physics and merely complicate the mathematics. The numerical factors in front of all the above expressions have been adjusted so that Eq. (7.42) is satisfied. The negative signs in front of certain terms are merely the result of a convention which simplifies algebraic manipulation.

7.8 THE SPECIFICATION OF STATES IN TERMS OF THEIR QUANTUM NUMBERS

One frequently wishes to refer to particular states or wavefunctions, similar to those we have been discussing, without there being any need to refer to their explicit form. This will be particularly true when we come to the electron wavefunctions in more complex atoms. In these cases the potential deviates from the simple $1/r$ law, although its shape is similar. In such cases the radial portion of the wavefunction cannot be expressed in any simple algebraic form. One would still like to be able to specify the state and have some feel for its properties.

The angular portion of the wavefunction is straightforward as it is completely specified by giving the values of the orbital angular momentum quantum number l and the z-component of angular momentum m, and this is true in any spherically symmetric potential. In the hydrogen atom the wavefunction is completely specified if, in addition to l and m, we specify the integer n that occurs in the expressions for the energy (Eqs. (7.32), and (7.41)). The three integers n, l and m are constrained by the relations

$$|m| \leqslant l \leqslant n - 1 \ . \tag{7.43}$$

l of course is a positive integer or zero, and n a positive integer. As the energy is solely dependent on n it is referred to as the *principal quantum number*.

The principal quantum number is of more significance than just being the number that determines the energy in Eq. (7.41). Together with l it determines the number of times the radial function crosses the axis, as we now indicate.

In section 7.5 we argued that, starting with the ground state solution, one could generate a whole family of s-wave solutions by letting the wavefunction cross the axis once, twice, three times, etc., each one being the solution with next lowest energy. Thus the nth solution crosses the axis $(n - 1)$ times, so in this case the principal quantum number n, in addition to specifying the energy, also specifies that $u(r)$ crosses the axis $(n - 1)$ times.

The lowest state with $l = 1$ has $n = 2$, and the radial function for this state is given by Eq. (7.38). We see that this, like the ground state function, does not cross the axis. The next lowest state in energy with $l = 1$, namely that with $n = 3$, must cross the axis once in order to be orthogonal to the lowest $l = 1$ state. Proceeding as for s-states we conclude that p-states have radial functions that cross the axis $(n - 2)$ times. If we make the assumption (which is true) that the lowest state for any given l value never crosses the axis, the next state crosses it once, and so on, then, as the lowest state for any given l value has $n = l + 1$ a state $u_{n,l}(r)$ specified by n and l crosses the axis $(n - l - 1)$ times.

Suppose now we change the potential from the Coulomb potential by small steps. At each stage we assume the wavefunction is adjusted so it is the eigenfunction of the new Hamiltonian which most resembles the initial eigenfunction. It is plausible that one thing this does not do is change the number of times $u(r)$ crosses the axis. Thus for potentials which differ from the Coulomb potential we can still define a principal quantum number n through the relation that $(n - l - 1)$ is the number of times $u(r)$ crosses the axis.

We can add further plausibility to the above argument as follows. The rate at which the wave function oscillates is a measure of the kinetic energy of the particle as can be seen from a study of the structure of the Schrödinger equation. For a given value of l the lowest state will be one for which $u(r)$ does not cross the axis at all. The next state must be orthogonal to the lowest state, and therefore will cross the axis once, and the third state, to be orthogonal to the first two must cross the axis twice, and so on. Hence for a given potential and angular momentum an ordering in energy is always equivalent to an ordering in terms of the number of times $u(r)$ crosses the axis. Hence we can always define a principle quantum number via $n = q + l + 1$ where q is the number of times $u(r)$ crosses the axis. Of course, for potentials other than the Coulomb potential and those similar to it we do not necessarily get the dominant dependence of energy on n.

7.9 ELECTRON SPIN IN THE HYDROGEN ATOM

So far we have not included the effects of electron spin. If we assumed that there was no interaction between the electron's spin and its translational motion the only effect would be to make the specification of an energy level by n, l and m incomplete. Such a level would be degenerate because the z-component of the electron spin could be either $+\frac{1}{2}\hbar$ or $-\frac{1}{2}\hbar$. Hence to specify a state completely we must specify n, l, m_l and m_s where m_l and m_s are respectively the z-components of the orbital and spin angular momentum.

This situation is very nearly true, the interaction between the spin and motion of the electron being very small. We have introduced electron spin in an *ad hoc* manner. The next stage in quantum mechanics is to introduce relativity which was done in 1927 by Dirac. When this is done the spin is found to arise quite naturally and a complete solution is possible. In heavier atoms it is found that the effect of electron spin is more significant, but of course an exact solution is not possible. We reserve discussion of this topic until later.

PROBLEMS 7

7.1 Find $\langle r \rangle$ and $\langle r^2 \rangle$ for the ground state of the hydrogen atom. (Note: $\int_0^\infty r^n \, e^{-\alpha r} \, dr = n!/\alpha^{n+1}$.)

7.2 Find $\langle V \rangle$ for the ground state of the hydrogen atom and compare it with the ground state energy. Hence determine $\langle T \rangle$ where T is the kinetic energy.

7.3 Show that $(x + iz)f(r)$ is an eigenfunction of $\hat{l}_y$ belonging to the eigenvalue $-\hbar$.

7.4 The radial wavefunction $u(r)$ for the state of the hydrogen atom with principal quantum number n and $l = n - 1$ is given by $u(r) = Ar^n \, e^{-r/na_0}$ when A is a normalization constant. Find $\langle r \rangle$ and $\langle r^2 \rangle$ and hence find the radial spread of the wavefunction as determined by

$$\left(\frac{\langle r^2 \rangle - \langle r \rangle^2}{\langle r \rangle^2} \right)^{1/2}$$

7.5 According to Eq. (7.23) the ground state energy of a single electron in the field of a nucleus of charge Ze is proportional to Z^2. Explain this in physical terms.

7.6 Show explicitly that the expressions for $u(r)$ given by Eq. (7.21b) and Eq. (7.29a) are orthogonal to each other.

7.7 Show that a wave function of the form $\psi = xf(r)$ is orthogonal to a spherically symmetric wavefunction, and hence that all p-states are orthogonal to all s-states.

7.8 Show that the minimum value for the effective radial potential for an electron of orbital angular momentum l in a hydrogen atom is given by

$$V = -\frac{1}{2}\frac{e^2}{4\pi\varepsilon_0 l(l+1)a_0}.$$

CHAPTER 8

The periodic table and the Pauli exclusion principle

8.1 INTRODUCTION

In the last chapter we discussed the problem of a single electron moving in a Coulomb field. We found we could specify the state completely by specifying the values of the four quantum numbers n, l, m_l and m_s. We would now like to extend the application of quantum mechanics, at least qualitatively, to cover atoms containing several electrons. We could then hope to account, at least in general terms, for the energy level spectra of more complicated atoms, and explain such things as the periodic table of the elements and the occurrence of the noble gases. We shall find that quantum mechanics, as we have developed it so far, is inadequate for these purposes, and we shall have to introduce an additional postulate in the form of the Pauli exclusion principle.

First, however, we must consider how we are going to describe a state of an atom possessing several electrons. We shall see that this can be done to a fair approximation by giving a set of quantum numbers for *each* electron similar to the set quoted above for the single electron case.

8.2 THE DESCRIPTION OF AN ATOM CONTAINING SEVERAL ELECTRONS

It is, of course, possible in principle to write down the Schrödinger equation for an atom containing several electrons and to attempt to solve it using a large electronic computer. Such a procedure would not normally prove very illuminating. For a start the wavefunction would be a complicated expression in $3Z$ dimensions, where Z is the number of electrons. Secondly there is no guarantee that the wavefunction would be presented in such a way as to enable one to pick out the main physical features of the problem and it would almost certainly prove very difficult to use in the calculation of other properties of the atom (that is, other than the energy). In complex problems in physics one is always seeking an approximate 'model' which gives the main features of the exact solution and at the same time gives one a physical 'feeling' for the nature of the solution, so that it is possible to see qualitatively what will happen under various sets of circumstances. Of course, in the majority of cases one does not know the exact solution, and in these cases one's faith in the approximate correctness of the model is to be founded on the number of physical quantities it predicts with an acceptable degree of accuracy when compared with experimental results.

Models are, in fact, essential in the calculation of all but the simplest problems. Where an exact analytical solution does not exist and numerical methods have to be employed there is bound to be approximation and truncation. In such circumstances a model is essential as a guide to which terms may be treated crudely and which should be preserved with the highest possible accuracy.

Thus we seek a simple model of a many electron atom which will enable us to describe its states in a way which will make it almost intuitively obvious what their main properties will be. As a first step in this direction we write down the Schrödinger equation for a many-electron atom so that we can see where the difficulty arises.

The Schrödinger equation for a neutral atom with Z electrons is

$$\left(\sum_{i=1}^{Z}\left(-\frac{\hbar^2}{2m}\nabla_i^2\right)-\sum_{i=1}^{Z}\frac{Ze^2}{4\pi\varepsilon_0 r_i}+\sum_{\substack{i,j\\ i>j}}^{Z}\frac{e^2}{4\pi\varepsilon_0|\mathbf{r}_i-\mathbf{r}_j|}\right)\Psi=E\Psi \tag{8.1}$$

where the individual electrons are numbered from 1 to Z.

In Eq. (8.1) the first term is the kinetic energy operator, there being one such operator for each electron present, the total kinetic energy operator being the sum of these operators. ∇_i^2 means that the operator ∇^2 is to act on the coordinates $\mathbf{r}_i$ of the ith electron only. Similarly the second term is the sum of the potential energies of the electrons in the field of the nucleus. The third term is the potential energy due to the electrostatic repulsion between the pairs of electrons. Unlike the previous terms, each term in this sum depends on the coordinates of *two* particles; the restriction $i > j$ is to avoid counting the contribution of a pair twice. It is this last term that causes all the trouble. If this term did not exist, Eq. (8.1) would be the Schrödinger equation for Z electrons, each moving in the field of a nucleus with charge Ze, but not interacting with each other. Each electron would then have a wavefunction of the type discussed in Chapter 7, and the total wavefunction Ψ would then simply be the product of the individual electron wavefunctions. Mathematically this result occurs because under these circumstances there would then be no term in Eq. (8.1) which referred to the coordinates of more than one electron. This gives us a clue as to the direction in which we should try and proceed. We should try and approximate to the third term of Eq. (8.1) so that it becomes a sum of terms, *each of which refers to the coordinates of one electron only.* If we can do this then we shall be able to refer not only to the total wavefunction Ψ *but to individual wavefunctions ϕ_i for each of the electrons.* The total wavefunction Ψ will merely be the product of these wavefunctions i.e.

$$\Psi = \prod_{i=1}^{Z} \phi_i(\mathbf{r}_i) \ . \tag{8.2}$$

Each of the wavefunctions ϕ_i can be specified by a set of quantum numbers n, l, m_l and m_s and the full set of Z of these will give us a specification of the state of the atom. Clearly this would give us a nice simple way of looking at things and we must pursue this matter further to see if it can be reasonably done, and if so, how.

Consider a singly ionized helium atom in its ground state. The single electron will have a wavefunction

$$\phi_0 = A\,\mathrm{e}^{-2r_1/a_0} \tag{8.3}$$

where A is a constant. Suppose now the helium ion captures another electron into a weakly bound state of high angular momentum. Because of its high angular momentum this electron has a negligible probability of being close to the nucleus. This means two things.

(i) Because it is always well away from the nucleus, and indeed outside the region where the first electron can appear with reasonable probability, it is essentially always moving in the Coulomb field of a single charge. Its wavefunction will be the appropriate hydrogen wavefunction $\phi_{nlm_l}(\mathbf{r}_2)$

(ii) Also because the second electron never gets close to the region occupied by the first electron it does not perturb the wavefunction for the first electron, which is still well represented by Eq. (8.3).

Hence it is clear that a very good approximation to the exact wavefunction will be

$$\left.\begin{aligned}\Psi &\approx A\,\mathrm{e}^{-2r_1/a_0}\phi_{nlm_l}(\mathbf{r}_2)\\ &= \phi_0(\mathbf{r}_1)\phi_{nlm_l}(\mathbf{r}_2) \ .\end{aligned}\right\} \tag{8.4}$$

In arriving at this wavefunction, what we have done, effectively, is replace the electron–electron interaction in Eq. (8.1) for each electron by the potential due to the charge distribution given by the wavefunction for the other electron, and then, because the second electron is so far away, contract the charge distribution of the first electron to a point. This suggests that we could get good approximate solutions to Eq. (8.1) by replacing the electron–electron repulsion terms in Eq. (8.1) by a series of potentials $V_i(\mathbf{r}_i)$, one for each electron, which is to be calculated from the charge distribution implied by the wavefunctions for the other electrons. The method of solution is then iterative. One starts by guessing a potential for each electron as we have done for the excited helium atom we have just discussed. This gives us separate wavefunctions for each electron. We now calculate a new potential for each electron from the nuclear charge and the charge distribution given by the wavefunctions for the other electrons, and use these new potentials to recalculate the wavefunctions. Providing our original guess was not too bad, after several steps we shall find the wavefunctions reproduce the charge distributions from which they were derived. For this reason this procedure is known as the 'Self-consistent field' Method. It was first introduced by D. R. Hartree in 1928.

The self-consistent field method turns out to be an extremely valuable approximation and it enables us to ascribe a wavefunction to each electron, and hence describe the state of an atom by giving the quantum numbers of the electron wavefunctions involved. Further, these electron wavefunctions are not too far removed from hydrogen wavefunctions. The angular portion is of course the same, as this does not depend on the radial potential, and the general behaviour for a given value of n (which specifies the number of times it crosses the axis, see section 7.8) is much the same as for hydrogen. While not good enough for quantitative estimates the hydrogen wavefunctions can frequently be used to get a fair qualitative picture.

To summarize this section then we assert that

(i) The electron–electron interaction in many-electron atoms can be replaced to a good approximation by a potential which is that due to the mean charge distribution of the other electrons.

(ii) The individual electrons can be regarded approximately as moving independently in the combination of the potential due to the nuclear charge and that due to the other electrons. This is called 'a screened Coulomb potential'.

(iii) The total wavefunction is now just a product of the individual wavefunctions, and can be specified by their quantum numbers.

(iv) The individual electron wavefunctions are not too different in *character* from the corresponding hydrogen wavefunctions.

8.3 THE PROPERTIES OF THE ELEMENTS AND THE PAULI EXCLUSION PRINCIPLE

In the last section we devised an approximate model for many electron atoms which we have good reason to hope will enable us to predict qualitatively the properties of the atoms. One reason we have for expecting this model to work well is that the electron–electron interaction is a slowly varying function of distance. Consequently replacing the actual interaction by the average interaction should not be too violent a simplification. We must now see how these qualitative predictions square with the observed experimental facts.

The most remarkable feature of the chemical elements, when taken as a whole, is that they can be arranged in the form of a periodic table, as was first done by Mendeleev in 1869. A modern version of the periodic table is shown in Fig. 8.1. When the elements are arranged in this way all the elements in the same column show similar chemical properties, but there is a rapid change of properties as we go across the table. Thus all the elements in the first group are monovalent, those in the second group are divalent, while those in group 0 on the right hand side are inert gases.

A moment's thought will make it clear that our model of the atom, as we have developed it so far, will not give us even approximately correct answers. According to our theory the lowest state of an atom will be the one in which each individual electron is in its lowest state in the field provided by the combination of the nuclear charge and the charge distribution of the other electrons. The lowest state in any potential is always an s-state, and therefore the individual electron wavefunctions will all be very similar. We see that the self-consistency requirement is satisfied if all the electron wavefunctions are identical. The potentials in which each electron moves will then be identical, leading to identical wavefunctions as required. We thus have a picture of the ground state of a many electron atom in which all the electrons have identical wavefunctions, this wavefunction being the lowest s-state wavefunction that can occur in the field provided by the nucleus and the remaining electrons. The nature of this wavefunction will change very little as we go from one element to the next, as the wavefunctions remain constant in type, and the addition of one unit to the nuclear charge and of one extra electron clearly cannot make any substantial change to the potential.

Thus on the picture we have developed one could visualize that there could possibly be a rapid change in chemical properties for the first few elements, but thereafter we would expect things to settle down with very little change as we go from element to element. Quite certainly our model will produce none of the regular repeating features of the periodic table, like the recurrence of the alkali metals, and the noble gases, and the steady progression of valency as we go through a period. Our model, therefore, fails to account for the most obvious qualitative features of the chemistry of the elements.

A clue to what is wrong with our picture is obtained if we look at the *first ionization potentials* of the atoms. The first ionization potential of an atom, when multiplied by the magnitude of the electron charge e, gives the difference in energy between the ground state of the neutral atom and the ground state of the singly charged positive ion.

Group		I	II	III	IV	V	VI	VII	VIII			0
Period	Series											
1	1	1 H 1.0080										2 He 4.003
2	2	3 Li 6.940	4 Be 9.013	5 B 10.82	6 C 12.011	7 N 14.008	8 O 16.0000	9 F 19.00				10 Ne 20.183
3	3	11 Na 22.991	12 Mg 24.32	13 Al 26.98	14 Si 28.09	15 P 30.975	16 S 32.066	17 Cl 35.357				18 **A** 39.944
4	4	19 K 39.100	20 Ca 40.08	21 Sc 44.96	22 Ti 47.90	23 V 50.95	24 Cr 52.01	25 Mn 54.94	26 Fe 55.85	27 Co 58.94	28 Ni 58.71	
	5	29 Cu 63.54	30 Zn 65.38	31 Ga 69.72	32 Ge 72.60	33 As 74.91	34 Se 78.96	35 Br 79.916				36 Kr 83.80
5	6	37 Rb 85.48	38 Sr 87.63	39 Y 88.92	40 Zr 91.22	41 Nb 92.91	42 Mo 95.95	43 Tc [99]	44 Ru 101.1	45 Rh 102.91	46 Pd 106.4	
	7	47 Ag 107.880	48 Cd 112.41	49 In 114.82	50 Sn 118.70	51 Sb 121.76	52 Te 127.61	53 I 126.91				54 Xe 131.30
6	8	55 Cs 132.91	56 Ba 137.36	57–71 Lanthanide series*	72 Hf 178.50	73 Ta 180.95	74 W 183.86	75 Re 186.22	76 Os 190.2	77 Ir 192.2	78 Pt 195.09	
	9	79 Au 197.0	80 Hg 200.61	81 Tl 204.39	82 Pb 207.21	83 Bi 209.00	84 Po 210	85 At [210]				86 Rn 222
7	10	87 Fr [223]	88 Ra 226.05	89–Actinide series†								

* Lanthanide series	57 La 138.92	58 Ce 140.13	59 Pr 140.92	60 Nd 144.27	61 Pm [147]	62 Sm 150.35	63 Eu 152.0	64 Gd 157.26	65 Tb 158.93	66 Dy 162.51	67 Ho 164.94	68 Er 167.27	69 Tm 168.94	70 Yb 173.04	71 Lu 174.99
† Actinide series:	89 Ac 227	90 Th 232.05	91 Pa 231	92 U 238.07	93 Nb [237]	94 Pu [242]	95 Am [243]	96 Cm [245]	97 Bk [249]	98 Cf [249]	99 E [253]	100 Fm [255]	101 Md [256]	102 No	103

Fig. 8.1. Periodic table of the elements. Each entry consists of the atomic number Z, followed by the chemical symbol, and below these the mean atomic weight. Exceptions are the rare earth or lanthanide series from $Z = 57$ to $Z = 71$ which form a single entry, and the actinide series beginning at $Z = 89$. Numbers in brackets are approximate values for unstable elements.

Consider the potential seen by one electron in a neutral atom containing Z electrons. At large distances from the nucleus the electron will have a potential energy of $-e^2/4\pi\varepsilon_0 r$ as the positive charge Ze on the nucleus is screened by the remaining $(Z - 1)$ electrons surrounding it. At very small distances the remaining electrons have negligible screening effect, and the potential energy of the electron will be very close to $-Ze^2/4\pi\varepsilon_0 r$. The result is an effective potential of the form shown in Fig. 8.2 where we have drawn the potential

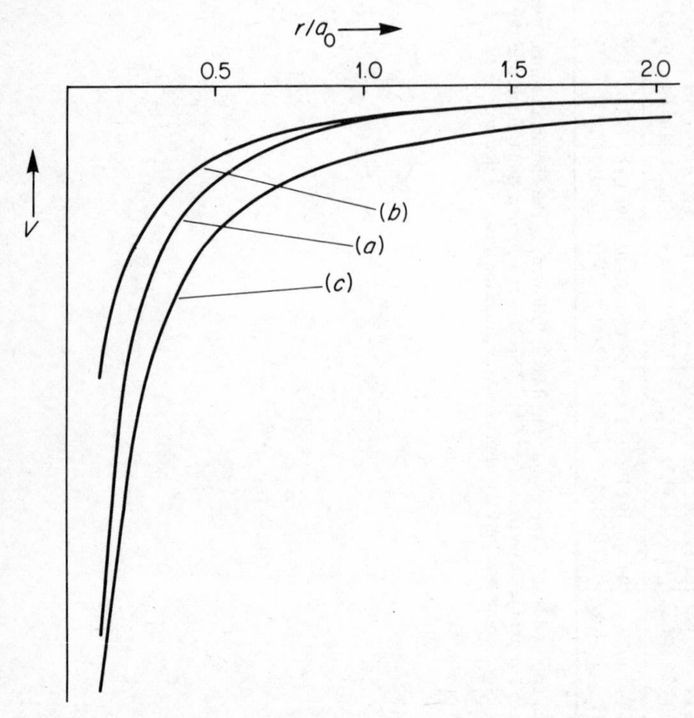

Fig. 8.2. (a) The effective potential seen by one of the electrons in the ground state of the helium atom, in arbitrary units, and (b) and (c): the potentials $-e^2/4\pi\varepsilon_0 r$ and $-2\,e^2/4\pi\varepsilon_0 r$ respectively.

for helium. If the potential were everywhere $-e^2/4\pi\varepsilon_0 r$ we should get an ionization potential of $e/4\pi\varepsilon_0 2a_0 = 13.6$ volts, while if the potential were $-Ze^2/4\pi\varepsilon_0 r$ everywhere we should get an ionization potential of $Z^2 e/4\pi\varepsilon_0 2a_0 = Z^2 \times 13.6$ volts. The lowering of the potential below $-e^2/4\pi\varepsilon_0 r$ for small values of r means the ionization potential must be greater than 13.6 volts, but it will be less than $Z^2 \times 13.6$ volts as the potential of $-Ze^2/4\pi\varepsilon_0 r$ applies only very close to the nucleus. We make a crude guess, and suggest it is likely to be around $Z \times 13.6$ volts.

The measured values of the ionization potential are shown in Fig. 8.3. We see that far from showing the smooth increase with Z we have suggested

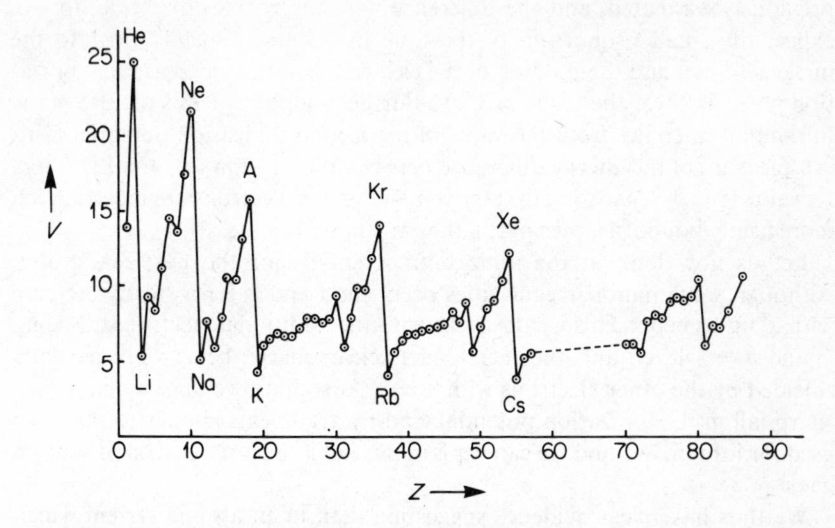

Fig. 8.3. The first ionization potentials of the elements, in volts, plotted in order of their atomic number. The lines joining the points are merely to guide the eye. The dotted region is the rare earths.

the ionization potentials show the same sort of periodic structure as do the chemical properties. However a study of the values for the first few elements will show us what is happening. The ionization potential for helium is 24.5 V in line with the behaviour we have proposed. Moving on to lithium there is a catastrophic fall to a value of 5.4 V, below that for hydrogen, instead of a value of between 30 V and 40 V as we should have expected. There is no way of getting such a low ionization potential if all the electrons have a principal quantum number $n = 1$. If, however, one of the electrons has a principal quantum number $n = 2$, then things are much more reasonable. Let us assume that in lithium two electrons go into the lowest state and have $n = 1$ but the third electron goes into the next available state which we assume to be the 2s state. The reason for preferring this over the 2p state is that the results of a Stern–Gerlach experiment on neutral lithium atoms show that the ground state is an s-state. Reference to Fig. 7.10 shows that the 2s wavefunction in the field of a single charge is relatively small for $r < 2a_0$ and the electron spends most of its time outside this radius. The two 1s electrons in lithium have a probability of greater than 99% of being within $2a_0$ of the nucleus due to the increased nuclear charge. Thus we expect the third electron in lithium to have a wavefunction very close to the 2s wavefunction for hydrogen, and to have an ionization potential similar to that for hydrogen in its first excited state, namely 3.4 V. This is indeed much closer to the measured value of 5.4 V than is the value of between 30 V and 40 V we previously estimated, and the difference is in the correct direction, for we expect the small proportion of the time the electron spends close to the nucleus to increase the binding of the electron, and thus increase the ionization potential over the value of 3.4 V. Further support for this picture of the lithium atom comes from the value of the second ionization potential. This is a measure of the energy difference between the Li^+ ion and the Li^{++} ion. Its value is 75.3 V, showing that the remaining two electrons are indeed much more firmly bound, as we expect if they are in the 1s state.

Let us now look at the ionization potentials for the next few atoms. Although some minor irregularities occur the trend is for a steady increase with Z up to neon. This is consistent with the additional electrons all being in the $n = 2$ level, and the increased nuclear charge being only partially shielded by the other electrons with $n = 2$. At sodium we once again have a sharp fall in the ionization potential, and by arguments similar to those we used for lithium, we find we can explain this fall if the last electron in sodium has $n = 3$.

We thus have clear evidence suggesting that, in an atomic system which we describe by electrons moving in a self-consistent field with wavefunctions described by the quantum numbers n, l, m_l and m_s, there can be *at most two electrons with* $n = 1$, and *eight electrons with* $n = 2$.

Why these particular numbers? We note that for $n = 1$ the remaining quantum numbers can have the values $l = 0$, $m_l = 0$, $m_s = \pm\frac{1}{2}$ i.e. *two* states, and for $n = 2$ we can have either $l = 0$ with $m_s = \pm\frac{1}{2}$, or $l = 1$ with $m_l = -1$, 0, 1 and $m_s = \pm\frac{1}{2}$, i.e. a total of eight states.

We put this in tabular form in Table 8.1.

Table 8.1. Possible quantum numbers of states with $n = 1$ and 2

n	l	m_l	m_s
1	0	0	$+\frac{1}{2}$
1	0	0	$-\frac{1}{2}$
2	0	0	$+\frac{1}{2}$
2	0	0	$-\frac{1}{2}$
2	1	-1	$+\frac{1}{2}$
2	1	-1	$-\frac{1}{2}$
2	1	0	$+\frac{1}{2}$
2	1	0	$-\frac{1}{2}$
2	1	$+1$	$+\frac{1}{2}$
2	1	$+1$	$-\frac{1}{2}$

Thus we can account for our observations to date if we make the following postulate:

Postulate

In a system containing many electrons no two electrons can be in states with identical quantum numbers.

This is the Pauli exclusion principle, and is the form in which it was first enunciated by Pauli in 1925. Pauli deduced the principle from spectroscopic data prior to the advent of quantum mechanics. Later in the chapter we shall give a more general formulation of the Pauli exclusion principle which can be applied in all situations even though it is not possible to describe the particles by individual states possessing specific quantum numbers.

8.4 THE EFFECT OF THE PAULI EXCLUSION PRINCIPLE ON THE DESCRIPTION OF ATOMS

Because of the exclusion principle the electrons in the ground state of an atom do not all have the same energy and wavefunction as we originally supposed. Instead different electrons have different wavefunctions distinguished by the quantities n, l, m_l and m_s. As the values of m_l and m_s are arbitrary, depending on the direction chosen for the z-axis, electrons with the same values of n and l but different values of m_l and m_s will have the same energy. Electrons with different values of n and l, however, will have different energies, as it is only in the pure Coulomb field that electrons with the same n but different l have the same energy. Now electrons with different energies have wavefunctions of substantially different extension. In Fig. 8.4. we show

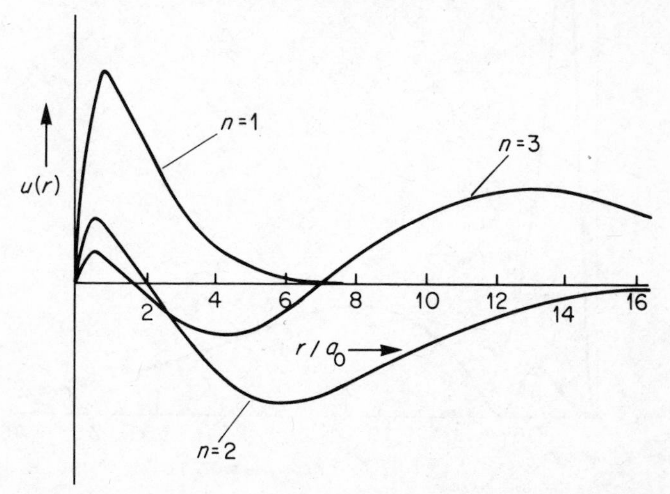

Fig. 8.4. The 1s, 2s and 3s radial wavefunctions $u(r)$ for hydrogen. The largest maxima occur at r approximately equal to a_0, $5a_0$ and $13a_0$ respectively.

the radial wavefunctions of hydrogen for the first three s-states. The considerable difference in their extension is immediately obvious. This difference is exaggerated when we consider atoms with several electrons.

The electrons with the highest value of n are largely shielded from the full nuclear charge by the electrons with small n-values, while the reverse is of course not true. Thus the innermost electrons, the 1s electrons, move in a field very close to $-Ze^2/4\pi\varepsilon_0 r$ and hence can be fairly well represented by the wavefunction

$$\phi = \text{constant} \times e^{-Zr/a_0}$$

which has an extent of $1/Z$ times the corresponding hydrogen wavefunction. The outermost electrons, with the highest value of n move in a field which varies from $-Ze^2/4\pi\varepsilon_0 r$ close to the nucleus to $-e^2/4\pi\varepsilon_0 r$ a long way away from the nucleus. The effect of this is to contract the early portion of the wavefunction but to leave the bulk of the wavefunction only moderately affected. The net result of this is that the outermost electrons always have about the same extent, namely two or three times a_0, as is indicated by the approximate constancy of experimentally determined atomic radii.

The wide differences in the extension of the wavefunctions for different electrons are most important. We illustrate these differences in Fig. 8.5 by showing approximate radial probability densities for the 1s and 2s electrons in lithium.

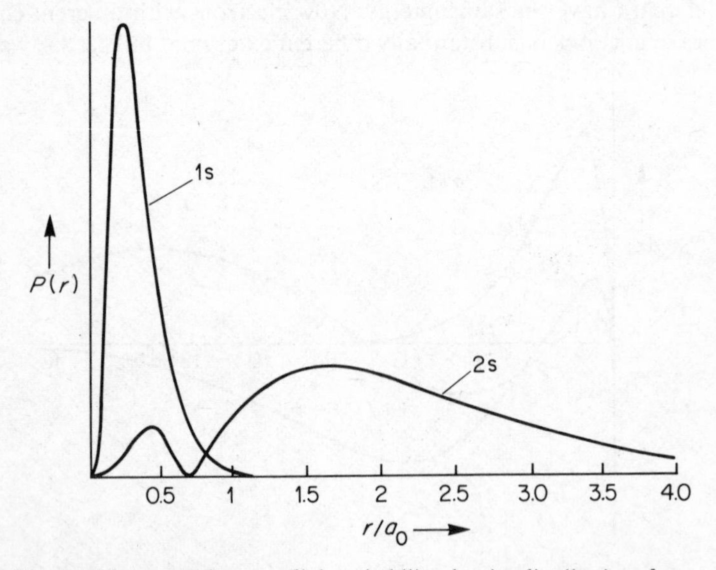

Fig. 8.5. The approximate radial probability density distributions for a 1s electron and a 2s electron in lithium.

Electrons with the same values of n and l and differing only in their values for m_l and m_s have the same radial wavefunction in the self-consistent field and therefore the same radial probability distribution. Every wavefunction has one peak that is bigger than the rest (the final one) and hence the electron spends most of its time in the vicinity of this peak. Thus electrons with the same values of n and l mainly occupy a 'shell' in space surrounding the nucleus whose mean radius is very strongly dependent on n. For this reason the description we have given is referred to as the *shell model of the atom.*

The word 'shell' is used extensively in the terminology of atomic physics. Thus the complete set of states with a given principal quantum number n is referred to as a shell, and we say for example, that the outer electrons in the atoms between lithium and neon are 'filling the $n = 2$ shell'. If all the states with a given n are occupied we say that the shell is full, or, that the electrons form 'a closed shell'. States with a given value of n and l are said to form a sub-shell, and if they are all occupied we say the electrons form 'a closed sub-shell'. Terminology is often a bit loose in this matter and the word 'shell' is often used in place of 'sub-shell' when the context makes it clear what is meant. Also, we shall find that major breaks in properties occur when the p-wave sub-shells are full, and it is almost universal practice to say that the electrons in these elements form a closed shell, even though states with the same n but higher l values are not occupied.

The effect, then, of the Pauli exclusion principle in requiring different electrons to have different quantum numbers is to cause different electrons, particularly those with different n values, to have different energies in the atom, and substantially different spatial extents; in fact a shell model.

Many properties of an atom are determined by the outermost electrons. This is obviously true of the chemical properties as it is the outer electrons that will be affected most when two atoms approach each other. The same is true of the optical spectra of the atoms. Not until we come to discuss characteristic X-rays will we be interested in the properties of the inner electrons. Thus for most properties we are only interested in the energies and wavefunctions of the outermost electrons, and to find these, or at least sufficient approximations to them, we do not have to solve the complete self-consistent field problem which is well beyond the scope of this book. Instead because the outer electrons spend most of their time outside the region of space occupied by the inner electrons we can treat the influence of the inner electrons by considering qualitatively how their presence affects the potential experienced by the outer electrons and in what way this affects the energies and wavefunctions of the outer electrons.

8.5 THE PERIODIC TABLE ON THE ATOMIC SHELL MODEL

We have said that the chemical properties of an element are determined by the electrons in the outermost shell, and have justified this in section 8.4

in terms of the varying extent of the wavefunctions for electrons in different shells. We now wish to account in more detail for the properties of the elements as we go through the periodic table, using the ideas of the atomic shell model we have developed so far. Before we can carry this through we need to do three things.

(i) Determine the maximum number of electrons there can be in a shell or sub-shell.

(ii) Find out qualitatively the order in energy of the possible states in terms of the quantum numbers n and l.

(iii) Discuss briefly the nature of chemical combination.

(i) The number of electrons in a shell or sub-shell

For a particular value of l there are $(2l + 1)$ magnetic substates with values of the z-component m_l ranging from $-l$ to $+l$. Each of these can be associated with two values of the z-component of spin m_s, namely $m_s = +\frac{1}{2}$ and $m_s = -\frac{1}{2}$. Hence in a particular l sub-shell we can accommodate up to $2(2l + 1)$ electrons.

For any particular value of n we can have any value of l from 0 to $n - 1$. Hence the number of states in a shell of given n is

$$\sum_{l=0}^{n-1} 2(2l + 1) = 4\left(\sum_{l=0}^{n-1} l\right) + 2n$$

$$= 4\frac{n}{2}(n - 1) + 2n$$

$$= 2n^2 .$$

(ii) The order in energy in terms of n and l

The potential energy of any electron in a neutral atom is $-e^2/4\pi\varepsilon_0 r$ at large distances, and deviates from this as we move in towards the nucleus, approaching $-Ze^2/4\pi\varepsilon_0 r$ as we get very close to the nucleus. As we are interested in the energies of the outermost electrons which spend a substantial fraction of their time in the field of a single charge we start by assuming that, to the lowest order of approximation, the level ordering will be the same as in hydrogen. We now consider the effect of the deviation from the pure Coulomb potential. The deviation is largest near the origin, so we expect the effect to be biggest on those states whose wavefunctions are biggest near the origin. These are the s-states, followed by the p-states, and so on, the effect decreasing with increasing l-value. To illustrate this we plot, in Fig. 8.6, the 3s, 3p and 3d wavefunctions for small r, and the deviation from the Coulomb potential for sodium. It is clear that there will be a substantial effect on the

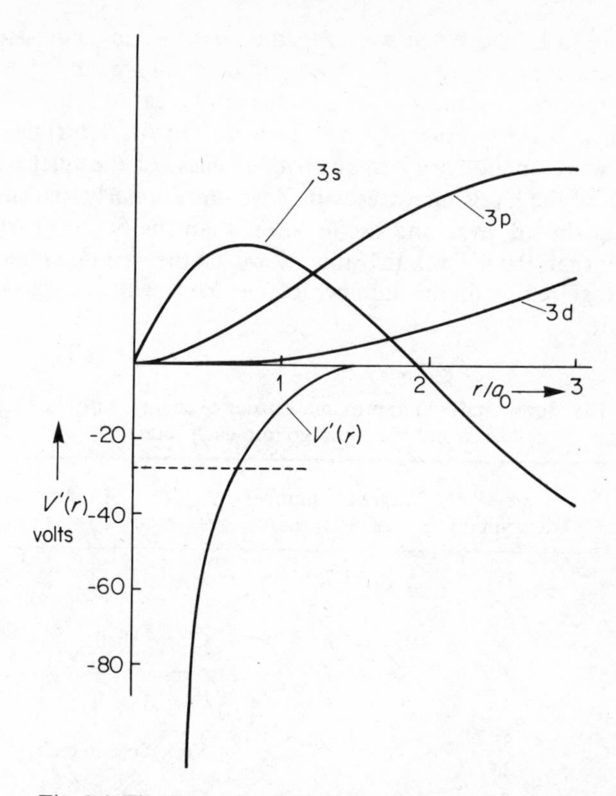

Fig. 8.6. The 3s, 3p, 3d wavefunctions for hydrogen close to the origin and the approximate deviation $V'(r)$ of the potential energy for the valence electron in sodium from the pure Coulomb potential. The horizontal dotted line is the value of the potential energy of an electron in the hydrogen atom at a radial distance a_0.

3s state, a moderate effect on the 3p state and negligible effect on the 3d state. Thus the effect of the deviation from the Coulomb potential is to destroy the degeneracy of the levels with the same value of n, but differing in l, that occurs in hydrogen. As the potential near the origin is always more attractive than the pure Coulomb potential the effect is always to lower the energies of the levels as compared with atomic hydrogen.

Our preliminary conclusion therefore is that the electrons will go into the state with the lowest available value of n, and within that n value, the lowest available value of l. This simple state of affairs will be spoiled however when the effect becomes large enough to be comparable with the energy spacing between successive values of n. This can best be illustrated by concrete examples. At argon with $Z = 18$ all the levels have been filled up to and including the 3p levels. When we go to the next nucleus, potassium, with $Z = 19$, we find that, due to the finite amplitude of all s-wave functions near the nucleus, the effect on the 4s level is sufficiently great to bring it *below* the 3d level which is not perturbed very much due to the centrifugal barrier keeping its wavefunction very small in the region of the nucleus. As the perturbation of the levels increases with Z we find similarly that the 5s level is lower than the 4d level, and the 6s lower than the 5d. Similarly the 5p level is lower than the 4f level. In Table 8.2 we list the various states in order of energy, together with the number of electrons which each state can accommodate.

Table 8.2. The atomic states in approximate order of energy, with the number of electrons which can go into each state

n	l	Level description	Maximum number of electrons	Comments
1	0	1s	2	
2	0	2s	2	
2	1	2p	6	Closed shell
3	0	3s	2	
3	1	3p	6	Closed shell
4	0	4s	2 }	Very close in energy
3	2	3d	10 }	
4	1	4p	6	Closed shell
5	0	5s	2 }	Very close in energy
4	2	4d	10 }	
5	1	5p	6	Closed shell
6	0	6s	2	
4	3	4f	14 }	These two are very close in energy. The filling of the 4f shell accounts for the rare earths
5	2	5d	6 }	
6	1	6p	10	

The details of the electron arrangements for the elements are shown in Table 8.3.

(iii) Chemical combination

Two or more atoms bind together to form a chemical compound due to the fact that, by sharing their outer electrons between them, they form a state whose total energy is lower than the sum of their individual ground state energies when they are separate. When one or more of the atoms involved has several electrons in its outermost shell the situation clearly becomes very complex and a simple description is not possible. Indeed whole books have been written on the subject. Even when there is only one electron in the outer shell the situation may be complicated by the fact that the electrons in the next lowest shell may be only slightly higher in energy. For example copper has one electron in the 4s shell, but due to the fact that the electrons in the 3d shell have nearly the same energy copper forms both monovalent and divalent compounds. Clearly in this section we can discuss only the simplest possibilities.

The reader is reminded that at the most elementary level valency is defined by the number of atoms of hydrogen that are displaced by, or will combine with, one atom of the element under consideration. As hydrogen has only one electron we expect that atoms with only one electron outside a tightly bound closed shell will be monovalent. This is indeed the case; they are the alkali metals. Similarly one might expect that atoms which require one more electron to complete a shell will be monovalent. These elements, the halogens, have high ionization potentials (see the discussion on fluorine which follows) and so do not readily share their electrons with other atoms, but they are capable of binding an additional electron to complete the shell, but not more than one, as a second one would have to go into a higher orbit. In the same way atoms with two electrons outside a closed shell, or requiring two more electrons to complete a closed shell are normally divalent. Beyond this point the variety of ways in which the electrons can arrange themselves usually means more than one valency can and does occur. For example there are three different oxides of nitrogen and two of carbon, and the valency can depend on the nature of the other atoms with which the element is combining. Nevertheless we find, as we should expect, that elements with the same numbers of outer electrons and with similar inner shell structures possess very similar chemical properties.

With these substantial preliminaries out of the way we are now in a position to discuss the periodic table. We treat the first few elements individually though not always in order, and thereafter treat the elements in groups, mentioning only particular features.

Hydrogen ($Z = 1$) With one electron in the 1s level this is our reference atom so to speak, used as the standard of comparison in valency, ionization potential etc.

Helium ($Z = 2$) With two electrons in the 1s shell this is the first of the noble gases. Electrons in the same shell are not very effective at shielding each other from the nuclear charge and the ionization potential at 24.6 volts is the highest of all the elements. This high binding prevents it sharing any of its electrons with other elements to form a compound. Equally it cannot take a share of an electron from another element, as this electron would

Table 8.3. Electron configurations and ionization potentials of atoms

Z	Element	1s	2s	2p	3s	3p	3d	4s	4p	4d	4f	5s	5p	5d	6s	6p	6d	7s	Ionization potential, eV
1	H	1																	13.595
2	He	2																	24.580
3	Li	2	1																5.390
4	Be	2	2																9.320
5	B	2	2	1															8.296
6	C	2	2	2															11.260
7	N	2	2	3															14.532
8	O	2	2	4															13.614
9	F	2	2	5															17.422
10	Ne	2	2	6															21·564
11	Na	2	2	6	1														5.138
12	Mg	2	2	6	2														7.644
13	Al	2	2	6	2	1													5.984
14	Si	2	2	6	2	2													8.149
15	P	2	2	6	2	3													10.486
16	S	2	2	6	2	4													10.357
17	Cl	2	2	6	2	5													12.967
18	Ar	2	2	6	2	6													15.759
19	K	2	2	6	2	6		1											4.339
20	Ca	2	2	6	2	6		2											6.111
21	Sc	2	2	6	2	6	1	2											6.540
22	Ti	2	2	6	2	6	2	2											6.280
23	V	2	2	6	2	6	3	2											6.740
24	Cr	2	2	6	2	6	5	1											6.764
25	Mn	2	2	6	2	6	5	2											7.432
26	Fe	2	2	6	2	6	6	2											7.870

27	Co	2	2	6	2	6	7	2							7.864
28	Ni	2	2	6	2	6	8	2							7.633
29	Cu	2	2	6	2	6	10	1							7.724
30	Zn	2	2	6	2	6	10	2							9.391
31	Ga	2	2	6	2	6	10	2	1						6.00
32	Ge	2	2	6	2	6	10	2	2						7.88
33	As	2	2	6	2	6	10	2	3						9.81
34	Se	2	2	6	2	6	10	2	4						9.75
35	Br	2	2	6	2	6	10	2	5						11.84
36	Kr	2	2	6	2	6	10	2	6						13.996
37	Rb	2	2	6	2	6	10	2	6		1				4.176
38	Sr	2	2	6	2	6	10	2	6		2				5.692
39	Y	2	2	6	2	6	10	2	6	1	2				6.377
40	Zr	2	2	6	2	6	10	2	6	2	2				6.835
41	Nb	2	2	6	2	6	10	2	6	4	1				6.881
42	Mo	2	2	6	2	6	10	2	6	5	1				7.131
43	Tc	2	2	6	2	6	10	2	6	(5)	(2)?				7.23
44	Ru	2	2	6	2	6	10	2	6	7	1				7.365
45	Rh	2	2	6	2	6	10	2	6	8	1				7.461
46	Pd	2	2	6	2	6	10	2	6	10					8.33
47	Ag	2	2	6	2	6	10	2	6	10	1				7.574
48	Cd	2	2	6	2	6	10	2	6	10	2				8.991
49	In	2	2	6	2	6	10	2	6	10	2	1			5.785
50	Sn	2	2	6	2	6	10	2	6	10	2	2			7.332
51	Sb	2	2	6	2	6	10	2	6	10	2	3			8.639
52	Te	2	2	6	2	6	10	2	6	10	2	4			9.01
53	I	2	2	6	2	6	10	2	6	10	2	5			10.44
54	Xe	2	2	6	2	6	10	2	6	10	2	6			12.127
55	Cs	2	2	6	2	6	10	2	6	10	2	6	1		3.893
56	Ba	2	2	6	2	6	10	2	6	10	2	6	2		5.210

(*Continued*)

Table 8.3 (*Continued*)

Z	Element	1s	2s	2p	3s	3p	3d	4s	4p	4d	4f	5s	5p	5d	5f	6s	6p	6d	7s	Ionization potential, eV
57	La	2	2	6	2	6	10	2	6	10		2	6	1		2				5.61
58	Ce	2	2	6	2	6	10	2	6	10	1	2	6	1		2				6.91
59	Pr	2	2	6	2	6	10	2	6	10	3	2	6			2				5.70
60	Nd	2	2	6	2	6	10	2	6	10	4	2	6			2				6.31
61	Pm	2	2	6	2	6	10	2	6	10	5	2	6			2				
62	Sm	2	2	6	2	6	10	2	6	10	6	2	6			2				5.6
63	Eu	2	2	6	2	6	10	2	6	10	7	2	6			2				5.67
64	Gd	2	2	6	2	6	10	2	6	10	7	2	6	1		2				6.16
65	Tb	2	2	6	2	6	10	2	6	10	8	2	6	1		2				6.74
66	Dy	2	2	6	2	6	10	2	6	10	9	2	6	1		2	?			6.32
67	Ho	2	2	6	2	6	10	2	6	10	10	2	6	1		2	?			
68	Er	2	2	6	2	6	10	2	6	10	11	2	6	1		2	?			
69	Tm	2	2	6	2	6	10	2	6	10	13	2	6			2				
70	Yb	2	2	6	2	6	10	2	6	10	14	2	6			2				6.22
71	Lu	2	2	6	2	6	10	2	6	10	14	2	6	1		2				6.15
72	Hf	2	2	6	2	6	10	2	6	10	14	2	6	2		2				5.5
73	Ta	2	2	6	2	6	10	2	6	10	14	2	6	3		2				7.7
74	W	2	2	6	2	6	10	2	6	10	14	2	6	4		2				7.98
75	Re	2	2	6	2	6	10	2	6	10	14	2	6	5		2				7.87
76	Os	2	2	6	2	6	10	2	6	10	14	2	6	6		2				8.7
77	Ir	2	2	6	2	6	10	2	6	10	14	2	6	7		2				9.2
78	Pt	2	2	6	2	6	10	2	6	10	14	2	6	9		1	?			9.0
79	Au	2	2	6	2	6	10	2	6	10	14	2	6	10		1				9.22
80	Hg	2	2	6	2	6	10	2	6	10	14	2	6	10		2				10.434
81	Tl	2	2	6	2	6	10	2	6	10	14	2	6	10		2	1			6.106
82	Pb	2	2	6	2	6	10	2	6	10	14	2	6	10		2	2			7.415
83	Bi	2	2	6	2	6	10	2	6	10	14	2	6	10		2	3			7.287

84	Po	2	2	6	2	6	10	2	6	10	14	2	6	10		2	4			8.43
85	At	2	2	6	2	6	10	2	6	10	14	2	6	10		2	5			9.2
86	Rn	2	2	6	2	6	10	2	6	10	14	2	6	10		2	6			10.745
87	Fr	2	2	6	2	6	10	2	6	10	14	2	6	10		2	6		1	4.0
88	Ra	2	2	6	2	6	10	2	6	10	14	2	6	10		2	6		2	5.277
89	Ac	2	2	6	2	6	10	2	6	10	14	2	6	10		2	6	1	2?	6.9
90	Th	2	2	6	2	6	10	2	6	10	14	2	6	10		2	6	2	2?	
91	Pa	2	2	6	2	6	10	2	6	10	14	2	6	10	2	2	6	1	2?	
92	U	2	2	6	2	6	10	2	6	10	14	2	6	10	3	2	6	1	2	4
93	Np	2	2	6	2	6	10	2	0	10	14	2	6	10	4	2	6	1	2?	
94	Pu	2	2	6	2	6	10	2	6	10	14	2	6	10	5	2	6	1	2?	
95	Am	2	2	6	2	6	10	2	6	10	14	2	6	10	7	2	6		2	
96	Cm	2	2	6	2	6	10	2	6	10	14	2	6	10	7	2	6	1	2?	
97	Bk	2	2	6	2	6	10	2	6	10	14	2	6	10	8	2	6	1	2?	
98	Cf	2	2	6	2	6	10	2	6	10	14	2	6	10	9	2	6	1	2?	

have to spend its time in states of n equal to two or greater. The wavefunctions for these states are in regions where the nuclear charge is completely shielded by the two 1s electrons and there is no field to bind it. Helium is thus inert.

Lithium ($Z = 3$) This consists of a closed 1s shell and one electron in the 2s state. The binding of the 1s electrons is even greater than in helium, and the properties are therefore determined by the single 2s electron. The ionization potential is low as the 2s electron spends most of its time in a region where the potential is that of a single charge. It thus shares its electron readily and is chemically a very active single valency element, the first of the alkalis. All the alkali metals have a single s-electron outside a closed shell.

Beryllium ($Z = 4$) has two 1s electrons forming an inert core and two 2s electrons. The ionization potential at 9.32 volts is higher than that of lithium due to the poor screening provided by electrons in the same state, but it can readily combine, and with two outer electrons has a valency of two.

Boron ($Z = 5$) has three outer electrons; two 2s electrons and one 2p electron. Two factors affect the ionization potential. There is the additional nuclear charge, which is poorly screened by the 2s electrons. Thus the potential in which the last electron moves is stronger than in beryllium. The last electron in this case however is in a p-state, and hence there is only a small probability of finding it close to the origin, where the potential is largest. The net result is a small drop in the ionization potential. With three outer electrons boron forms trivalent compounds, e.g. boron trifluoride.

Neon ($Z = 10$) We now take a jump to $Z = 10$. Table 8.2 shows that the $n = 1$ and $n = 2$ shells can just hold ten electrons between them. The outer electrons ($n = 2$) are poor at shielding each other from the nuclear charge, and the ionization potential is consequently high (21.56 V). The next electron has to go into the $n = 3$ shell, and so we have an inert noble gas, as in helium.

Fluorine ($Z = 9$) The same arguments as used for neon predict a high ionization potential (17.42 V). However it can take another electron in the $n = 2$ shell, which due to the poor shielding is readily bound with an energy of approximately 4 eV. It therefore easily forms negative ions and easily accepts *one* electron from other elements. It is therefore monovalent and is the first of the halogens. All the halogens require one more electron to complete a p-shell.

In combination with the alkali metals the halogens form ionic compounds in which the alkali metal has essentially lost an electron and is thus positively charged, and the halogen gained an electron and is negatively charged, the two ions being held together by electrostatic attraction. Taking lithium fluoride as an example a positively charged lithium ion and a negatively charged fluorine ion together have a lower energy than the two neutral atoms for separations of less than 10 Å.

Oxygen ($Z = 8$) By similar arguments oxygen is expected to be divalent, as it is.

Carbon ($Z = 6$) *and Nitrogen* ($Z = 7$) both show more than one valency.

Sodium ($Z = 11$) *to Argon* ($Z = 18$) The elements in this sequence fill the 3s and 3p subshells in just the same way as the $n = 2$ shell was filled, and so we get an extremely similar sequence of properties, including the drop in ionization potential when the s sub-shell is filled. At argon we get an inert gas in spite of the fact that the 3d levels are not yet filled. This is a case where one has to consider the change in wavefunction as well as the change in energy. The perturbation of the p-waves at the origin is sufficiently strong to substantially lower the energy, and cause the range of the 3p wavefunction to be so much less than that of the 3d wavefunction that the p-wave particles effectively provide full shielding of the nuclear charge for the 3d particle. We therefore have all the conditions for an inert gas.

Potassium ($Z = 19$) *to Zinc* ($Z = 30$) The strong perturbation acting on s-wave levels is sufficiently strong at this point to make the 4s level slightly lower than the 3d level. The first two elements, potassium and calcium, have respectively 1 and 2 electrons in the 4s subshell, and potassium is a typical alkali, and calcium a typical alkaline earth. From this point up to zinc i.e. ($Z = 21$ to 30) the ten vacant 3d levels are being filled. The range of the 3d and 4s wavefunctions is very similar and a complex valency situation arises with many of the elements (they are all metals) showing several different valencies. At zinc the shells from $n = 1$ to $n = 3$ are completely full, and there are two 4s electrons.

Gallium ($Z = 31$) *to Krypton* ($Z = 36$) This is a straightforward region in which the 4p shell is filling ending in the noble gas krypton. It is similar to the region from aluminium to argon.

Rubidium ($Z = 37$) *to Xenon* ($Z = 54$) These elements follow a precisely similar path to the one followed in going from potassium to krypton. The two regions have the same number of elements. In this case the sequence is 5s, 4d, 5p ending in a noble gas. Note that the 4f level is completely empty.

Cesium ($Z = 55$) *to Mercury* ($Z = 80$) This region starts off in the conventional way following a noble gas with an alkali (cesium) and an alkaline earth (barium) with one and two 6s electrons respectively. The following elements, from lanthanum ($Z = 57$) to lutetium ($Z = 71$), are known as the rare-earths or lanthanides. They have extremely similar chemical properties due to the fact that it is the 4f shell that is filling here. The situation is quite different from the case of the 3d and 4s shell discussed above. In that case there were only two electrons in shells with higher principal quantum numbers. In the present case there are ten such electrons. It turns out that the 4f electrons that are being added have the bulk of their wavefunction well inside the atom, and thus do not seriously affect the chemical properties. Although this means the 4f electrons have a lower energy than the 6s electrons it does not mean that the 6s electrons ought to make a transition to the 4f shell. In the most elementary terms, if they did, although they themselves would have a lower energy the screening effect they would have as 4f electrons on the electrons with higher principal quantum numbers would result in a net raising of the energy of the atom.

The filling of the 4f shell is followed by the filling of the 5d shell and at mercury the first four shells are full, the $n = 5$ shell is full up to and including the 5d sub-shell, and there are two 6s electrons.

Thallium ($Z = 81$) *and upwards* For the first six elements the 6p shell is filling to form a noble gas at radon ($Z = 86$). We then have an alkali at francium ($Z = 87$), an alkaline earth at radium ($Z = 88$) followed by the actinides, a series similar to the rare earths in which the 5f shell is filling.

8.6 THE INDISTINGUISHABILITY OF IDENTICAL PARTICLES AND THE SYMMETRY OF WAVEFUNCTIONS

We have introduced the Pauli exclusion principle in a purely empirical way, as a necessary principle to enable us to account for the periodic table of the elements. We were able to use it in a simple and convenient manner because we had a model, the self consistent field model, in which the particles moved independently and could therefore have their states labelled by a set of quantum numbers, n, l, m_l and m_s. However, this is an approximation to what actually happens and it is not in the least clear how to apply the exclusion principle to a more complex description. In some problems it might not be easy to find identifying quantum numbers and this too would put us at a loss. It is clear that we should look for a more general formulation of the exclusion principle, applicable in all circumstances.

Let us consider how we have been describing atoms. In our discussion in section 8.2 leading up to the formulation of the self-consistent field we considered the case of an excited helium atom in which one electron was in its lowest state and the other in an excited state, and said that a very good approximation to the exact wavefunction would be

$$\Psi = A\,\mathrm{e}^{-2r_1/a_0}\phi_{nlm_l}(\mathbf{r}_2)\ . \tag{8.4}$$

In writing Eq. (8.4) in that particular form we have done more than assert that two particular states are involved, for we have ascribed one electron to one particular state and the other electron to the other state; we have identified the electrons by their wavefunctions. We have implicitly taken this attitude throughout, and we should enquire whether this is something that can be done.

Electrons are, of course, indistinguishable. There is nothing that will enable us to tell one electron from another. This, however, has nothing to do with quantum mechanics, and there are many imaginable classical systems which consist of indistinguishable particles and we have no hesitation in identifying the particles in these by their states of motion. Let us consider the way we would set about doing this.

Suppose we have any collection of particles forming a classical gas and further assume that electromagnetic waves behave purely classically too. We now take a pair of photographs of the particles with a very short time interval between them, so that each particle in the second picture has moved only very slightly from its position in the first picture. We illustrate what we mean in Fig. 8.7(a) which represents a superposition of the two pictures with an arrow linking the position of each particle in the first picture to its position in the second picture. From the two pictures together and the time interval between them we may deduce the position and velocity of each of the particles in the second picture. We may label each particle in the second picture by a number, and make a table giving, opposite each number, the position and velocity of the particle to which each number refers. It is now possible, in principle, to work out the positions and velocities of the particles as a function of time, given the law of force between the particles. If, at some subsequent time, a third picture is taken we can relate the positions the particles have in this picture to their positions in the previous pictures, and hence identify the particles as shown in Fig. 8.7(*b*). This is what we mean by identifying particles by their states of motion.

The experiment we have just described is not possible even in principle in quantum mechanics. When we take quantum phenomena into account the uncertainty principle tells us that the first two pictures cannot provide us with sufficiently precise information to enable us to be sure of our later identification. Experimentally what would happen is that if we used light of a fairly long wavelength to take our pictures they would be fuzzy and it would not be possible to determine the particle positions very accurately, whereas if we use a short wavelength so that we get sharp pictures, then the velocities determined from the first two pictures will *not* correspond to the particle velocities after the second picture has been taken due to the unknown momenta imparted by the photons to the particles when we took the second picture. Hence we can never obtain sufficiently precise information on both the positions and velocities of the particles to enable us to identify the particles at a later stage, and we *cannot* ascribe a specific state of motion to a specific particle. We therefore make a new postulate as follows:

Postulate

The wavefunction for a system containing two or more identical particles must be written in such a way that it is impossible to distinguish between the particles by their states of motion.

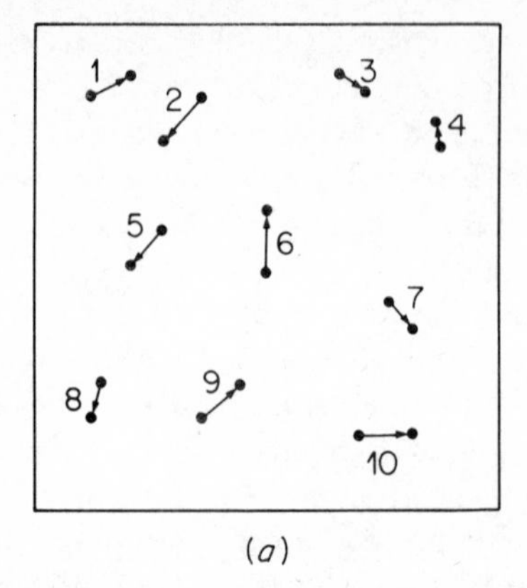

(a)

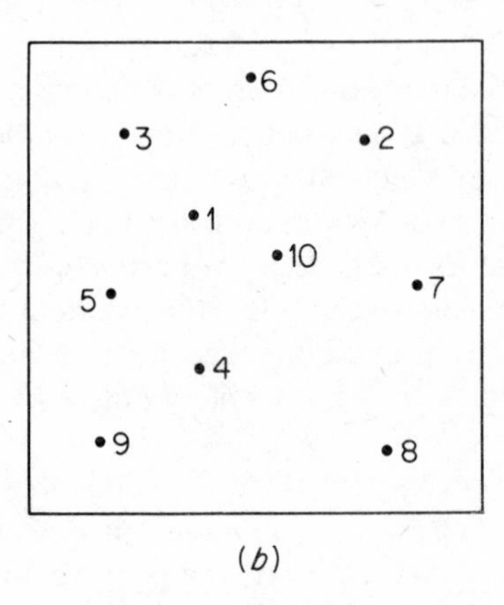

(b)

Fig. 8.7. (*a*) Represents the superposition of two photographs of ten classical atoms which were taken with a very short time interval between them. The positions of each atom in the two photographs are joined by an arrow which proceeds from the position in the first photograph to the position in the second photograph. The lengths and directions of the arrows, therefore, represent the velocities of the atoms. From the initial positions and velocities we can, in principle, calculate the positions of the atoms at all subsequent times, and hence identify the atoms in a later picture, as in (*b*).

We must now devise a mathematical way of expressing this postulate, and investigate its consequences.

A wavefunction for a two particle system must contain the coordinates of the particles as variables and a set of descriptive parameters labelling the functions. In the simple, approximate, cases we have discussed in this chapter where we can write the total wavefunction as a product of individual wavefunctions, one for each particle as in Eq. (8.2) and Eq. (8.4), the sets of quantum numbers n, l, m_l, m_s form such descriptive parameters. As we have already pointed out the wavefunctions of Eq. (8.2) and Eq. (8.4) associate the coordinates of a *particular particle* with a *particular set* of descriptive parameters. Our new postulate says this is not permissible. We have to find a way of writing our wavefunction so that no form of *measurement* will distinguish between the particles.

Let us illustrate this last point with an example. Suppose the function ϕ_{nlm_l} in Eq. (8.4) has $n = 3, l = 2$. The probability of finding a particle in this state within, say, $a_0/2$ of the origin is less than one part in ten million, whereas for the 1s electron it is approximately 30%. Eq. (8.4) implies that if, on two separate occasions, we find an electron within $a_0/2$ of the origin we can be virtually certain it is the same electron on each occasion. Our new postulate asserts that this is precisely what we *cannot* say.

We start with a wavefunction of the form we have been using before introducing our new postulate and which does make a distinction between the particles, and write it as

$$\Phi(1, 2)$$

where the numbers 1 and 2 label the particles, and the *positions* of the numbers signify the set of wavefunction parameters to be associated with the particle. Let us label these sets of parameters by the labels m and n respectively so that $\Phi(1, 2)$ means particle 1 is associated with the set of parameters denoted by m, and particle 2 with the set denoted by n.

We assume that $\Phi(1, 2)$ is a solution of the Schrödinger equation

$$H(1, 2)\Phi(1, 2) = E\Phi(1, 2) \tag{8.5}$$

where

$$H(1, 2) = H_0(1) + H_0(2) + V(1, 2) \tag{8.6}$$

and $H_0(1)$ and $H_0(2)$ are the Hamiltonians for particles 1 and 2 in the external field, and $V(1, 2)$ is the interaction between the particles. In the problem of the helium atom, for example, we would have

$$H_0(1) = -\frac{\hbar^2}{2m}\nabla_1^2 - \frac{2e^2}{4\pi\varepsilon_0 r_1}$$

$$H_0(2) = -\frac{\hbar^2}{2m}\nabla_2^2 - \frac{2e^2}{4\pi\varepsilon_0 r_2}$$

$$V(1, 2) = \frac{e^2}{4\pi\varepsilon_0|\mathbf{r}_1 - \mathbf{r}_2|} .$$

The Hamiltonians $H_0(1)$ and $H_0(2)$ are identical in form as the two electrons have the same mass and charge and interact with the same nucleus. This is always true as we have defined H_0 as the Hamiltonian containing the interaction with the external field, and if the particles are identical they must each have the same interaction with outside bodies, by definition. In the case of the helium atom the expression for $V(1, 2)$ is symmetrical with respect to the particles, and again this is always true, as if it weren't we should have a way of distinguishing between the particles and they could not be considered identical. Hence $V(1, 2) = V(2, 1)$ from which it follows, with the identity of form of $H_0(1)$ and $H_0(2)$, that the Hamiltonian is symmetrical, i.e.

$$H(1, 2) \equiv H(2, 1) . \tag{8.7}$$

The way we labelled the particles in $\Phi(1, 2)$ was quite arbitrary, and we could equally well have labelled the particles the other way round, so that the wavefunction becomes $\Phi(2, 1)$ with particle 2 now associated with the parameters m, and particle 1 associated with the parameters n. In terms of the approximate wavefunction Eq. (8.4) for the helium atom what we are saying is it does not matter if we write it as

$$\Psi = A\,\mathrm{e}^{-2r_1/a_0}\phi_{nlm_l}(\mathbf{r}_2)$$

or as

$$\Psi = A\,\mathrm{e}^{-2r_2/a_0}\phi_{nlm_l}(\mathbf{r}_1) .$$

$\Phi(2, 1)$ will be a solution of the corresponding equation to Eq. (8.5)

$$H(2, 1)\Phi(2, 1) = E\Phi(2, 1)$$

which, as $H(1, 2) \equiv H(2, 1)$, can be written

$$H(1, 2)\Phi(2, 1) = E\Phi(2, 1) . \tag{8.5a}$$

Thus if $\Phi(1, 2)$ is an eigenfunction of the Hamiltonian $H(1, 2)$, belonging to the eigenvalue E, then so is $\Phi(2, 1)$. This is physically obvious in the case of helium. If we started with both electrons in the ground state, and excited one to a higher state, leaving the other behind, it clearly does not matter which one we excite. But if $\Phi(1, 2)$ and $\Phi(2, 1)$ are both solutions belonging to the same eigenvalue, then so is any linear combination

$$a\Phi(1, 2) + b\Phi(2, 1) \; . \tag{8.8}$$

But in Eq. (8.8) particle 1 is partly associated with the parameters n, and partly with the parameters m, and it appears reasonable that it is possible to choose the ratio a/b so that it is impossible to say which particle is associated with which state of motion. If so we shall have a wavefunction that satisfies our new postulate.

If a wavefunction $\Psi(1, 2)$ is to satisfy our postulate it must have the property that if we interchange the labels associated with the particles we must get no observable change in any measurable quantity, i.e. it does not matter whether we take $\Psi(1, 2)$ as our wavefunction, or $\Psi(2, 1)$; everything we measure will be the same.

One quantity we can measure is the probability that one of the particles is in a small volume around $\mathbf{r}_1$ and the other is in a small volume around $\mathbf{r}_2$. If $\Psi(1, 2)$ and $\Psi(2, 1)$ are to give the same answer we must have

$$|\Psi(1, 2)|^2 \, d^3\mathbf{r}_1 \, d^3\mathbf{r}_2 \equiv |\Psi(2, 1)|^2 \, d^3\mathbf{r}_1 \, d^3\mathbf{r}_2 \; . \tag{8.9}$$

From Eq. (8.9) it follows that

$$\Psi(2, 1) = e^{i\delta}\Psi(1, 2) \tag{8.10}$$

where δ is a real number.

If we exchange the labels again we get

$$\begin{aligned} \Psi(1, 2) &= e^{i\delta}\Psi(2, 1) \\ &= e^{2i\delta}\Psi(1, 2) \end{aligned}$$

whence

$$e^{2i\delta} = 1$$

and therefore

$$e^{i\delta} = \pm 1$$

where the symbol $\pm$ is to be interpreted as 'plus or minus'.

Wavefunctions which satisfy our new postulate therefore satisfy the condition

$$\Psi(1, 2) = \pm \Psi(2, 1) \; . \tag{8.11}$$

Wavefunctions which stay the same are said to be symmetric under exchange and those that change sign are said to be antisymmetric under exchange.

If we take Eq. (8.8) as our $\Psi(1, 2)$ we may use Eq. (8.11) to determine the ratio of the constants a and b. Thus

$$\{a\Phi(1, 2) + b\Phi(2, 1)\} = \pm\{a\Phi(2, 1) + b\Phi(1, 2)\} .$$

We see that this equation is satisfied with the positive sign if $a = b$ and with the negative sign if $a = -b$. Thus from $\Phi(1, 2)$ we can construct two wavefunctions

$$\Psi^+(1, 2) \equiv \Phi(1, 2) + \Phi(2, 1) \tag{8.12a}$$

and

$$\Psi^-(1, 2) \equiv \Phi(1, 2) - \Phi(2, 1) \tag{8.12b}$$

where $\Psi^+(1, 2)$ is symmetric under exchange and $\Psi^-(1, 2)$ is antisymmetric.

In the simple case where $\Phi(1, 2)$ can be expressed as a product

$$\Phi(1, 2) = \phi_m(1)\phi_n(2)$$

the functions that satisfy our postulate are

$$\Psi^+(1, 2) = \phi_m(1)\phi_n(2) + \phi_m(2)\phi_n(1) \tag{8.13a}$$

and

$$\Psi^-(1, 2) = \phi_m(1)\phi_n(2) - \phi_m(2)\phi_n(1) \tag{8.13b}$$

Before we can use Eq. (8.4) as an example of this procedure we must make one small modification. The labels n and m in Eq. (8.13a,b) each must include a specification of the spin of the particle. This is necessary, as if we did not exchange these labels too we would be left with the possibility of identifying the particle through its spin direction. Eq. (8.4) makes no reference to the spins of the particles, and these must be included to give us a complete set of labels. We introduce labels for the spin of the electron as follows. If the result of measuring the z-component of the electron spin is always $+\frac{1}{2}\hbar$ we say the electron is in the state α i.e.

$$\hat{s}_z\alpha = \tfrac{1}{2}\hbar\alpha \tag{8.14a}$$

and if the result is always $-\frac{1}{2}\hbar$ we say the electron is in the state β, i.e.

$$\hat{s}_z\beta = -\tfrac{1}{2}\hbar\beta . \tag{8.14b}$$

The states α and β are the analogues, for electron spin, to the wavefunctions for spatial motion. For this reason they are often referred to as 'spin wavefunctions'. However it should be emphasized that they are not functions of position. Equally the operator $\hat{s}_z$ is *not* a differential function of position, as is $\hat{l}_z$, and cannot be expressed in some other more familiar form.

Let us assume that particle 1 has spin up, i.e. a spin wavefunction $\alpha(1)$ and particle 2 has spin down, i.e. a spin wavefunction $\beta(2)$, so that Eq. (8.4) now reads

$$\Psi = A\,e^{-2r_1/a_0}\phi_{nlm_l}(\mathbf{r}_2)\alpha(1)\beta(2)\ . \tag{8.4a}$$

From Eq. (8.4a) we may construct both a symmetric and an antisymmetric wavefunction satisfying the principle of indistinguishability:

$$\Psi^+(1,2) = A\,e^{-2r_1/a_0}\phi_{nlm_l}(\mathbf{r}_2)\alpha(1)\beta(2) + A\,e^{-2r_2/a_0}\phi_{nlm_l}(\mathbf{r}_1)\alpha(2)\beta(1) \tag{8.4b}$$

$$\Psi^-(1,2) = A\,e^{-2r_1/a_0}\phi_{nlm_l}(\mathbf{r}_1)\alpha(1)\beta(2) - A\,e^{-2r_2/a_0}\phi_{nlm_l}(\mathbf{r}_1)\alpha(2)\beta(1)\ . \tag{8.4c}$$

Eqs. (8.4b,c) illustrate a quite general point. We have constructed the equations by demanding a certain behaviour when we interchange the labels 1 and 2 specifying the particles. We get exactly the same result by demanding the same behaviour when we interchange the labels specifying the wavefunctions, including the spin wavefunctions. We may therefore talk about the behaviour of the wavefunction under the exchange of a pair of particles, and not worry about whether we mean the labels specifying the wavefunctions (including spin) of the particles, or the coordinates (including spin) of the particles, as these are entirely equivalent.

The extension to systems of more than two identical particles is quite straightforward. We must make no observable change when we exchange any pair of particles. We can always choose to label the pair to be exchanged by the labels 1 and 2 and everything we have said up to now will apply.

We may summarize the situation then as follows. In order to satisfy the postulate of indistinguishability the wave function must either remain the same (symmetric) or change sign (antisymmetric) under the exchange of any pair of particles.

In the general case we write this as

$$\Psi(1,2,\ldots,n) = \pm\Psi(2,1,\ldots,n)\ . \tag{8.15}$$

In the special case where the wavefunction can be written as a product, either exactly because there is no interaction between the particles, or approximately as in the self-consistent field model, we may write the wavefunction in the form of Eq. (8.13) for two particles. There is a corresponding form for more than two particles.

8.7 THE CONSERVATION OF SYMMETRY

Is it possible for the symmetry of a system of identical particles to change with time? The answer is, no it is not, the symmetry of a system is conserved however one perturbs it. One result of this is that all systems of a particular type of particle have the same symmetry under exchange. In particular all systems of electrons are antisymmetric under exchange as is shown by the fact that they obey the Pauli exclusion principles (see section 8.8).

We may demonstrate the conservation of symmetry as follows. We have already shown that the Hamiltonian for identical particles is symmetric under exchange, and this remains true, however we vary the external field, from the definition of what is meant by identical. Consider now how the wavefunctions change with time. From Schrödinger's time-dependent equation we have

$$i\hbar \frac{\partial \Psi(1,2)}{\partial t} = H(1,2)\Psi(1,2)$$

and

$$i\hbar \frac{\partial \Psi(2,1)}{\partial t} = H(2,1)\Psi(2,1) \ .$$

As $H(1,2) \equiv H(2,1)$ it follows that $\partial\Psi(1,2)/\partial t$ has the same symmetry under exchange as $\Psi(1,2)$. Now

$$\Psi(t + \delta t) = \Psi(t) + \frac{\partial \Psi(t)}{\partial t}\delta t \ . \tag{8.16}$$

As $\partial\Psi(t)/\partial t$ has the same symmetry as $\Psi(t)$ it follows that $\Psi(t + \delta t)$ has the same symmetry as $\Psi(t)$. Thus once a certain symmetry has been established, it will continue for *all* subsequent time, whatever happens. We may change the external potential, bombard the system with other particles, and do whatever we please, the symmetry will be preserved.

Consider two systems each of the same sort of identical particles coming together and interacting. The combination of the two now makes a bigger system whose wavefunction must have a definite symmetry. The symmetry of the two original systems cannot be changed, from which it follows that the two original systems and their combination must all have the same symmetry. It follows by extension that *all* systems in the universe of a particular type of particle must have the same symmetry. Thus *all* electrons, we find, belong to antisymmetric systems.

We shall show in the next section how the exchange symmetry for electrons can be determined from the structure of atoms, but while this is obviously very important, the simplest and most accurate experiments on the properties of a particle can usually be best done on single specimens. It would thus be very useful if there were to be a relation between the exchange symmetry and the properties of the particle. It was shown by Pauli in 1940, using relativistic quantum field theory, that all particles with half integral spin ($\frac{1}{2}, \frac{3}{2}, \ldots$ in units of $\hbar$) have antisymmetric exchange properties, and all particles with integral spin ($0, 1, 2, \ldots$ in units of $\hbar$) have symmetric exchange properties. Thus electrons, μ-mesons, protons and neutrons always have wavefunctions which are antisymmetric under the exchange of identical particles, while photons, π-mesons and alpha particles have symmetric wavefunctions.

8.8 EXCHANGE SYMMETRY AND THE PAULI EXCLUSION PRINCIPLE

We now show that particles which have antisymmetric exchange properties obey the Pauli exclusion principle, while those that have symmetric exchange properties do not. The Pauli exclusion principle, as we formulated it at the end of section 8.3, applied to the approximate description of electrons in an atom where we have a product of wavefunctions one, for each electron, each one being labelled by the quantum numbers n, l, m_l, m_s. Thus we are interested in the formulation of the exchange property as given by Eq. (8.13) for product wavefunctions. For antisymmetric particles we require the formulation of Eq. (8.13b)

$$\Psi^{-}(1,2) = \phi_m(1)\phi_n(2) - \phi_m(2)\phi_n(1) \ . \tag{8.13b}$$

If the labels m and n are the same, i.e. $\phi_m \equiv \phi_n$, Eq. (8.13b) becomes

$$\Psi^{-}(1,2) \equiv 0 \ .$$

Thus we cannot write down an antisymmetric wavefunction for two particles in the same state; to preserve the antisymmetric property they must be in different states. Thus no more than one electron may occupy any particular state, the requirement of the Pauli exclusion principle.

If we consider Eq. (8.13a) for symmetric particles

$$\Psi^{+}(1,2) = \phi_m(1)\phi_n(2) + \phi_m(2)\phi_n(1) \tag{8.13a}$$

and let $\phi_m \equiv \phi_n$ we get, leaving out normalization factors,

$$\Psi^{+}(1,2) = \phi_m(1)\phi_m(2) \ .$$

This can be extended to any number of symmetric identical particles, the wavefunction

$$\Psi = \phi_m(1)\phi_m(2)\phi_m(3)\ldots\phi_m(j) \tag{8.17}$$

clearly being symmetric under the exchange of any pair of particles. Thus, any number of symmetric particles can be in the same state.

The fact that there is no restriction on the number of particles occupying a given state if they are symmetric under exchange (and therefore have integral spin), while they must all be in different states if they are antisymmetric (and hence have half-integral spin) will clearly lead to very different behaviour for assemblies of such particles. In fact they obey different laws of statistical mechanics, those with antisymmetric exchange properties obeying Fermi–Dirac statistics, and those with symmetrical exchange properties obeying Bose–Einstein statistics. For this reason particles with half-integral spin are frequently referred to as fermions, and those of integral spin as bosons. (See F. Mandl, *Statistical Physics*, Manchester Physics Series, Wiley, London 1972.)

PROBLEMS 8

8.1 List the angular momenta that can occur for a hydrogen atom for $n = 5$. List the number of states in each subshell and show the sum equals $2n^2$

8.2 Estimate the mean radius of the electrons occupying the $n = 1$ and $n = 2$ shells in zinc ($Z = 30$).

8.3 Show that states with different exchange symmetries are orthogonal to each other.

8.4 Which do you think is the smallest neutral atom? Why?

8.5 $\psi(1, 2, 3)$ is a wave function for identical particles which changes sign under the exchange of any pair. Show that

$$\psi(3, 1, 2) = \psi(2, 3, 1) = \psi(1, 2, 3)$$

$$\text{and } \psi(2, 1, 3) = \psi(1, 3, 2) = \psi(3, 2, 1) = -\psi(1, 2, 3) \ .$$

8.6 At what values of Z would the $n = 1$ and $n = 2$ shells be full if electrons had a spin of $\frac{3}{2}\hbar$?

CHAPTER 9

One and two electron spectra

9.1 INTRODUCTION

In Chapter 8 we were concerned with the arrangement of the electrons in the ground states of atoms. For any particular atom there are many other arrangements possible, and these will lead to possible excited states of the atom. There are an enormous number of these, and clearly we cannot consider all of them. For the moment, we shall only be interested in the levels that arise from excitations of the outer electrons. Let us make this point more explicit: if we study the arrangement of the electrons in the atoms given in Table 8.3 we see that it consists of a number of electrons in sub-shells that are full, plus a small number in unfilled sub-shells. It is these electrons in unfilled subshells which we mean when we refer to the outer electrons, with two exceptions. The first, and most important, is when we have two electrons in an s-shell and the p-subshell with the same principal quantum is not completely full; in this case, the two s-electrons are regarded as members of the outer electron group. The second exception is in the case of the rare earth elements where the electrons in the unfilled 4f shell are not normally regarded as members of the outer electron group.

Any particular arrangement of electrons, specified by the quantum numbers n and l for each electron, is called a *configuration*. The arrangement corresponding to the ground state is called *the ground state configuration.* The configurations we are going to consider are those in which the electrons in closed sub-shells, which we call the *core*, are undisturbed from the ground state configuration, and only the quantum numbers of the outer electrons are changed. These configurations have excitation energies of a few electron volts above the ground state, and transitions between the levels resulting from these configurations are responsible for the spectral lines emitted by atoms in the visible and near ultra-violet regions. Configurations in which one of the core electrons is excited are much higher in energy, and are responsible for the X-ray spectra of the elements.

The complexity of the spectrum of energy levels rises very sharply with the increase of the number of electrons in the outer group. For this reason, we shall confine ourselves to the simplest cases, namely the alkalis where there is only one outer electron and to helium which may be taken as the model for the cases with two electrons outside the core.

The levels we are to discuss will be characterized by their energy, and such other labels as we find are necessary, including, in particular, their angular momentum. The significance of the division of the electrons into the core and the outer electrons arises from the fact that, as we shall show in the next section, an electron configuration consisting of closed shells or subshells has zero angular momentum and provides a spherically symmetric charge distribution. Consequently, the angular momentum properties of the levels will be determined by the angular momentum properties of the outer electrons, and the electrostatic interaction between the outer electrons and the core electrons will be well represented by the field due to the spherically symmetric charge distribution of the core.

9.2 THE ANGULAR MOMENTUM AND CHARGE DISTRIBUTION OF CLOSED SHELLS

One may conclude that closed shells have zero angular momentum from the experimental observation that atoms with closed shell ground state configurations (i.e. the noble gases) only show a single undeviated line in a Stern–Gerlach experiment. In sections 6.9 and 6.10 we pointed out that if we have an assembly of atoms, all in the same state, with angular momentum which we denote by J, then there are $(2J + 1)$ degenerate magnetic substates, each with a different magnetic moment, with the result that we get $(2J + 1)$ beams emerging from a Stern–Gerlach apparatus. Thus the existence of only one line in the case of noble gases implies they have angular momentum $J = 0$.

We can take this further. The existence of $(2J + 1)$ different magnetic substates for an angular momentum J implies that there are $(2J + 1)$ different degenerate wavefunctions. The angular momentum of the atom as a whole must surely arise from the motions and intrinsic spins of the electrons comprising the atom. Hence, without knowing how the electron spins and motions combine to form a total angular momentum for the atom as a whole, we can say that if a particular configuration, specified by an n value and l

value for each electron, gives rise to a certain total angular momentum J, then this configuration must have *at least* a degeneracy of $(2J + 1)$. It could have a greater degeneracy, as it is possible (indeed probable) that the chosen configuration can give rise to other values of the angular momentum besides the one specified. The degeneracy of a configuration specified by a set of n, l values is determined by the number of different ways the magnetic quantum numbers m_l and m_s can be assigned, taking into account the effect of the Pauli exclusion principle. An example will illustrate the point best. Suppose we have a configuration of two electrons, each with $n = 2, l = 1$. As m_l can equal $1, 0$, or -1, and m_s can be $+\frac{1}{2}$ or $-\frac{1}{2}$, there are six different possible combinations of the quantum numbers m_l and m_s which can be assigned to each of the electrons. According to the Pauli exclusion principle, the two electrons must have different combinations of m_l and m_s. This means the number of degenerate states that can be formed is equal to the number of ways there are of selecting two out of six distinguishable objects, namely fifteen. The configuration of two electrons both with $n = 2$, $l = 1$, gives rise to fifteen different degenerate wavefunctions.

We now turn the argument the other way round and say that if a configuration has a degeneracy g, then the *maximum* value, J_{max}, of the angular momentum to which it can give rise, must satisfy the condition:

$$(2J_{max} + 1) \leqslant g \tag{9.1a}$$

or

$$J_{max} \leqslant \frac{g - 1}{2}. \tag{9.1b}$$

A full sub-shell has an electron for every possible combination of m_l and m_s and hence there is only one state; g is therefore equal to one and Eq. (9.1b) tells us that the angular momentum J must be equal to zero.

We now demonstrate that a state with zero angular momentum is always spherically symmetric. We do this using the uncertainty principle. So far, we have always expressed the uncertainty principle in cartesian components of position and momentum, but there is no reason why we should not express it in polar coordinates. Some care is needed in this case as θ and ϕ are cyclic coordinates. However, the essential result is that the uncertainty in *angular position* about some axis multiplied by the uncertainty in *angular momentum* about the same axis is of the order of $\hbar$ or greater. The important point is that if we have a *definite* angular momentum about a particular axis the orientation of the system about that axis is completely undetermined and it is thus cylindrically symmetrical about that axis. This is explicitly displayed by the wavefunction for a single particle with a definite component m of angular momentum about the z-axis. The wavefunction is of the form

$$\psi = A\,e^{im\phi}$$

and

$$\psi^*\psi = A^*A \text{ is independent of } \phi \ .$$

A state with zero angular momentum is peculiar in that it is the *only* state with definite components of angular momentum in all three cartesian directions, each of them being zero. It is therefore cylindrically symmetrical about all three axes. This can only be achieved if it is spherically symmetrical.

Hence a full subshell has zero angular momentum and is spherically symmetric. It therefore gives rise to a spherically symmetric potential.

9.3 ELEMENTARY PERTURBATION THEORY

In this chapter we are going to be discussing problems for which there are no exact solutions, but for which approximate solutions are fairly obvious. A good example is the ground state of the lithium atom, which has two 1s electrons and one 2s electron, the problem being to calculate the energy of the 2s electron.

The two 1s electrons have a probability of greater than 99% of being within $2a_0$ of the nucleus, and so outside this distance the potential energy of the third electron is $-e^2/4\pi\varepsilon_0 r$. Indeed the potential energy does not differ from this by more than 5% until r becomes less than a_0. Within this distance the potential falls more rapidly, approaching $-3e^2/4\pi\varepsilon_0 r$ as one gets close to the nucleus. An electron in a 2s state in hydrogen spends only 8% of its time within a_0 of the nucleus, so it is reasonable to suppose that the 2s hydrogen wavefunction is a reasonable approximation for the 2s wavefunction of the third electron in lithium and that the energy of the third electron will not be too different from that of a 2s electron in hydrogen. As an analytical solution does not exist, it would appear sensible to start by assuming that the energy and wavefunction are given by the solution for hydrogen and try to find the true energy and wavefunction by a process of successive approximation.

A systematic procedure has been developed for doing this which is known as perturbation theory.

In perturbation theory, we approximate to the solution of the problem

$$H\psi = E\psi \tag{9.2}$$

by starting from the problem

$$H_0\psi_0 = E_0\psi_0 \tag{9.3}$$

whose solutions are known. We call E_0 and ψ_0 the zero order approximation to the energy and wavefunction respectively. We start by writing

$$H = H_0 + U \tag{9.4}$$

where U is called the perturbation. Not surprisingly, the procedure only works well if U is small. To introduce the idea of U being small we write

$$U = \lambda u \tag{9.5}$$

where λ is a measure of the size of the perturbation. This procedure allows us to keep track of the magnitudes of the terms that occur. Perturbation theory then gives us successively, first corrections of order λ, then of order λ^2 and so on. In this book we are only interested in the correction to the energy of order λ. It can be shown that to this order the energy E in Eq. (9.2) is given by

$$E \approx E_0 + \lambda \int \psi_0^* u \psi_0 \, \mathrm{d}^3\mathbf{r} \tag{9.6}$$

where ψ_0 is normalized. Remembering that λ is a constant and using Eq. (9.5) we can express our approximation to E in terms of E_0 and U as

$$E \approx E_0 + \int \psi_0^* U \psi_0 \, \mathrm{d}^3\mathbf{r} \ . \tag{9.7}$$

Eq. (9.7) is known as *the first order perturbation theory* approximation to the energy. We note that to find the energy to this order we only need to know the zero order wavefunction.

A difficulty with Eq. (9.7) is that it gives us very little idea whether the perturbation is small or not. It can be misleading to compare the correction with E_0 as this depends on the base level from which E_0 is measured. Unless we are prepared to go to the next order and calculate the change in the wavefunction, we must rely on physical intuition: usually a fairly reliable guide.

★ 9.4 FORMAL DERIVATION OF THE FIRST ORDER PERTURBATION THEORY RESULT

To find an approximate solution to Eq. (9.2), starting from the known solutions of Eq. (9.3), we start by assuming that because the effect of the perturbation is small, the wavefunction ψ consists largely of ψ_0, but also contains a small component ψ_1, orthogonal to ψ_0. It is easier to work with an unnormalized wavefunction which has its magnitude adjusted so that the coefficient of the normalized wavefunction ψ_0 is unity. Thus we assume the correct unnormalized wavefunction ψ in Eq. (9.2) can be written

$$\psi = \psi_0 + \lambda\psi_1 + \text{terms of order } \lambda^2 \text{ and of higher order} \tag{9.8}$$

where ψ_0 and ψ_1 are normalized and orthogonal. We drop the terms in λ^2 and higher in Eq. (9.8) and insert the remainder and Eqs. (9.4) and (9.5) in (9.2) to obtain

$$(H_0 + \lambda u)(\psi_0 + \lambda\psi_1) \approx E(\psi_0 + \lambda\psi_1) \ . \tag{9.9}$$

We multiply out the left hand side of Eq. (9.9), drop terms in λ^2 and replace $H_0\psi_0$ from Eq. (9.3), leaving us with

$$E_0\psi_0 + \lambda u\psi_0 + \lambda H_0\psi_1 \approx E(\psi_0 + \lambda\psi_1) \ . \qquad (9.10)$$

We multiply both sides on the left by ψ_0^* and integrate over all space to obtain

$$E_0 + \int \psi_0^* \lambda u \psi_0 \, \mathrm{d}^3\mathbf{r} + \lambda \int \psi_0^* H_0 \psi_1 \, \mathrm{d}^3\mathbf{r} \approx E$$

where we have used the fact that ψ_0 is normalized and orthogonal to ψ_1. We show below that

$$\int \psi_0^* H_0 \psi_1 \, \mathrm{d}^3\mathbf{r} = \int \psi_1 H_0 \psi_0^* \, \mathrm{d}^3\mathbf{r} \ . \qquad (9.11)$$

Using Eq. (9.3) and the orthogonality of ψ_0 and ψ_1, we see that the last expression equals

$$E_0 \int \psi_1 \psi_0^* \, \mathrm{d}^3\mathbf{r} = 0 \ .$$

Hence we have

$$E \approx E_0 + \int \psi_0^* \lambda u \psi_0 \, \mathrm{d}^3\mathbf{r}$$

$$\approx E_0 + \int \psi_0^* U \psi_0 \, \mathrm{d}^3\mathbf{r} \ . \qquad (9.7)$$

To complete the proof we require to show that

$$\int \psi_0^* H_0 \psi_1 \, \mathrm{d}^3\mathbf{r} = \int \psi_1 H_0 \psi_0^* \, \mathrm{d}^3\mathbf{r}$$

as stated in Eq. (9.11).

Let f and g be any two functions of $\mathbf{r}$ with continuous first and second derivatives, and which go to zero sufficiently quickly as $\mathbf{r}$ goes to infinity. (How quickly we shall see in a minute.) If $H_0 = -\dfrac{\hbar^2}{2m}\nabla^2 + V(\mathbf{r})$, then

$$\int f H_0 g \, \mathrm{d}^3\mathbf{r} = -\frac{\hbar^2}{2m} \int f \nabla^2 g \, \mathrm{d}^3\mathbf{r} + \int f g V(\mathbf{r}) \, \mathrm{d}^3\mathbf{r}$$

and

$$\int gH_0 f\, d^3\mathbf{r} = -\frac{\hbar^2}{2m}\int g\,\nabla^2 f\, d^3\mathbf{r} + \int g f\, V(\mathbf{r})\, d^3\mathbf{r}\ .$$

Hence

$$\int (fH_0 g - gH_0 f)\, d^3\mathbf{r} = -\frac{\hbar^2}{2m}\int (f\,\nabla^2 g - g\,\nabla^2 f)\, d^3\mathbf{r}$$

$$= -\frac{\hbar^2}{2m}\int \nabla\cdot(f\nabla g - g\nabla f)\, d^3\mathbf{r}\ .$$

We convert the volume integral to a surface integral using the relation that for any vector function of position $\mathbf{F}(\mathbf{r})$

$$\int_V \nabla\cdot\mathbf{F}(\mathbf{r})\, d^3\mathbf{r} = \int_S \mathbf{F}\cdot d\mathbf{S}$$

where the volume V is bounded by the surface S. We take $(f\nabla g - g\nabla f)$ for $\mathbf{F}$, and as the integral is over all space the surface must be taken at infinity. Thus we get

$$\int (fH_0 g - gH_0 f)\, d^3\mathbf{r} = -\frac{\hbar^2}{2m}\int_{\substack{S\\ r\to\infty}} (f\,\nabla g - g\,\nabla f)\cdot d\mathbf{S}\ .$$

As r tends to infinity, the bounding surface increases as $4\pi r^2$. Hence if $(f\,\nabla g - g\,\nabla f)$ decreases more rapidly than $1/r^2$ the surface integral tends to zero as $\mathbf{r}$ tends to infinity and we get

$$\int fH_0 g\, d^3\mathbf{r} = \int gH_0 f\, d^3\mathbf{r}\ .$$

If we take ψ_0^* for f and ψ_1 for g, being bound state wavefunctions they and their gradients go to zero exponentially and the surface integral vanishes as r tends to infinity. Hence we have proved that

$$\int \psi_0^* H_0 \psi_1\, d^3\mathbf{r} = \int \psi_1 H_0 \psi_0^*\, d^3\mathbf{r}\ . \tag{9.11}$$

★ 9.5 TWO EXAMPLES OF PERTURBATION CALCULATIONS

We now give two examples of a perturbation calculation, the first of which is to a problem where we know the exact solution and it demonstrates that the error is indeed of order λ^2.

9.5.1 Increase in the nuclear charge

Suppose we have a nucleus of charge Ze and a single electron. The energy of the ground state of this electron is $-Z^2e^2/(4\pi\varepsilon_0 a_0 \times 2)$. Suppose also that the nucleus undergoes β-decay, emitting an electron, thereby raising its charge to $(Z + 1)e$. The energy of the ground state is now $-(Z + 1)^2e^2/(4\pi\varepsilon_0 a_0 \times 2)$. Let us calculate the energy we obtain using perturbation theory. The Hamiltonian for the original state is

$$H_0 = -\frac{\hbar^2}{2m}\nabla^2 - \frac{Ze^2}{4\pi\varepsilon_0 r} \tag{9.12}$$

with a ground state solution

$$u(r) = \left(\frac{4Z^3}{a_0^3}\right)^{1/2} r\,\mathrm{e}^{-Zr/a_0} \tag{9.13}$$

where the normalization is chosen so that

$$\int_0^\infty |u(r)|^2\,\mathrm{d}r = 1 \ .$$

The new Hamiltonian is

$$\begin{aligned} H &= -\frac{\hbar^2}{2m}\nabla^2 - \frac{(Z+1)e^2}{4\pi\varepsilon_0 r} \\ &= H_0 - \frac{e^2}{4\pi\varepsilon_0 r} \ . \end{aligned}$$

The perturbation is just $1/Z$ times the original potential everywhere, so we may take λ as equal to $1/Z$. According to Eq. (9.7) the perturbed energy is

$$\begin{aligned} E &\approx E_0 + \int \psi^* U \psi\,\mathrm{d}^3\mathbf{r} \\ &\approx E_0 + \int_0^\infty u(r)\left\{-\frac{e^2}{4\pi\varepsilon_0 r}\right\}u(r)\,\mathrm{d}r \\ &\approx E_0 - \frac{4Z}{a_0^3}\frac{e^2}{4\pi\varepsilon_0}\int_0^\infty r\,\mathrm{e}^{-2Zr/a_0}\,\mathrm{d}r \\ &\approx E_0 - \frac{4Z}{a_0^3}\frac{e^2}{4\pi\varepsilon_0}\frac{a_0^2}{4Z^2} \end{aligned}$$

where the integral has been evaluated using Eq. (7.27). Hence

$$E \approx E_0 - \frac{Ze^2}{4\pi\varepsilon_0 a_0}$$

$$\approx -\frac{(Z^2 + 2Z)e^2}{4\pi\varepsilon_0 \cdot 2a_0} \, . \qquad (9.14)$$

The correction we have calculated is $2/Z$ times the original energy, i.e. $2\lambda E_0$. The correct answer for E is

$$E = -\frac{(Z+1)^2 e^2}{4\pi\varepsilon_0 \cdot 2a_0}$$

$$= -\frac{(Z^2 + 2Z + 1)e^2}{4\pi\varepsilon_0 \cdot 2a_0}$$

and the difference between this and the perturbation calculation is

$$-e^2/4\pi\varepsilon_0 \cdot Z = \lambda^2 E_0 \, .$$

9.5.2 The effect of the finite size of the nucleus

Our calculations of the energies of the states of a single electron in the field of a charge Ze have assumed this charge to be a point charge. Actual nuclei have a finite extent and hence the potential is $-Ze/4\pi\varepsilon_0 r$ only up to the surface of the nucleus, deviating from this inside the nucleus. We calculate the effect of this using perturbation theory.

We assume the nucleus to be a uniformly charged sphere of radius R. Using Gauss' theorem in electrostatics (see I. S. Grant and W. R. Phillips, *Electromagnetism* (Manchester Physics Series), Wiley, London, 1975), it is easy to show that inside the nucleus the field rises linearly from zero at the origin to $Ze/4\pi\varepsilon_0 R^2$ at the surface.

Thus

$$E_{\text{interior}} = \frac{Zer}{4\pi\varepsilon_0 R^3} \, .$$

From

$$E = -\frac{\mathrm{d}V'}{\mathrm{d}r}$$

where V' is the potential we have

$$\int_r^R E(r')\,\mathrm{d}r' = -\int_r^R \mathrm{d}V'$$

$$= V'(r) - V'(R) \, .$$

Therefore

$$V'(r) = V'(R) + \int_r^R E(r')\,dr'$$

$$= \frac{Ze}{4\pi\varepsilon_0 R} + \frac{Ze}{4\pi\varepsilon_0 R^3}\int_r^R r'\,dr'$$

$$= \frac{Ze}{4\pi\varepsilon_0 R} + \frac{Ze}{4\pi\varepsilon_0 R^3}\left(\frac{R^2}{2} - \frac{r^2}{2}\right)$$

$$= \frac{3}{2}\frac{Ze}{4\pi\varepsilon_0 R} - \frac{Zer^2}{4\pi\varepsilon_0 \cdot 2R^3}\,. \tag{9.15}$$

The potential energy for an electron inside the nucleus is therefore

$$V(r) = -\frac{3}{2}\frac{Ze^2}{4\pi\varepsilon_0 R} + \frac{Ze^2r^2}{4\pi\varepsilon_0 \cdot 2R^3}\,.$$

The potential energy of the electron in the case of a point charge is $-Ze^2/4\pi\varepsilon_0 r$ and hence the perturbation is given by

$$\left.\begin{aligned} U(r) &= \frac{1}{4\pi\varepsilon_0}\left(\frac{Ze^2}{r} - \frac{3}{2}\frac{Ze^2}{R} + \frac{Ze^2r^2}{2R^3}\right), \quad r < R \\ &= 0, \qquad\qquad\qquad\qquad\qquad\qquad r > R\,. \end{aligned}\right\} \tag{9.16}$$

The correction to the energy of the ground state due to the finite size of the nucleus is, from Eq. (9.13),

$$\delta E = \frac{4Z^3}{a_0^3}\int_0^R r^2\,e^{-2Zr/a_0}U(r)\,dr\,. \tag{9.17}$$

The radius of a nucleus is given to a good approximation by

$$R = r_0 A^{1/3} \tag{9.18}$$

where $r_0 = 1.3 \times 10^{-13}$ cm and A is the mass number. Thus, even for the heavier nuclei (e.g. lead, $A = 208$, $Z = 82$) R does not exceed 10^{-12} cm, while the radius parameter a_0/Z is always in excess of 10^{-10} cm. Thus, the exponential term in Eq. (9.17) can be put equal to unity, and Eq. (9.17) becomes

$$\begin{aligned}\delta E &= \frac{4Z^3}{a_0^3}\int_0^R r^2 U(r)\,\mathrm{d}r \\ &= \frac{4Z^4e^2}{4\pi\varepsilon_0 a_0^3}\int_0^R \left(r - \frac{3}{2}\frac{r^2}{R} + \frac{1}{2}\frac{r^4}{R^3}\right)\mathrm{d}r \\ &= \frac{4Z^4e^2}{4\pi\varepsilon_0 a_0^3}\left[\tfrac{1}{2}R^2 - \tfrac{1}{2}R^2 + \tfrac{1}{10}R^2\right] \\ &= \frac{4}{5}\left(\frac{ZR}{a_0}\right)^2 \frac{Z^2e^2}{4\pi\varepsilon_0 \cdot 2a_0}\,. \end{aligned} \tag{9.19}$$

Using Eq. (9.18) for R we obtain the following results. For neon ($Z = 10$, $A = 20$)

$$\delta E = 4.9 \times 10^{-4}\,\text{eV} = 3.6 \times 10^{-7}\,E_0$$

where E_0 is the unperturbed ground state energy. For lead ($Z = 82$, $A = 208$)

$$\delta E = 10.4\,\text{eV} = 1.2 \times 10^{-4}\,E_0\,.$$

For light elements the effect is totally negligible, but for heavy elements the effect, though small, can be detected.

The situation is quite different if we consider the system formed when an atom captures a negatively charged μ-meson. Such a μ-mesic atom has an energy level system similar in every way to that of an atom with a single electron, except that the radius parameter $a_0(\mu)$ is reduced, compared to that for an electron by the ratio of their masses. The μ-meson has a mass of 207 electron masses. Hence

$$\begin{aligned}a_0(\mu) &= \frac{4\pi\varepsilon_0\hbar^2}{m_\mu e^2} = \left(\frac{4\pi\varepsilon_0\hbar^2}{m_e e^2}\right)\cdot\frac{m_e}{m_\mu} \\ &= \frac{0.529 \times 10^{-8}}{207}\,\text{cm} \\ &= 2.55 \times 10^{-11}\,\text{cm}\,. \end{aligned}$$

Thus for Z greater than about 50 the point nucleus ground state wave-function has a radius parameter *smaller* than the nuclear radius. The correction for neon is about 1.5 %.

9.6 THE ENERGY LEVELS OF THE ALKALI METALS

The ground state configuration of the alkali metals (Li, Na, K, Rb, Cs) consists of a single valence electron in an s-state outside a closed shell, as may be seen by reference to Table 8.3. The excited states we are going to consider are those in which the valence electron is promoted to higher states. To make our arguments specific we shall discuss a particular one of the alkali metals, namely sodium, but the arguments we use can be equally applied to all the alkali metals with appropriate changes in numerical values and quantum numbers.

Sodium has eleven electrons. Ten of these fill the $n = 1$ and $n = 2$ shells, and the valence electron in the ground state is a 3s electron. The excited states we are going to consider are those in which the $n = 1$ and $n = 2$ shells remain full, and the valence electron is promoted to the 3p and 3d states, the 4s, 4p, 4d and 4f states, and so on.

As a first approximation we shall assume that the wavefunction for the valence electron in any of the above states (including the ground state) is the same as for an electron with the same quantum numbers in the hydrogen atom. We must first justify this assumption.

Our assumption will be justified if we can show that over the bulk of the region occupied by the valence electron wavefunction the potential seen by this electron is just the coulomb potential. In Fig. 9.1 we plot the square of the radial wavefunction $|u(r)|^2$ for the 3s, 3p and 3d states in hydrogen. We see that the bulk of the probability lies beyond $r = 4a_0$, where a_0 is the Bohr radius $4\pi\varepsilon_0\hbar^2/(me^2)$. Higher states, being more weakly bound, will extend even further out.

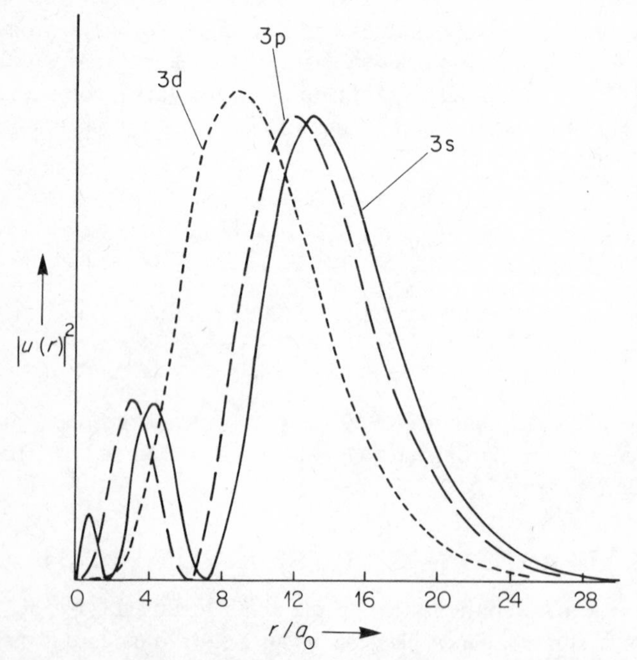

Fig. 9.1. The square of the radial wavefunction $u(r)$ for the 3s, 3p and 3d states in hydrogen.

At large distances, the potential seen by the valence electron will clearly be the coulomb potential due to a single charge as the ten electrons in the core will completely shield all but one of the eleven charges on the nucleus. We wish to estimate at what distance this shielding becomes incomplete. We may do this as follows. In section 7.3 we pointed out that at large distances *all* electron wavefunctions fall off as $e^{-\alpha r}$ where $\alpha = (2mE/\hbar^2)^{1/2}$ and E is the binding energy of the electron. The energy of the electrons in the inner shell will be of the order of the second ionization potential, which is 47 volts for sodium. Thus we have $E \approx 47$ eV, and hence α is approximately $2/a_0$. The *charge* density due to the inner electrons falls off as e^{-4r/a_0} as it is proportional to the square modulus of the wavefunction.

As there are ten inner electrons we have to let the factor e^{-4r/a_0} get fairly small, say of the order of 10^{-3}, before we can safely say the amount of charge likely to be found at greater distances is negligible and the screening effect of the inner electrons is complete. This condition is achieved at $r = 2a_0$ ($e^{-8} = 0.3 \times 10^{-3}$). Thus the bulk of the valence electron wavefunction is indeed in a region where the potential it sees is that due to a single charge.

Thus to find the energy level spectrum of sodium we start with wavefunctions and energies as given by the hydrogen atom, and use perturbation theory to calculate the change in energy due to the deviation of the potential from the pure coulomb form at small distances. We have, in fact, obtained the essential qualitative results in section 8.5 under sub-heading (iii). In Fig. 8.6, we showed the initial portion of the 3s, 3p and 3d wavefunctions, together with the perturbation potential U due to the decreasing effectiveness of the screening of the nuclear charge by the core electrons.

The qualitative features of what happens are immediately obvious. The d-wave with $l = 2$ is very small throughout the whole of the region in which the perturbation is significant, and consequently the energy of this level is hardly shifted from the value for hydrogen. The p-wave, with $l = 1$, has rather more of its wavefunction inside the perturbing potential, due to the peak that occurs at $r = 3a_0$ (see Fig. 9.1), while the s-wave solution has the whole of the first peak inside the perturbing potential.

Thus, the effect of the deviation from the pure coulomb field is to split the degeneracy of the $n = 3$ levels with different orbital angular momenta. Not surprisingly, the high angular momentum state is hardly affected, as the centrifugal force due to its angular momentum keeps it away from the region where the perturbation exists. We illustrate this point in Fig. 9.2 where we plot the total effective radial potential

$$V_{\text{eff}} = V(r) + \frac{l(l+1)\hbar^2}{2mr^2}$$

for the pure coulomb potential and the potential in sodium, in both cases with $l = 2$. We see that the two cases do not deviate from each other significantly until we are well into the classically forbidden region.

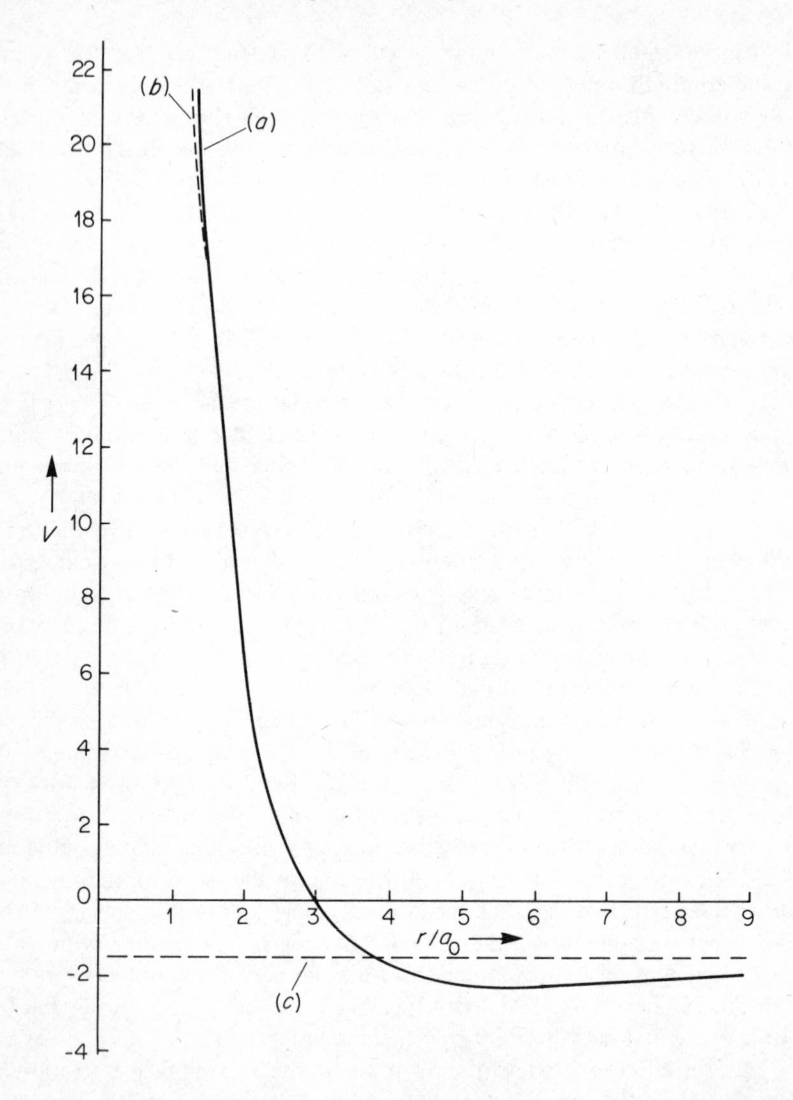

Fig. 9.2. The potential energy in electron volts versus the radius in units of a_0 for (*a*) an electron with $l = 2$ in the hydrogen atom, and (*b*) an electron with $l = 2$ in the sodium atom. The horizontal dotted line (*c*) represents the energy of the 3d state in hydrogen.

The low angular momentum states have a high probability of the electron being near the nucleus, where it will be strongly affected by the high nuclear charge, with a consequent lowering of its energy level.

These qualitative features apply with equal force to the higher levels of sodium. We have shown in section 7.6, Eq. (7.36), that the radial function $u(r)$

behaves for small r as

$$u(r) \propto r^{l+1}$$

and the way the wavefunction starts off near the origin is therefore determined by the value of the angular momentum. Hence all the s-wave levels are substantially shifted, all the p-wave levels are moderately shifted, and the levels with $l = 2$ are hardly affected at all.

The energy levels of sodium are shown in Fig. 9.3 together with the

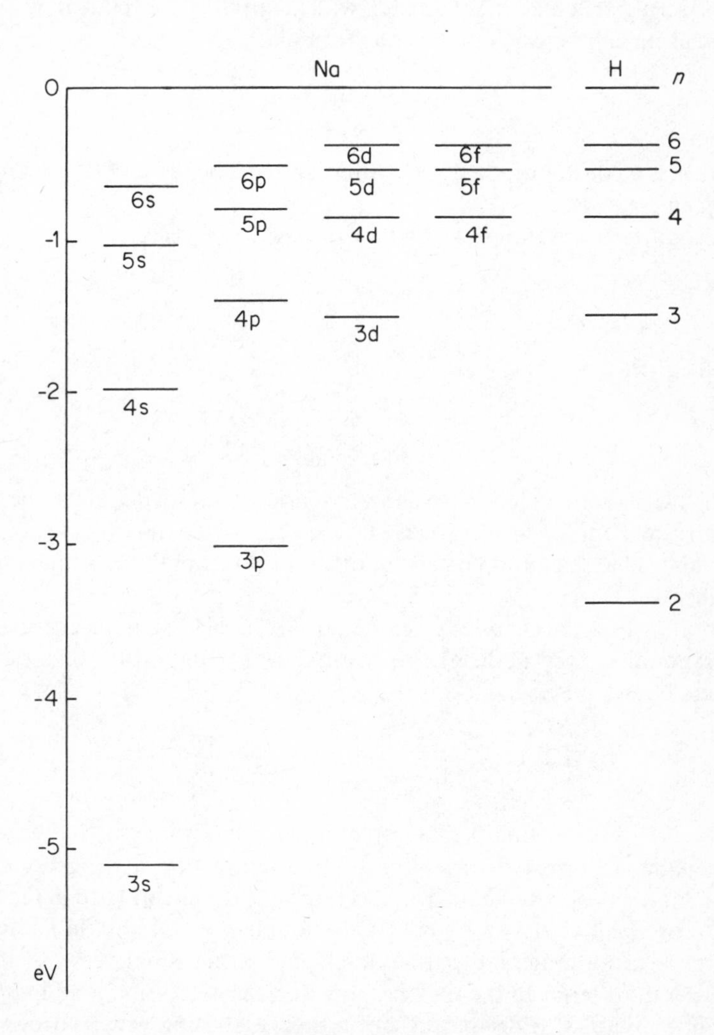

Fig. 9.3. The energy levels of the sodium atom in electron volts for $l = 0, 1, 2$ and 3 with, on the right, the levels of the hydrogen atom for $n = 2$ to 6.

positions of the corresponding levels in hydrogen. The situation in the other alkali metals is similar, except that in rubidium and caesium, we observe fairly large shifts of the d-wave levels. The f-wave levels remain essentially unshifted throughout the whole of the alkali metal series.

★ 9.7 THE QUANTUM DEFECT

In practice, it is found that, to a very good approximation, the energy levels in a particular alkali metal with a given value of orbital angular momentum can be expressed by the formula

$$E_{n,l} = -\frac{e^2}{4\pi\varepsilon_0 \cdot 2a_0} \cdot \frac{1}{(n-\delta_l)^2} \tag{9.20}$$

where δ_l is a constant for a given l value, and independent of n. δ_l is known as the *quantum defect*.

In sodium, for example, we find for the s-wave levels

$$E_{n,0} = -\frac{e^2}{4\pi\varepsilon_0 \cdot 2a_0} \frac{1}{(n-1.38)^2}$$

and for p-wave levels

$$E_{n,1} = -\frac{e^2}{4\pi\varepsilon_0 \cdot 2a_0} \frac{1}{(n-0.88)^2} \cdot$$

The discussion in section 9.6 shows why the quantum defect for the p-wave levels is smaller than for the s-wave levels, but provides no explanation for the constancy of the quantum defect for different values of the principal quantum number.

To show how the energies given by Eq. (9.20) differ from the energies of the corresponding levels in the hydrogen atom, we expand Eq. (9.20) in ascending powers of δ_l

$$E_{n,l} = -\frac{e^2}{4\pi\varepsilon_0 \cdot 2a_0}\left(\frac{1}{n^2} + \frac{2\delta_l}{n^3} + \frac{3\delta_l^2}{n^4}\cdots\right) . \tag{9.20a}$$

The first term in Eq. (9.20a) is the unperturbed value for hydrogen, and the remainder form the correction to be applied. For large values of n the second term, proportional to $1/n^3$, is clearly the dominant term in the correction. For small values of n and l, for example $n = 3$, $l = 0$, in sodium, the second term, although larger than the higher terms, is not very much bigger than the third term. In these cases, however, the correction is so big that we cannot expect first order perturbation theory, starting with hydrogen atom wavefunctions, to give more than a qualitative indication of the correction. Thus we shall have achieved some understanding of why Eq. (9.20) is a good representation of the energies of the levels if we can show that perturbation theory gives a correction approximately proportional to $1/n^3$.

In Fig. 9.4 we plot unnormalized versions of the radial functions $u(r)$ for the 3s, 4s and 5s levels of hydrogen in which the coefficient of the leading

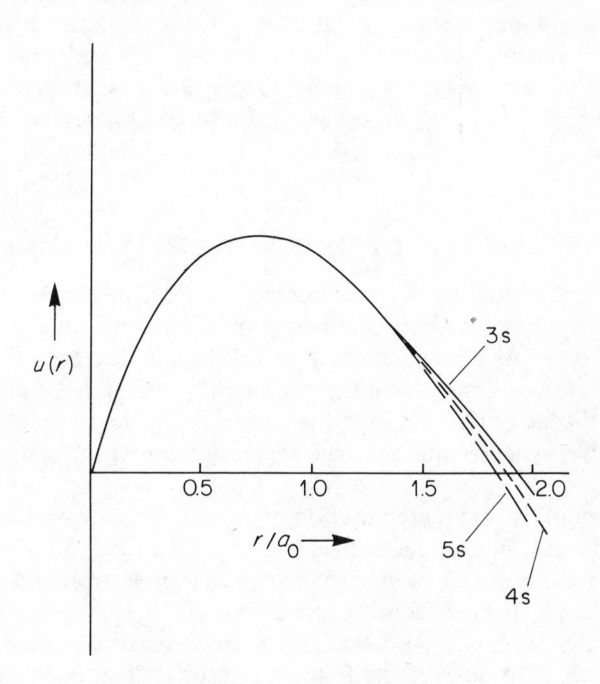

Fig. 9.4. The 3s, 4s and 5s radial wavefunctions of hydrogen with the leading coefficient equal to unity.

term is unity. We see that in the relevant region of up to $r = 2a_0$ (compare Fig. 8.6) the wavefunctions are very nearly identical. The reason for this is that in this region the energy of the state E is small compared with the potential energy $V(r)$ and hence the value of E has very little effect on the shape of the wavefunction. Let $Au(r)$ be the normalized wavefunction of the s-state we are considering. Then it follows that the perturbation correction is

$$A^2 \int_0^\infty u^*(r) U u(r)\, \mathrm{d}r \propto A^2 \ . \tag{9.21}$$

A^2 is given by

$$A^2 = 4/n^3 a_0^3 \tag{9.22}$$

and hence the correction is proportional to $1/n^3$. To obtain the result of Eq. (9.22) we need to integrate the hydrogen wavefunctions but we can see that we expect a result of this form as follows. A level with principal quantum number n has a range parameter na_0 as its energy is proportional to $-1/n^2$ and the range parameter is proportional to $1/\sqrt{(-E)}$ (see Eq. (7.20)). Consequently the volume in which an electron is likely to appear is proportional to n^3. As the shapes of all the s-wave functions are very similar near the origin, we therefore expect their square amplitudes to be proportional to $1/n^3$.

We can follow a similar line of argument for the p-wave solutions.

9.8 THE STRUCTURE OF THE 3p STATE IN SODIUM

The energies of the levels in sodium, as described in section 9.6, depended on the values of the quantum numbers n and l, but not on the quantum numbers m_l and m_s. As a result, each level will have a degeneracy of $2(2l + 1)$, the factor of two coming from the two possible orientations of the electron spin, and the factor $(2l + 1)$ from the $(2l + 1)$ possible values of m_l. We now ask ourselves whether this is a true representation of the state of affairs in sodium.

It is certainly a true representation of the ground state, which is split into two components in a magnetic field. Indeed, it was this splitting in a Stern–Gerlach experiment which led us to the existence of electron spin in Chapter 6. What about the first excited state, the 3p state?

If the description of section 9.6 is a true representation of the first excited state we expect to observe the following results. First, in the absence of a magnetic field, all six levels with different values of m_s and m_l will have the same energy, and a single level will be observed. Secondly, in the presence of a magnetic field each line will be displaced from its original position by an amount proportional to its z-component of magnetic moment. There are two contributions to the magnetic moment, one from the orbital motion determined by the value of m_l and one from the spin determined by m_s. As the gyromagnetic ratio is one for orbital motion and two for the electron spin (see section 6.11) we get the result that the z-component of magnetic moment μ_z is given by

$$\mu_z = -\frac{e\hbar}{2m}(m_l + 2m_s) \ . \tag{9.23}$$

In Table 9.1 we show the possible values of $(m_l + 2m_s)$ and hence the magnetic moments in Bohr magnetons.

Table 9.1

m_l	m_s	$m_l + 2m_s$
1	$\frac{1}{2}$	2
0	$\frac{1}{2}$	1
1	$-\frac{1}{2}$	0
-1	$\frac{1}{2}$	0
0	$-\frac{1}{2}$	-1
-1	$-\frac{1}{2}$	-2

Thus we would expect to see the 3p level split into five levels in a magnetic field, the central level with $\mu_z = 0$ being doubly degenerate. In fact, because the gyromagnetic ratio of the electron is not exactly two, there would be a small splitting of these two levels. The expected situation is shown in Fig. 9.5.

Let us now compare these expected situations with what is observed experimentally.

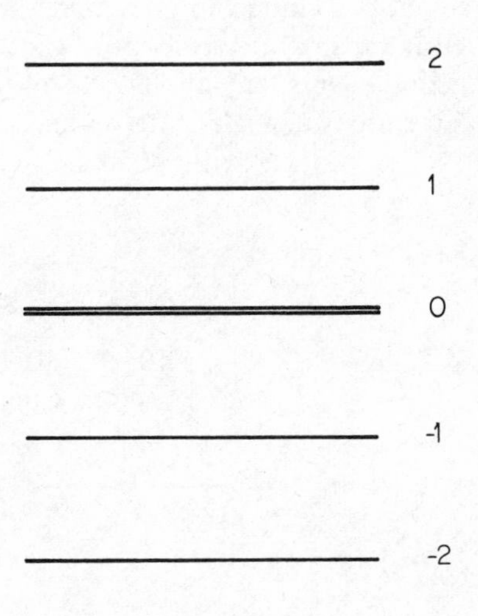

Fig. 9.5. The expected energy level system for a p-wave electron in a magnetic field assuming that the orbital angular momentum and electron spin contribute independently to the value of the magnetic moment. The levels are labelled by the value of $m_l + 2m_s$.

From Fig. 9.3 we see that the 3p state lies 2.11 eV above the 3s state. According to the selection rules for the emission of electromagnetic radiation, which we shall not discuss until the next chapter, the 3p state can decay to the 3s state by emitting photons of 2.11 eV.

From the relation

$$E = h\nu = hc/\lambda$$

we calculate the wavelength of the emitted light as

$$\begin{aligned}\lambda &= hc/E \\ &= \frac{6.625 \times 10^{-34} \times 3 \times 10^{8}}{2.11 \times 1.6 \times 10^{-19}} \\ &= 5.89 \times 10^{-7}\ \text{m} .\end{aligned}$$

We identify this light with the very strong line seen at 5893 Å in the emission spectrum of sodium. It is this line which is responsible for the characteristic yellow colour of sodium vapour lamps.

If we study this spectral line in a spectroscope of good resolution, we find that it consists of *two* components, one at 5890 Å and the other at 5896 Å. As we know the ground state is a single state we must assume that the 3p state, in fact, consists of *two* states with a small energy separation ΔE, given by

$$\begin{aligned}\Delta E &= \frac{hc}{\lambda_1} - \frac{hc}{\lambda_2} \\ &= 6.625 \times 10^{-34} \times 3 \times 10^{8} \left(\frac{10^7}{5.890} - \frac{10^7}{5.896} \right) \text{J} \\ &= 19.9 \times 10^{-26} \times \frac{6 \times 10^4}{5.890 \times 5.896} \times \frac{1}{1.60 \times 10^{-19}}\,\text{eV} \\ &= 2.15 \times 10^{-3}\ \text{eV}\end{aligned}$$

If we apply a magnetic field we find the lower of the two levels splits into two components implying it has angular momentum $\frac{1}{2}\hbar$, and the upper level splits into four components, implying an angular moment of $\frac{3}{2}\hbar$. We see that, in a magnetic field, we do get six levels, but the spacings are quite different from what we expected, as we illustrate in Fig. 9.6.

We wish to emphasize two points. The first is that, at zero field there are two levels, not one, and each of these behaves, in a magnetic field, as is expected of a level with a unique angular momentum. The second is that the expected splitting, as illustrated in Fig. 9.5, does *not* correspond to a unique value of the angular momentum as the centre level contains two degenerate

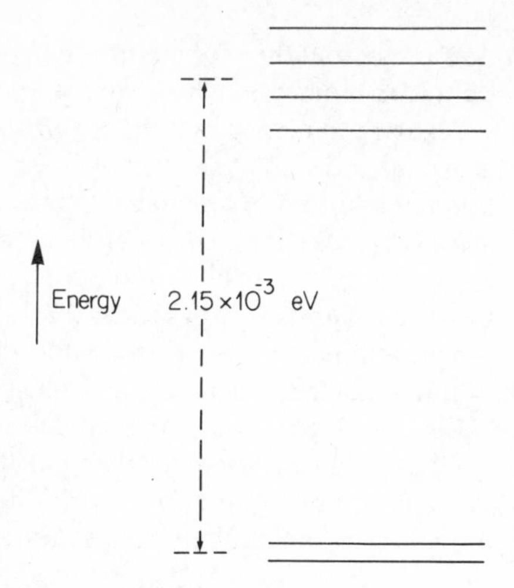

Fig. 9.6. The components (to scale) of the 3p level in sodium in a magnetic field of 2 tesla.

components. The sub-levels in a magnetic field (at least for fields below 1 tesla) of a state with unique angular momentum are always equally spaced and are not degenerate. From this we must conclude that there is a further interaction that we have hitherto neglected, that splits the 3p level (with non-unique angular momentum) into two levels, each with unique angular momentum. Before considering the source of this interaction, we must consider the addition of angular momentum.

9.9 THE ADDITION OF ANGULAR MOMENTUM

We remarked in section 6.7 that a free atom, not being subject to any external constraint, would be expected to have a definite value for its total angular momentum. (We use the phrase '*total* angular momentum' to indicate that its angular momentum may arise from several sources, i.e. the orbital angular momenta of the various electrons and their spin angular momenta.) However, if for a particular value of the energy, there exist solutions corresponding to several values of angular momentum, i.e. it is degenerate with respect to certain values of the angular momentum, then, as any linear combination of the solutions will also be a solution, we would not necessarily expect to observe a definite value for the total angular momentum. Conversely, we can say that if we observe a situation that does not obey the rules expected for a definite value of the angular momentum, then it must be a superposition of degenerate states corresponding to two, or more, values of the total angular momentum. Clearly the result we expected for the 3p state in sodium, and which was illustrated in Fig. 9.5 does not obey the rules, as was mentioned above, and therefore corresponds to a superposition of states of different angular momenta. The results actually found suggest very strongly that the angular momenta involved are $\frac{1}{2}\hbar$ and $\frac{3}{2}\hbar$, each created by the addition of the orbital and spin angular momenta.

Before embarking on the addition of angular momenta, we must make a few comments on *nomenclature*. We have consistently used the letters l and s to denote respectively values of the orbital and spin angular momentum of a single electron, measured in units of $\hbar$. In section 9.2 we used the symbol J to denote the total angular momentum of an atom, and this is the generally accepted symbol for total angular momentum (though sometimes I is used instead). The symbol j is used for the total angular momentum of a *single* electron. The symbol j is also frequently used for the total angular momentum of an atom when this angular momentum arises from the motion and spin of *one* electron only, as in the case of the alkali metals we are presently discussing.

With these preliminaries over, let us now consider the addition of the orbital angular momentum l to the spin angular momentum s to form a total angular momentum j. If the electron were a classical particle, the total angular momentum would be a vector $\mathbf{j}$ formed from the orbital angular momentum vector $\mathbf{l}$ and the spin vector $\mathbf{s}$, i.e.

$$\mathbf{j} = \mathbf{l} + \mathbf{s} \; . \tag{9.24}$$

If the magnitudes of the vectors $\mathbf{l}$ and $\mathbf{s}$ were held constant, but the angle between them allowed to vary, the resultant j could have any magnitude between $|l - s|$ and $l + s$, that is, classically

$$|l - s| \leqslant j \leqslant (l + s) \; . \tag{9.25}$$

This situation is illustrated in Fig. 9.7.

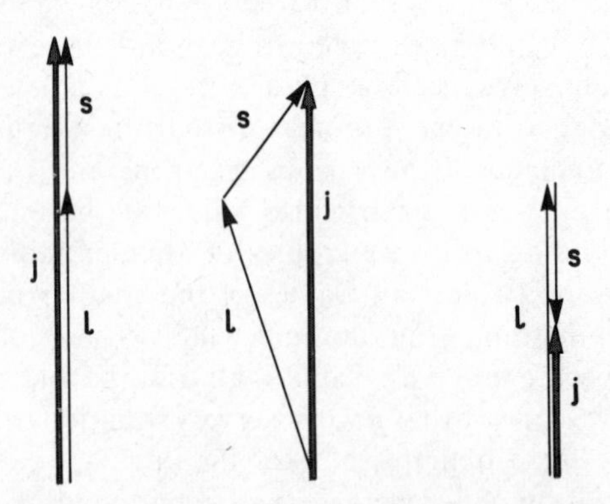

Fig. 9.7. Classically if two angular momentum vectors $\mathbf{l}$ and $\mathbf{s}$ are added together to form a resultant $\mathbf{j}$ the magnitude of the resultant must lie between $|l - s|$ and $l + s$.

We expect the quantum mechanical situation to be analogous with this but modified because of the quantization of angular momentum.

At every stage in our development of quantum mechanics so far we have proceeded by looking at the classical situation, and replacing the corresponding classical quantities by operators according to the prescription under sub-heading 7 of the summary at the end of Chapter 5. Doing the same on this occasion means that the classical relation of Eq. (9.24) becomes

$$\hat{\mathbf{j}} = \hat{\mathbf{l}} + \hat{\mathbf{s}} \tag{9.26}$$

or in components

$$\hat{j}_x = \hat{l}_x + \hat{s}_x \tag{9.26a}$$

$$\hat{j}_y = \hat{l}_y + \hat{s}_y \tag{9.26b}$$

$$\hat{j}_z = \hat{l}_z + \hat{s}_z \; . \tag{9.26c}$$

As the components of $\hat{\mathbf{l}}$ and $\hat{\mathbf{s}}$ operate on different portions of the wavefunction, it follows that the order in which they are written in a product is of no significance.

For example, if ψ is *any* wavefunction

$$\hat{l}_x \hat{s}_y \psi = \hat{s}_y \hat{l}_x \psi$$

independent of the wavefunction ψ. We denote this by writing the relation in the form

$$\hat{l}_x \hat{s}_y \equiv \hat{s}_y \hat{l}_x \tag{9.27}$$

and we say the operators $\hat{l}_x$ and $\hat{s}_y$ *commute.*

This identity holds for any product formed with one component from $\hat{\mathbf{l}}$ and the other from $\hat{\mathbf{s}}$. Thus each component of $\hat{\mathbf{l}}$ commutes with every component of $\hat{\mathbf{s}}$.

Using Eq. (9.27) it is readily shown that the components of $\hat{\mathbf{j}}$, as defined in Eq. (9.26), obey the algebraic relations of Eq. (6.24) in section 6.11 which we asserted defined an angular momentum operator. The reader is reminded that in section 6.11 we also stated (but did not prove) that the algebraic relations (6.24) were sufficient to demonstrate that the eigenvalues of the square of the angular momentum are given by $j(j + 1)\hbar^2$ where j is integral or half-integral, and the eigenvalues of the z-component range from $-j$ to $+j$ in units of $\hbar$.

Our problem, therefore, is to determine the eigenvalues of $\hat{j}^2$ and $\hat{j}_z$ knowing the values of l and s, and in particular that $s = \frac{1}{2}$. The eigenfunctions of $\hat{j}^2$ and $\hat{j}_z$ in this situation will, of course, be linear combinations of the eigenfunctions specified by l, m_l and m_s. We shall not need to know the exact form of these eigenfunctions.

From Eq. (9.26c) we note that, in units of $\hbar$, the eigenvalues m_j of $\hat{\jmath}_z$ are given by

$$m_j = m_l + m_s \ . \tag{9.28}$$

As m_l is always integral and $m_s = \pm\frac{1}{2}$ it follows that m_j is always *half-integral*, and hence the quantum numbers j are *half-integral.* The maximum value of j is equal to the maximum value of m_j, which from Eq. (9.28) is $(l + \frac{1}{2})$.

Hence the maximum allowed value of j is given by

$$j_{\max} = l + \tfrac{1}{2} \ . \tag{9.29}$$

As j must be half-integral the next highest value of j is given by $j = (l - \frac{1}{2})$. We now show that this is the only other allowed value of j, as would be expected from the classical relation, Eq. (9.25).

We have frequently pointed out that if we have two degenerate orthogonal solutions to a problem, ϕ_1 and ϕ_2, then any linear combination of ϕ_1 and ϕ_2

$$a\phi_1 + b\phi_2 \tag{9.30}$$

is also a solution to the problem. Within a normalization factor there is only one linear combination of ϕ_1 and ϕ_2 that is orthogonal to Eq. (9.30), namely the solution

$$b^*\phi_1 - a\phi_2 \ .$$

Thus the *number* of orthogonal solutions is fixed. This argument is easily extended to the case where there are any number of orthogonal degenerate solutions to show that the number of orthogonal solutions is fixed whatever the linear combinations used.

The number of orthogonal degenerate states for a given value of l and $s(= \frac{1}{2})$ is

$$2(2l + 1) \ .$$

As the states with definite values of j and m_j are linear superpositions of the states with definite l, m_l and m_s, there must be exactly the same number of them.

The number of states with a given value of j is $(2j + 1)$, hence the *total* number of states with $j = l + \frac{1}{2}$ or $j = l - \frac{1}{2}$ is

$$\begin{aligned} &\{2(l + \tfrac{1}{2}) + 1\} + \{2(l - \tfrac{1}{2}) + 1\} \\ &= \{2l + 2\} + 2l \\ &= 2(2l + 1) \ . \end{aligned}$$

Hence these two values of j exhaust the total possible number of states.

To summarize the situation, then, suppose we have an electron in a state specified by the quantum numbers n and l. This has $2(2l + 1)$ magnetic substates specified by the quantum numbers m_l and m_s. The states described by the quantum numbers n, l, m_l, m_s do not possess a specific value of the total angular momentum. We can add together the orbital angular momentum l and the spin angular momentum s to form states with specified total angular momentum j, with j equal to $l + \frac{1}{2}$ or $l - \frac{1}{2}$. These states are now completely specified by the quantum numbers n, l, j, m_j, and are linear combinations of the states specified by n, l, m_l, and m_s. Because of Eq. (9.28) the combinations are limited to those in which $(m_l + m_s)$ equals m_j. The situation is illustrated in Fig. 9.8.

A level specified by the quantum numbers n, l and j is referred to as the nl_j level; e.g. the two components of the 3p level are referred to as the $3p_{1/2}$ and $3p_{3/2}$ levels, as they have $j = \frac{1}{2}$ and $j = \frac{3}{2}$ respectively.

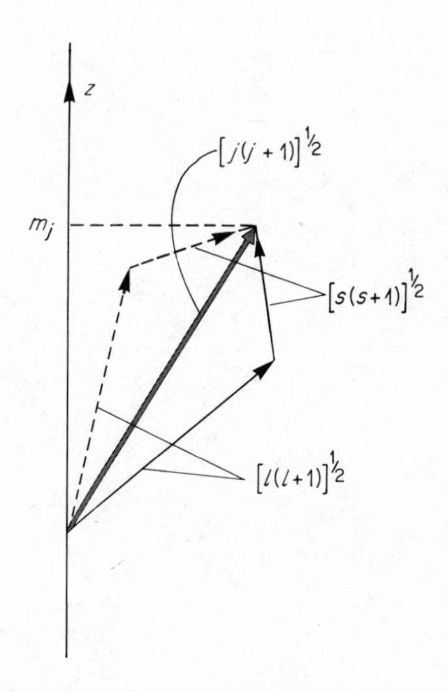

Fig. 9.8. If quantum angular momentum vectors of magnitude $\hbar[l(l + 1)]^{1/2}$ and $\hbar[s(s + 1)]^{1/2}$ are added to form a vector of magnitude $\hbar[j(j + 1)]^{1/2}$ then j can only take the values $|l - s|$, $|l - s| + 1$, etc. in integral steps up to $(l + s)$. If $s = \frac{1}{2}$ this means $j = l + \frac{1}{2}$ or $l - \frac{1}{2}$. The z-component of j, denoted by m_j, is well defined but not the components m_l and m_s.

9.10 THE SPIN–ORBIT INTERACTION

As was pointed out at the end of section 9.8 the experimental results on the first excited state of sodium imply that there is an interaction in the atom we have hitherto neglected which splits the degenerate sub-levels of this state into two groups, one with total angular momentum $j = \frac{3}{2}$ and one with $j = \frac{1}{2}$. In Fig. 9.9 we show how the spin and orbital angular momentum

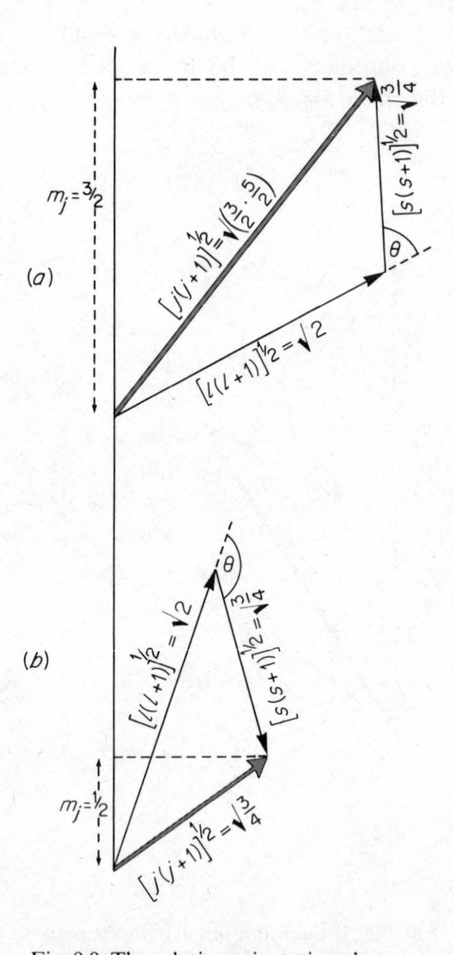

Fig. 9.9. The relative orientations between the spin, orbital and total angular momentum for the cases (*a*) $l = 1$, $s = \frac{1}{2}$, $j = \frac{3}{2}$ and (*b*) $l = 1$, $s = \frac{1}{2}$, $j = \frac{1}{2}$. Note the considerable difference in the value of θ, the angle between the orbital and spin vectors in the two cases.

are added together in the two cases. We see that they correspond to different angles between the spin and the orbital angular momentum. We will get a difference in the energy of the two j-values if the neglected interaction is one depending on the angle between the two component angular momenta, i.e. a spin–orbit interaction.

The source of such an interaction is readily seen if we consider a single electron moving around a nucleus of charge Ze. If we adopt a coordinate system in which the electron is at rest, then the nucleus is apparently moving round the electron, and due to its motion will produce a magnetic field at the electron at right angles to the plane of the orbit (see Fig. 9.10). This magnetic

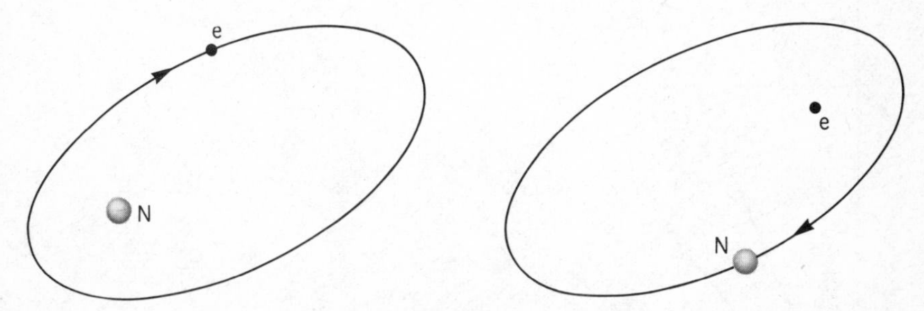

Fig. 9.10. If an electron e moves round a nucleus N, then from the electron's point of view the nucleus is moving round it in a similar orbit. As the nucleus is positively charged this motion will produce a magnetic field B at the position of e into the paper.

field will interact with the magnetic moment of the electron, the energy being dependent on the angle between the magnetic moment of the electron, and hence its spin direction, and the normal to the orbit. But the normal to the orbit points in the same direction as the angular momentum of the electron in the usual coordinate frame. Not surprisingly the interaction energy turns out to be of the form

$$W = \eta(r)\hat{\mathbf{l}} \cdot \hat{\mathbf{s}} \,. \tag{9.31}$$

We defer discussion of $\eta(r)$ for the present, except to note that the smallness of the separation between the two levels constituting the 3p state in sodium indicates it must be small, and hence we will be fully justified in using perturbation theory to evaluate the effect of Eq. (9.31).

Let us first consider what would be the effect on a classical system of a term like Eq. (9.31) in the expression for the energy. Such a term arises from the existence of a couple acting between the spin and the orbital angular momentum, the line of action of the couple being perpendicular to the directions of both the spin and orbital angular momentum as may be seen by differentiating the energy with respect to the angle between the two vectors.

The effect of such a couple is to make the angular momentum vectors precess. As there is no external couple acting the total angular momentum must remain constant in magnitude and direction. Hence the classical effect of a term like Eq. (9.31) is to cause the spin and orbital angular momenta to precess round the total angular momentum as illustrated in Fig. 9.11. As a result the *components* of both the orbital and spin angular momenta will be functions of time.

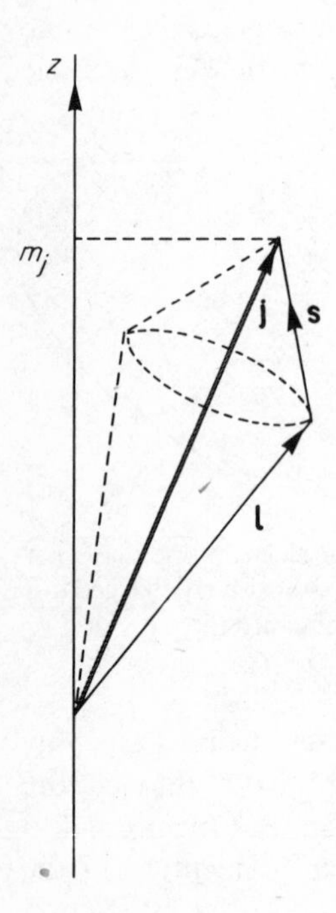

Fig. 9.11. If **l** and **s** precess round their resultant **j**, the components of **l** and **s** along z are not well defined. Quantum mechanically m_l and m_s are not good quantum numbers.

The quantum mechanical analogue of a fixed total angular momentum vector is a state with a definite value of j and m_j. One would not expect definite values quantum mechanically for the z-components of the orbital angular momentum and the spin angular momentum, however, when they do not even possess definite values classically. This presents no difficulty, however, as we have already pointed out that a state with definite values of j and m_j does not possess definite values for m_l and m_s.

From Fig. 9.9 it appears that because the magnitudes $\sqrt{(j(j+1))}\hbar$, $\sqrt{(l(l+1))}\hbar$ and $\sqrt{(s(s+1))}\hbar$ are all fixed in a state with a given value of l, s and j, the *angle* between **l** and **s** is determined. Hence we do not expect the eigenvalues of the operator $\hat{\mathbf{l}} \cdot \hat{\mathbf{s}}$ to depend on the value of m_j. We may show that this is so, and evaluate the eigenvalues as follows. We start with Eq. (9.26) for the operator $\hat{\mathbf{j}}$

$$\hat{\mathbf{j}} = \hat{\mathbf{l}} + \hat{\mathbf{s}} \ . \tag{9.26}$$

Squaring both sides we have

$$\begin{aligned} \hat{j}^2 &= (\hat{\mathbf{l}} + \hat{\mathbf{s}})^2 \\ &= \hat{l}^2 + 2\hat{\mathbf{l}}\cdot\hat{\mathbf{s}} + \hat{s}^2 \ . \end{aligned}$$

Therefore

$$\hat{\mathbf{l}}\cdot\hat{\mathbf{s}} = \tfrac{1}{2}(\hat{j}^2 - \hat{l}^2 - \hat{s}^2) \ . \tag{9.32}$$

Hence the effect of the operator acting on a wavefunction $\psi_{jls}(\mathbf{r})$ which is an eigenfunction of $\hat{j}^2$, $\hat{l}^2$ and $\hat{s}^2$ with eigenvalues $\hbar^2 j(j+1)$, $\hbar^2 l(l+1)$ and $\hbar^2 s(s+1)$ is

$$\begin{aligned} \hat{\mathbf{l}}\cdot\hat{\mathbf{s}}\psi_{jls}(\mathbf{r}) &= \tfrac{1}{2}(\hat{j}^2 - \hat{l}^2 - \hat{s}^2)\psi_{jls}(\mathbf{r}) \\ &= \tfrac{1}{2}[j(j+1) - l(l+1) - s(s+1)]\hbar^2\psi_{jls}(\mathbf{r}) \ . \end{aligned} \tag{9.33}$$

Remembering that $s = \frac{1}{2}$ we find that for the two cases $j = l + \frac{1}{2}$ and $j = l - \frac{1}{2}$ we get

$$\hat{\mathbf{l}}\cdot\hat{\mathbf{s}}\psi_{jls}(\mathbf{r}) = \tfrac{1}{2}l\hbar^2\psi_{jls}(\mathbf{r}), \qquad j = l + \tfrac{1}{2} \tag{9.34a}$$

and

$$\hat{\mathbf{l}}\cdot\hat{\mathbf{s}}\psi_{jls}(\mathbf{r}) = -\tfrac{1}{2}(l+1)\hbar^2\psi_{jls}(\mathbf{r}), \qquad j = l - \tfrac{1}{2} \ . \tag{9.34b}$$

According to the perturbation theory of section 9.3 the effect of the spin-orbit interaction is to change the energy of the level by the expectation value $\int \psi^* W \psi \, \mathrm{d}^3\mathbf{r}$ where W is given by Eq. (9.31). Hence we have, using Eq. (9.7),

$$\langle W \rangle = \int \psi^*_{njls}(\mathbf{r})\eta(r)\hat{\mathbf{l}}\cdot\hat{\mathbf{s}}\psi_{njls}(\mathbf{r})\,\mathrm{d}^3\mathbf{r}$$

$$\begin{cases} = \tfrac{1}{2}l\hbar^2 \displaystyle\int \psi^*_{njls}(\mathbf{r})\eta(r)\psi_{njls}(\mathbf{r})\,\mathrm{d}^3\mathbf{r}, \qquad j = l + \tfrac{1}{2} & \text{(9.35a)} \\ = -\tfrac{1}{2}(l+1)\hbar^2 \displaystyle\int \psi^*_{njls}(\mathbf{r})\eta(r)\psi_{njls}(\mathbf{r})\,\mathrm{d}^3\mathbf{r}, \qquad j = l - \tfrac{1}{2}. & \text{(9.35b)} \end{cases}$$

As the radial dependence of the wavefunctions in Eqs. (9.35) is determined by the quantum numbers n and l and is independent of the values of m_l, m_s or m_j the integrals occurring in Eqs. (9.35) are the same for both members of the pair of levels that occur for specified values of n and l.

The evaluation of the integrals is well beyond the scope of this book, but we can estimate their order of magnitude. The effective magnetic field acting on the electron spin due to the motion of the electron through the electric field of the atom is

$$\mathbf{B} = -\frac{\mathbf{v} \wedge \mathbf{E}}{c^2} = \frac{E\mathbf{l}}{mrc^2}$$

remembering that $\mathbf{E}$ points in the direction of $\mathbf{r}$. The magnetic moment of the electron is given by $\boldsymbol{\mu} = -(e/m)\mathbf{s}$. If we express $\mathbf{l}$ and $\mathbf{s}$ in units of $\hbar$ and substitute the quoted expressions for $\boldsymbol{\mu}$ and $\mathbf{B}$ we obtain

$$\langle W \rangle = -\langle \boldsymbol{\mu} \cdot \mathbf{B} \rangle = \frac{\hbar^2}{m^2c^2}\frac{eE}{r}\mathbf{l} \cdot \mathbf{s} .$$

E varies from $Ze/4\pi\varepsilon_0 r^2$ close to the nucleus to a very much smaller value at a distance of a few times a_0. Hence we write eE/r as

$$\frac{eE}{r} = \frac{1}{a_0^2}\beta\frac{e^2}{4\pi\varepsilon_0 a_0}$$

where β is a number and $e^2/(4\pi\varepsilon_0 a_0)$ is, of course, the magnitude of the expectation value of the potential energy in the ground state of the hydrogen atom. The magnitude of β depends on the state involved, but we should expect it to have an order of magnitude close to unity. Thus we obtain

$$\langle W \rangle = \left(\frac{\hbar^2}{m^2c^2a_0^2}\right)\left(\frac{\beta e^2}{4\pi\varepsilon_0 a_0}\right)(\mathbf{l} \cdot \mathbf{s}) .$$

On substituting $a_0 = 4\pi\varepsilon_0\hbar^2/me^2$ in the first term we obtain

$$\langle W \rangle = \left(\frac{e^2}{4\pi\varepsilon_0\hbar c}\right)^2\left(\frac{\beta e^2}{4\pi\varepsilon_0 a_0}\right)(\mathbf{l} \cdot \mathbf{s}) .$$

A proper relativistic calculation gives precisely half the value we have obtained, namely

$$\langle W \rangle = \frac{1}{2}\left(\frac{e^2}{4\pi\varepsilon_0\hbar c}\right)^2\left(\frac{\beta e^2}{4\pi\varepsilon_0 a_0}\right)(\mathbf{l} \cdot \mathbf{s}) .$$

The quantity $\alpha = e^2/(4\pi\varepsilon_0\hbar c)$ is a dimensionless number known as the fine structure constant. The most accurate recent value is given in Appendix B, from which we see that it is equal to $\frac{1}{137}$ to better than 0.1% accuracy.

Thus we find that the spin orbit splitting is of order α^2 times $e^2/4\pi\varepsilon_0 a_0$ remembering that $(\mathbf{l} \cdot \mathbf{s})$ is of order unity as l and s are measured in units of $\hbar$. The dependence of the splitting on the parameters of the individual levels has, of course, been absorbed into β. In general β will be relatively large for orbits which penetrate into regions of high electric field, i.e. orbits with low values of n and l in atoms with high Z, and vice versa. For the case of the 3p level in sodium which we have been discussing we find β equals 1.98. $\langle W \rangle$ is of course zero for s-states as $(\mathbf{l} \cdot \mathbf{s})$ is zero.

The quantity α is of fundamental significance in quantum electrodynamics as it is a measure of the interaction between an electron and a photon. It has already occurred in Eq. (6.31) for the gyromagnetic ratio of the electron. The reader will readily verify that the ground state energy of the hydrogen atom is $-\frac{1}{2}\alpha^2 mc^2$, where mc^2 is the rest energy of the electron.

Summary

Due to the motion of an electron through the electric field in an atom there is a magnetic field acting on the electron in the frame of reference which moves with the electron. The interaction between this magnetic field and the magnetic moment of the electron causes the spin and orbital angular momenta to precess round the total angular momentum j. As a result the set of sub-levels with a given value of n and l separate into two groups, one with $j = l + \frac{1}{2}$ and the other with $j = l - \frac{1}{2}$. The sub-levels within these groups do not possess definite values of m_l and m_s, but *do* possess definite values of m_j. The changes in energy due to the spin–orbit interaction are given by Eqs. (9.35), and the integrals in Eqs. (9.35) decrease with increasing n and l, and increase with increasing Z.

9.11 THE HELIUM ATOM—INTRODUCTION

The atoms we have discussed so far have been those with one electron outside a closed shell, the alkali metals. These atoms have the simplest energy level schemes. The next simplest atoms are those with two electrons outside a closed shell, the alkaline earths together with helium. In this case we find it necessary to take into account the coulomb repulsion between the two valence electrons rather more explicitly, and not merely represent it by an average field. All the features of the energy level spectra of atoms with two valence electrons are brought out by the example of helium, and as in this case we have the additional simplification of knowing precisely what the central field is, namely just that due to the nucleus, we shall confine our attention to this case.

The essential difference between helium and the alkali metals is as follows. In the one-electron case the closed shells are spherically symmetric and provide no angular momentum. There is thus no preferred direction in the core and, as we have seen, the direction of the electron spin only affects the energy through the spin–orbit interaction. In the case of the helium atom we have to consider not only the direction of each electron spin with respect to its own orbital angular momentum, but the direction of the spin of one electron relative to that of the other. At first sight it might be thought that these two effects would be of the same order of magnitude. This is not so however, as it is not the direct interaction of the two electron spins that causes the relative orientation of the two spins to be important, but an indirect effect operating through the Pauli principle. As we shall see in section 9.14 the relative orientation of the two electron spins affects the probability of the two electrons being close together, and hence the average value of the coulomb repulsion energy between the electrons. This effect is much larger than the spin–orbit interaction. The spin–orbit interaction is complicated by the presence of two electrons, and as it is so small we shall neglect it for the rest of this chapter.

9.12 THE GROUND STATE OF HELIUM

As we have agreed to neglect the effects of spin–orbit interaction in the helium atom, the Hamiltonian in this approximation is

$$H = -\frac{\hbar^2}{2m}(\nabla_1^2 + \nabla_2^2) - \frac{2e^2}{4\pi\varepsilon_0 r_1} - \frac{2e^2}{4\pi\varepsilon_0 r_2} + \frac{e^2}{4\pi\varepsilon_0|\mathbf{r}_1 - \mathbf{r}_2|} . \tag{9.36}$$

The first term is the sum of the two kinetic energy operators, the second and third terms are the potential energies of the electrons in the field of the nucleus with nuclear charge $2e$ and the last term is the energy of electrostatic repulsion between the two electrons. The Schrödinger equation for the states of helium

$$H\psi = E\psi$$

where H is given by Eq. (9.36) cannot be solved analytically, so we are forced to use approximate methods. The simplest approach is to use first order perturbation theory. We split the Hamiltonian H of Eq. (9.36) into two parts,

$$H = H_0 + U$$

where

$$H_0 = -\frac{\hbar^2}{2m}(\nabla_1^2 + \nabla_2^2) - \frac{2e}{4\pi\varepsilon_0}\left(\frac{1}{r_1} + \frac{1}{r_2}\right) \tag{9.37}$$

and

$$U = \frac{e^2}{4\pi\varepsilon_0|\mathbf{r}_1 - \mathbf{r}_2|} . \tag{9.38}$$

H_0 is the Hamiltonian for two non-interacting electrons in the field of the nucleus, and the perturbation U is the electrostatic repulsion between the two electrons. We note that U is not substantially smaller than H_0 and therefore while we expect our solutions to be qualitatively correct we do not expect very great quantitative accuracy.

As H_0 contains only terms which apply either to one electron or the other, but not to both, it has solutions of the form

$$H_0\psi = E\psi$$

where

$$\psi = \phi_{n_1l_1}(\mathbf{r}_1)\phi_{n_2l_2}(\mathbf{r}_2) \tag{9.39}$$

and

$$E = E_{n_1l_1} + E_{n_2l_2} \; . \tag{9.40}$$

We have suppressed the magnetic quantum numbers in the labels specifying the wavefunctions as the result is independent of these, as there is no preferred direction in space.

The wavefunctions ϕ_{nl} and energies E_{nl} occurring in Eqs. (9.39) and (9.40) are solutions of

$$\left(-\frac{\hbar^2}{2m}\nabla_1^2 - \frac{2e^2}{4\pi\varepsilon_0 r_1}\right)\phi_{n_1l_1}(\mathbf{r}_1) = E_{n_1l_1}\phi_{n_1l_1}(\mathbf{r}_1) \tag{9.41a}$$

and

$$\left(-\frac{\hbar^2}{2m}\nabla_2^2 - \frac{2e^2}{4\pi\varepsilon_0 r_2}\right)\phi_{n_2l_2}(\mathbf{r}_2) = E_{n_2l_2}\phi_{n_2l_2}(\mathbf{r}_2) \; . \tag{9.41b}$$

These are, of course, just the solutions and energies we discussed in Chapter 7 i.e. the hydrogen-like one electron problem with $Z = 2$. The lowest energy for H_0 occurs when $n_1 = n_2 = 1$ and $l_1 = l_2 = 0$. Thus the solution of Eq. (9.39) which corresponds to the lowest energy is

$$\psi_{\text{gs}} = \phi_{10}(\mathbf{r}_1)\phi_{10}(\mathbf{r}_2) \; . \tag{9.42}$$

$\phi_{10}(\mathbf{r})$ is given by Eq. (7.13) with $\chi(\theta, \phi) = 1/\sqrt{(4\pi)}$ and $\psi(r) = u(r)/r$ where $u(r)$ is given by Eq. (7.21b) with $Z = 2$, that is

$$\phi_{10}(\mathbf{r}) = \left(\frac{8}{\pi a_0^3}\right)^{1/2} \mathrm{e}^{-2r/a_0} \tag{9.43}$$

and the ground state energy is given by

$$E_{0\,\text{gs}} = 2E_{10}$$

where

$$E_{10} = -4e^2/4\pi\varepsilon_0 2a_0$$

from Eq. (7.32) with $Z = 2$ and $n = 1$. Therefore

$$E_{0\,\text{gs}} = -8\left(\frac{e^2}{4\pi\varepsilon_0 2a_0}\right) = -8 \times 13.6\,\text{eV} = -108.8\,\text{eV}. \tag{9.44}$$

However Eq. (9.42) is unsatisfactory as the ground state wavefunction in two respects. It makes no reference to the electron spins, and it is not antisymmetric. We may remove both objections simultaneously by multiplying Eq. (9.42) by a spin wavefunction which is antisymmetric in the spins of the two particles. As the space wavefunctions of the two particles are identical the whole wavefunction will then be antisymmetric. There is only one spin wavefunction of two particles that is antisymmetric, namely

$$\chi(s) = \frac{1}{\sqrt{2}}[\alpha(1)\beta(2) - \alpha(2)\beta(1)] \tag{9.45}$$

where the functions α and β are defined by Eqs. (8.14). The factor of $1/\sqrt{2}$ outside is a normalization factor which ensures that the spin wavefunction is normalized.

We thus modify Eq. (9.42) for the ground state wavefunction to read

$$\Psi_{gs} = \phi_1(\mathbf{r}_1)\phi_1(\mathbf{r}_2)\chi(s) \tag{9.42a}$$

where $\phi_1(\mathbf{r})$ is given by Eq. (9.43) and $\chi(s)$ by Eq. (9.45). Eq. (9.42a) is now antisymmetric. We also note that as one of the electrons in Eq. (9.45) has $s_z = +\frac{1}{2}\hbar$ while the other has $s_z = -\frac{1}{2}\hbar$ the total z-component of spin angular momentum is zero. We shall give reasons in the next section for supposing that Eq. (9.45) corresponds to a state in which the total spin angular momentum equals zero.

As the Hamilton H_0 of Eq. (9.37) contains no reference to the spins of the electrons the extra factor $\chi(s)$ in the wavefunction (9.42a) makes no difference to the calculation of the unperturbed energy Eq. (9.44).

To calculate the perturbation theory result for the ground state energy of helium we have to calculate the expectation value of Eq. (9.38) in the state given by (9.42a)

$$\langle U \rangle = \int \phi_1^*(\mathbf{r}_1)\phi_1^*(\mathbf{r}_2)\chi^*(s)\frac{e^2}{4\pi\varepsilon_0|\mathbf{r}_1 - \mathbf{r}_2|}\phi_1(\mathbf{r}_1)\phi_1(\mathbf{r}_2)\chi(s)\,\mathrm{d}^3\mathbf{r}_1\,\mathrm{d}^3\mathbf{r}_2 \,. \tag{9.46}$$

As $\chi(s)$ is normalized and U has no term referring to the spins Eq. (9.46) becomes

$$\langle U \rangle = \int \phi_1^*(\mathbf{r}_1)\phi_1^*(\mathbf{r}_2)\frac{e^2}{4\pi\varepsilon_0|\mathbf{r}_1 - \mathbf{r}_2|}\phi_1(\mathbf{r}_1)\phi_1(\mathbf{r}_2)\,\mathrm{d}^3\mathbf{r}_1\,\mathrm{d}^3\mathbf{r}_2 \,. \tag{9.46a}$$

For readers who are interested we evaluate the integral of Eq. (9.46a) in section 9.13. The result is

$$\begin{aligned}\langle U\rangle &= \frac{5}{4}\frac{e^2}{4\pi\varepsilon_0 a_0} \\ &= \frac{5}{2}\left(\frac{e^2}{4\pi\varepsilon_0 2a_0}\right) \\ &= \frac{5}{2}\times 13.6\,\text{eV} \\ &= 34.0\,\text{eV}\ .\end{aligned} \tag{9.47}$$

$\langle U\rangle$ is the average potential energy of repulsion between the two electrons and is therefore positive.

Adding together Eq. (9.44) and Eq. (9.47) we get the ground state energy

$$\begin{aligned}E_{\text{gs}} &= E_{0\,\text{gs}} + \langle U\rangle \\ &= -108.8 + 34.0\,\text{eV} = -74.8\,\text{eV}\ .\end{aligned} \tag{9.48}$$

From experiment the ground state energy of helium is known to be -79.0 eV and in view of the fact the perturbation cannot be considered small the result of Eq. (9.48) is very satisfactory.

★ 9.13 EVALUATION OF THE INTEGRAL EQ. (9.46a)

We may re-write Eq. (9.46a) as

$$\langle U\rangle = \int |\phi_1(\mathbf{r}_1)|^2|\phi_1(\mathbf{r}_2)|^2\frac{e^2}{4\pi\varepsilon_0|\mathbf{r}_1-\mathbf{r}_2|}\,\mathrm{d}^3\mathbf{r}_1\,\mathrm{d}^3\mathbf{r}_2\ . \tag{9.49}$$

As $|\phi_1(\mathbf{r}_1)|^2$ and $|\phi_1(\mathbf{r}_2)|^2$ are just the probabilities of finding electron 1 in the region of $\mathbf{r}_1$ and electron 2 near $\mathbf{r}_2$ we see that Eq. (9.49) in fact represents the energy of electrostatic repulsion between two charge clouds each with a charge density distribution $-e|\phi_1(\mathbf{r})|^2$. We also note from Eq. (9.43) that the distribution is spherically symmetric.

Consider first the interaction between a spherical shell of the first charge distribution, of radius r_1, with an element of the second charge distribution of volume $\mathrm{d}^3\mathbf{r}_2$ situated at r_2 where r_2 is *greater* than r_1. The situation is illustrated in Fig. 9.12. For points outside r_1 the spherical shell of charge acts as if it were concentrated at the origin and the potential energy between the shell and the element of charge at r_2 is therefore

$$\mathrm{d}U = \frac{\rho_1 4\pi r_1^2\,\mathrm{d}r_1 \times \rho_2\,\mathrm{d}^3\mathbf{r}_2}{4\pi\varepsilon_0 r_2} \tag{9.50}$$

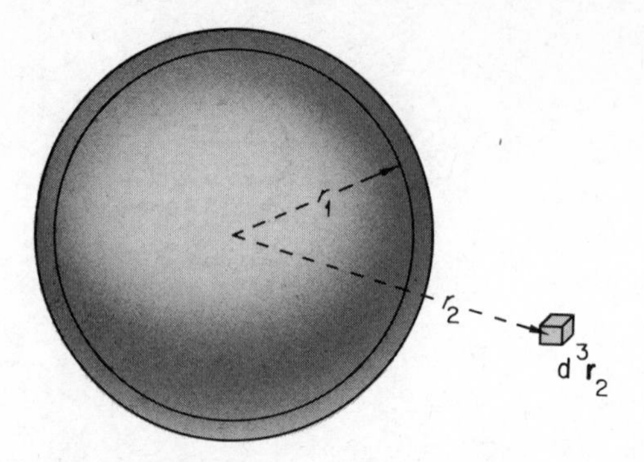

Fig. 9.12. From the point of view of an element of volume $d^3\mathbf{r}_2$ at $\mathbf{r}_2$ with charge density ρ_2 a thin spherical shell of charge density ρ_1 at radius r_1 acts as if its charge were concentrated at the origin. The electrostatic interaction between them is

$$dU = \frac{\rho_1 4\pi r_1^2\, dr_1 \rho_2\, d^3\mathbf{r}_2}{4\pi\varepsilon_0 r_2}\,.$$

where ρ_1 and ρ_2 are the densities of the two charge distributions. As Eq. (9.50) is independent of the angular position of the elementary volume $d^3\mathbf{r}_2$ the electrostatic potential acting between two shells of radius r_1 and r_2, where r_2 is *greater* than r_1 is

$$dU = \frac{\rho_1\rho_2 4\pi r_1^2 4\pi r_2^2\, dr_1\, dr_2}{4\pi\varepsilon_0 r_2}\,. \tag{9.51}$$

A similar expression with r_2 replaced by r_1 in the denominator of Eq. (9.51) clearly holds for the case when r_2 is *less* than r_1. We may put the two cases together by writing

$$dU = \frac{\rho_1\rho_2 4\pi r_1^2 4\pi r_2^2\, dr_1\, dr_2}{4\pi\varepsilon_0 r_>} \tag{9.52}$$

where $r_>$ is the greater of r_1 and r_2.

Remembering $\rho_1 = -e|\phi_1(\mathbf{r}_1)|^2$ and similarly for ρ_2, and using Eq. (9.43) for $\phi_1(\mathbf{r})$, we get

$$dU = \frac{e^2\left(\dfrac{32}{a_0^3}\right)^2 r_1^2 r_2^2\, e^{-4r_1/a_0}\, e^{-4r_2/a_0}\, dr_1\, dr_2}{4\pi\varepsilon_0 r_>} \tag{9.53}$$

Thus

$$\langle U\rangle = e^2\left(\frac{32}{a_0^3}\right)^2 \iint \frac{r_1^2 r_2^2\, e^{-4r_1/a_0}\, e^{-4r_2/a_0}\, dr_1\, dr_2}{4\pi\varepsilon_0 r_>} . \tag{9.54}$$

As the integral in Eq. (9.54) is symmetrical in r_1 and r_2 we shall clearly get equal contributions from r_1 less than r_2 and r_1 greater than r_2. We may evaluate Eq. (9.54) therefore by taking $r_2 = r_1 + r_0$ and integrating over r_1 from zero to infinity, and then integrating over r_0 from zero to infinity and doubling the final result. Making these substitutions we get

$$\begin{aligned}\langle U\rangle &= \frac{2e^2}{4\pi\varepsilon_0}\left(\frac{32}{a_0^3}\right)^2 \int_0^\infty dr_0 \int_0^\infty \frac{r_1^2(r_1+r_0)^2\, e^{-4r_1/a_0}\, e^{-4(r_1+r_0)/a_0}}{r_1+r_0}\, dr_1 \\ &= \frac{2e^2}{4\pi\varepsilon_0}\left(\frac{32}{a_0^3}\right)^2 \int_0^\infty e^{-4r_0/a_0}\, dr_0 \int_0^\infty (r_1^3 + r_1^2 r_0)\, e^{-8r_1/a_0}\, dr_1 .\end{aligned} \tag{9.55}$$

Using the result already quoted in Eq. (7.27)

$$\int_0^\infty x^n e^{-\alpha x}\, dx = \frac{n!}{\alpha^{n+1}}, \qquad (\alpha > 0, n \geqslant 0) \tag{7.27}$$

integration over r_1 leads to the result

$$\begin{aligned}\langle U\rangle &= \frac{2e^2}{4\pi\varepsilon_0}\left(\frac{32}{a_0^3}\right)^2 \int_0^\infty e^{-4r_0/a_0}\left(\frac{3!a_0^4}{8^4} + \frac{2!a_0^3 r_0}{8^3}\right) dr_0 \\ &= \frac{e^2}{4\pi\varepsilon_0}\int_0^\infty \left(\frac{3}{a_0^2} + \frac{8r_0}{a_0^3}\right) e^{-4r_0/a_0}\, dr_0\end{aligned}$$

which, again using Eq. (7.27) for the integral over r_0, gives us

$$\begin{aligned}\langle U\rangle &= \frac{e^2}{4\pi\varepsilon_0}\left(\frac{3}{a_0^2}\cdot\frac{a_0}{4} + \frac{8}{a_0^3}\left(\frac{a_0}{4}\right)^2\right) \\ &= \frac{e^2}{4\pi\varepsilon_0}\left(\frac{3}{4a_0} + \frac{1}{2a_0}\right) \\ &= \frac{5}{4}\frac{e^2}{4\pi\varepsilon_0 a_0}\end{aligned} \tag{9.56}$$

which was the result quoted in Eq. (9.47).

9.14 EXCITED STATES OF THE HELIUM ATOM AND THE EFFECT OF ANTISYMMETRY

There is an enormous number of possible excited states of the helium atom corresponding to the different sets of quantum numbers $n_1 l_1 m_{l_1} m_{s_1}$ and $n_2 l_2 m_{l_2} m_{s_2}$ which may be ascribed to the two electrons. Some of these will certainly be degenerate, but this still leaves a vast number of distinct energy levels. We shall concern ourselves almost entirely with a small sub-set of these possibilities, namely those in which one of the electrons remains in the ground state with $n = 1$ and $l = 0$. Our reason for this is that these levels bring out most of the physical principles involved in all of the levels, and transitions among them are responsible for almost all of the observed spectral lines in helium. The main point we shall be interested in is the influence of the total electron spin on the energy of the system. According to the discussion of section 9.9, if we add the two electron spins together we can form states with total spin $S = 1$, or $S = 0$, in units of $\hbar$. Although the Hamiltonian Eq. (9.36) has no term in it which depends on electron spin, we shall find that the value of the total spin S, has an important influence on the energy of the state through the requirement that the total wavefunction is antisymmetric under the exchange of the two electrons.

We start by considering the wavefunction, ignoring the antisymmetry requirement. In these circumstances we make a first estimate of both the wavefunctions and energies of the states of helium with one electron excited by repeating the arguments used in section 8.2. We argue that, as the excited electron spends most of its time well away from the nucleus, it is largely moving in the field due to a single charge, one of the two charges on the nucleus being screened by the unexcited electron. If it has quantum numbers n and l then its wave function will be the corresponding hydrogen wavefunction $\phi_{nl}(\mathbf{r}_2)$. Likewise the unexcited electron spends most of its time inside the excited electron, and hence experiences the full field of the nucleus. Thus its wavefunction will be closely approximated by the ground state wavefunctions for $Z = 2$:

$$\phi_0(\mathbf{r}_1) = \left(\frac{8}{\pi a_0^3}\right)^{1/2} \mathrm{e}^{-2r_1/a_0} .$$

We thus get an approximate total wavefunction

$$\psi = \phi_0(\mathbf{r}_1)\phi_{nl}(\mathbf{r}_2) \tag{8.4}$$

with an approximate value for the energy of

$$E = E_0 + E_{nl} \tag{9.57}$$

where E_0 is the ground state energy of the He^+ ion, namely -54.4 eV, and E_{nl} is the energy of an electron in a hydrogen atom with quantum numbers n and l.

We must now consider how to modify Eq. (8.4) to take account of the electron spin and antisymmetry. When we discussed the ground state of helium, the two space wavefunctions were identical, and hence we were forced to make the spin wavefunctions antisymmetric in Eq. (9.45) and the two electron states were distinguished by having different quantum numbers for their electron spins. In the present case the two space wavefunctions are different and so there is no need for the spin wavefunctions to be different. The only requirement is that the total wavefunction be antisymmetric under exchange of electrons. If both electrons have the same spin quantum numbers, the spin wavefunction is necessarily symmetric, and thus the space wavefunction must be antisymmetric. We may write the necessary modifications to Eq. (8.4) in this case as follows:

$$\Psi = \{\phi_0(\mathbf{r}_1)\phi_{nl}(\mathbf{r}_2) - \phi_0(\mathbf{r}_2)\phi_{nl}(\mathbf{r}_1)\}\alpha(1)\alpha(2) . \tag{9.58a}$$

This represents an antisymmetric wavefunction in which both electrons have $m_s = +\frac{1}{2}$. The antisymmetric wavefunction in which both electrons have $m_s = -\frac{1}{2}$ is clearly

$$\Psi = \{\phi_0(\mathbf{r}_1)\phi_{nl}(\mathbf{r}_2) - \phi_0(\mathbf{r}_2)\phi_{nl}(\mathbf{r}_1)\}\beta(1)\beta(2) . \tag{9.58b}$$

When we come to the case in which one electron has $m_s = +\frac{1}{2}$ and the other has $m_s = -\frac{1}{2}$ there are two ways of writing down antisymmetric wavefunctions, namely

$$\Psi = \{\phi_0(\mathbf{r}_1)\phi_{nl}(\mathbf{r}_2) - \phi_0(\mathbf{r}_2)\phi_{nl}(\mathbf{r}_1)\}\{\alpha(1)\beta(2) + \beta(1)\alpha(2)\} \tag{9.58c}$$

and

$$\Psi = \{\phi_0(\mathbf{r}_1)\phi_{nl}(\mathbf{r}_2) + \phi_0(\mathbf{r}_2)\phi_{nl}(\mathbf{r}_1)\}\{\alpha(1)\beta(2) - \beta(1)\alpha(2)\} . \tag{9.58d}$$

Clearly any linear combination of the wavefunctions of Eqs. (9.58) will also form an antisymmetric wavefunction. However Eqs. (9.58a,b) are unique states possessing a physically measurable quantity to distinguish them, namely, the z-component of their total spin which is $\hbar$ for Eq. (9.58a) and $-\hbar$ for Eq. (9.58b). As the space wavefunctions of these two states are identical they will be degenerate in the approximation of Eq. (9.36). The state represented by Eq. (9.58c) which has an antisymmetric space and a symmetric spin wavefunction will also be degenerate with them. The state represented by Eq. (9.58d) has a different space wavefunction from the other three, being symmetric, and hence is not necessarily degenerate with the others. We shall show that the energy is indeed dependent on the symmetry of the space wavefunction and hence the wavefunctions of Eq. (9.58) represent two energy levels, one of which has the symmetric spin combination and the other the antisymmetric spin combination.

Before we do this we must make a remark on the single electron wavefunctions $\phi_0(\mathbf{r})$ and $\phi_{nl}(\mathbf{r})$. ϕ_0 is the ground state wavefunction for the helium ion, He^+, and the ϕ_{nl} are the excited state wavefunctions for the hydrogen atom. For l equal to any value other than zero ϕ_0 and ϕ_{nl} are orthogonal to each other, as wavefunctions with different l-values are always orthogonal.

We shall show that, as a result, the expectation values of the Hamiltonian Eq. (9.36) for the wavefunctions of Eq. (9.58) differ only in the expectation value of the term $e^2/(4\pi\varepsilon_0|\mathbf{r}_1 - \mathbf{r}_2|)$ in Eq. (9.36), that is, in the average value of the coulomb repulsion between the two electrons. For l equal to zero $\phi_0(\mathbf{r})$ and $\phi_{n0}(\mathbf{r})$ are *not* orthogonal. This lack of orthogonality not only adds greatly to the complexity of the calculation, but actually leads to an answer of the wrong sign for the energy difference between the two states. Clearly we must make a different choice for our functions $\phi_{n0}(\mathbf{r})$. The simplest thing to do is to choose them to be eigenfunctions for the case of $Z = 2$, when they will be orthogonal to $\phi_0(\mathbf{r})$. Clearly such a choice will not be such a good approximation as our original choice when using total wavefunctions of the form of Eq. (8.4), but in Eqs. (9.58) it will avoid the problems arising from lack of orthogonality of the single electron wavefunctions. Further, we may write Eq. (9.36) as

$$H = H_1 + H_2 + H_{12} \tag{9.59}$$

where

$$H_1 = -\frac{\hbar^2}{2m}\nabla_1^2 - \frac{2e^2}{4\pi\varepsilon_0 r_1}$$

$$H_2 = -\frac{\hbar^2}{2m}\nabla_2^2 - \frac{2e^2}{4\pi\varepsilon_0 r_2}$$

$$H_{12} = \frac{e^2}{4\pi\varepsilon_0|\mathbf{r}_1 - \mathbf{r}_2|} \,.$$

With our new choice of wavefunctions for the case when l equals zero in Eqs. (9.58) we find that for all of them

$$(H_1 + H_2)\Psi = (E_0 + E_n)\Psi$$

where ϕ_0 belongs to the eigenvalue E_0 and ϕ_{n0} belongs to the eigenvalue E_n. Any difference in the expectation values of the total Hamiltonian now depends solely on the perturbing term $e^2/(4\pi\varepsilon_0|\mathbf{r}_1 - \mathbf{r}_2|)$, thus creating a situation identical to the cases when l is other than zero.

The source of the trouble when ϕ_{n0} is not orthogonal to ϕ_0 is clearly seen if we express ϕ_{n0} as

$$\phi_{n0} = \phi'_{n0} + \alpha\phi_0$$

where α is given by

$$\alpha = \int \phi_{n0}^*(\mathbf{r})\phi_0(\mathbf{r})\,\mathrm{d}^3\mathbf{r}$$

which ensures that ϕ'_{n0} is orthogonal to ϕ_0. If we substitute for ϕ_{n0} in Eqs. (9.58c) and (9.58d) we find, for the space portion of Eq. (9.58c)

$$\Psi = \{\phi_0(\mathbf{r}_1)\phi'_{n0}(\mathbf{r}_2) - \phi_0(\mathbf{r}_2)\phi'_{n0}(\mathbf{r}_1)\}$$

and for the space portion of Eq. (9.58d)

$$\Psi = \{\phi_0(\mathbf{r}_1)\phi'_{n0}(\mathbf{r}_2) + \phi_0(\mathbf{r}_2)\phi'_{n0}(\mathbf{r}_1)\} + 2\alpha\phi_0(\mathbf{r}_1)\phi_0(\mathbf{r}_2) .$$

Thus Eq. (9.58d) contains a term that does not occur in Eq. (9.58c), and further this term has the same form as our first approximation to the ground state wavefunction for helium! Not surprisingly this term lowers the expectation value of the Hamiltonian for Eq. (9.58d) and brings it below that of Eq. (9.58c), contrary to observation.

We can now show qualitatively that the expectation value of the Hamiltonian is different in two states with the same single electron wavefunctions, but which differ in their symmetry; as we have already stated, we shall show that, because of the orthogonality of the single electron wavefunctions, the expectation value of the Hamiltonian in the two states differs only in the expectation value of the electrostatic repulsion $e^2/4\pi\varepsilon_0|\mathbf{r}_1 - \mathbf{r}_2|$. This term gets larger the smaller the value of $|\mathbf{r}_1 - \mathbf{r}_2|$, becoming infinite when $\mathbf{r}_1 = \mathbf{r}_2$. Hence the expectation value of this term is dominated by the behaviour of the wavefunctions when $\mathbf{r}_1$ and $\mathbf{r}_2$ are equal or very nearly so. Inspection of Eqs. (9.58a,b,c) shows that for these cases the wavefunction Ψ *vanishes* when $\mathbf{r}_1$ and $\mathbf{r}_2$ are equal; that is the probability of the two electrons both being at the same place is zero. As the individual wavefunctions are slowly varying functions of position it follows that the probability of the two electrons being close together is small. The space symmetric wavefunction Eq. (9.58d) on the other hand has no special behaviour as $\mathbf{r}_1$ approaches $\mathbf{r}_2$ and there is a finite probability of finding the two electrons very close together. As a result the expectation value of the electrostatic repulsion term is greater for symmetric space wavefunctions than for antisymmetric space wavefunctions and hence the expectation value of the Hamiltonian for Eq. (9.58a,b,c) will be lower than for Eq. (9.58d) for a given pair of individual electron wavefunctions ϕ_0 and ϕ_{nl}.

Let us put this on a slightly more formal basis. We write the wavefunctions Eqs. (9.58) in the form

$$\Psi(\mathbf{r}_1, \mathbf{r}_2, s_1, s_2) = \psi(\mathbf{r}_1, \mathbf{r}_2)\chi(s_1, s_2) \tag{9.60}$$

where $\psi(\mathbf{r}_1, \mathbf{r}_2)$ represents the space portions of Eq. (9.58) and $\chi(s_1, s_2)$ the spin portions. As the Hamiltonian Eq. (9.36) is independent of the spins its expectation value will be given by

$$\langle H \rangle = \frac{\iint \psi^*(\mathbf{r}_1, \mathbf{r}_2) H \psi(\mathbf{r}_1, \mathbf{r}_2)\, \mathrm{d}^3\mathbf{r}_1\, \mathrm{d}^3\mathbf{r}_2}{\iint \psi^*(\mathbf{r}_1, \mathbf{r}_2)\psi(\mathbf{r}_1, \mathbf{r}_2)\, \mathrm{d}^3\mathbf{r}_1\, \mathrm{d}^3\mathbf{r}_2} . \tag{9.61}$$

If, as we have assumed, the individual particle wavefunctions ϕ_0 and ϕ_{nl} are normalized and orthogonal it is straightforward to show that the normalization integral in the denominator of Eq. (9.61) has the value 2 for both types of space wavefunction. Thus we get

$$\begin{aligned}\langle H\rangle_+ = \tfrac{1}{2}\iint &\{\phi_0(\mathbf{r}_1)\phi_{nl}(\mathbf{r}_2) + \phi_0(\mathbf{r}_2)\phi_{nl}(\mathbf{r}_1)\}^* \\ &\times H\{\phi_0(\mathbf{r}_1)\phi_{nl}(\mathbf{r}_2) + \phi_0(\mathbf{r}_2)\phi_{nl}(\mathbf{r}_1)\}\,\mathrm{d}^3\mathbf{r}_1\,\mathrm{d}^3\mathbf{r}_2\end{aligned} \tag{9.62a}$$

and

$$\begin{aligned}\langle H\rangle_- = \tfrac{1}{2}\iint &\{\phi_0(\mathbf{r}_1)\phi_{nl}(\mathbf{r}_2) - \phi_0(\mathbf{r}_2)\phi_{nl}(\mathbf{r}_1)\}^* \\ &\times H\{\phi_0(\mathbf{r}_1)\phi_{nl}(\mathbf{r}_2) - \phi_0(\mathbf{r}_2)\phi_{nl}(\mathbf{r}_1)\}\,\mathrm{d}^3\mathbf{r}_1\,\mathrm{d}^3\mathbf{r}_2\end{aligned} \tag{9.62b}$$

where $\langle H\rangle_+$ and $\langle H\rangle_-$ are the expectation values of the Hamiltonian in the symmetric and antisymmetric space wavefunction cases respectively.

If we multiply out the terms in Eq. (9.62a,b) we see that each expression contains four terms. In two of the terms the wavefunction in front of H is just the complex conjugate of the wavefunction following H; these are called the *direct* terms. In the other two terms the wavefunction in front of H is the complex conjugate of the wavefunction following H, but with the *particle coordinates exchanged.* These are called the *exchange terms.* Thus if we multiply out Eq. (9.62a) we get

$$\begin{aligned}\langle H\rangle_+ &= \tfrac{1}{2}\Big[\iint \phi_0^*(\mathbf{r}_1)\phi_{nl}^*(\mathbf{r}_2)H\phi_0(\mathbf{r}_1)\phi_{nl}(\mathbf{r}_2)\,\mathrm{d}^3\mathbf{r}_1\,\mathrm{d}^3\mathbf{r}_2 \\ &\quad + \iint \phi_0^*(\mathbf{r}_2)\phi_{nl}^*(\mathbf{r}_1)H\phi_0(\mathbf{r}_2)\phi_{nl}(\mathbf{r}_1)\,\mathrm{d}^3\mathbf{r}_1\,\mathrm{d}^3\mathbf{r}_2\Big] \\ &\quad + \tfrac{1}{2}\Big[\iint \phi_0^*(\mathbf{r}_1)\phi_{nl}^*(\mathbf{r}_2)H\phi_0(\mathbf{r}_2)\phi_{nl}(\mathbf{r}_1)\,\mathrm{d}^3\mathbf{r}_1\,\mathrm{d}^3\mathbf{r}_2 \\ &\quad + \iint \phi_0^*(\mathbf{r}_2)\phi_{nl}^*(\mathbf{r}_1)H\phi_0(\mathbf{r}_1)\phi_{nl}(\mathbf{r}_2)\,\mathrm{d}^3\mathbf{r}_1\,\mathrm{d}^3\mathbf{r}_2\Big] \\ &= \langle H\rangle_{\mathrm{D}} + \langle H\rangle_{\mathrm{Ex}}\end{aligned} \tag{9.63a}$$

where $\langle H\rangle_{\mathrm{D}}$ is the expectation value of the direct terms and $\langle H\rangle_{\mathrm{Ex}}$ the expectation value of the exchange terms.

Similarly if we multiply out Eq. (9.62b) we find

$$\langle H\rangle_- = \langle H\rangle_{\mathrm{D}} - \langle H\rangle_{\mathrm{Ex}}\,. \tag{9.63b}$$

Let us first look at the direct terms. As the Hamiltonian is symmetric in the coordinates of the two particles the two integrals inside the first bracket in Eq. (9.63a) are identical, and hence we get the result

$$\langle H\rangle_{\mathrm{D}} = \iint \phi_0^*(\mathbf{r}_1)\phi_{nl}^*(\mathbf{r}_2)H\phi_0(\mathbf{r}_1)\phi_{nl}(\mathbf{r}_2)\,\mathrm{d}^3\mathbf{r}_1\,\mathrm{d}^3\mathbf{r}_2 \tag{9.64}$$

which is just the result we would have got for $\langle H \rangle$ if we had ignored the antisymmetry requirement and treated the two electrons as being distinguishable.

Now let us look at the exchange terms. In this case we split up the Hamiltonian into three parts as before

$$H = H_1 + H_2 + H_{12} \tag{9.59}$$

where

$$H_1 = -\frac{\hbar^2}{2m}\nabla_1^2 - \frac{2e^2}{4\pi\varepsilon_0 r_1}$$

and

$$H_2 = -\frac{\hbar^2}{2m}\nabla_2^2 - \frac{2e^2}{4\pi\varepsilon_0 r_2}$$

depend on the coordinates of one of the particles only, and

$$H_{12} = \frac{e^2}{4\pi\varepsilon_0|\mathbf{r}_1 - \mathbf{r}_2|}$$

depends on the coordinates of both particles. Because ϕ_0 and ϕ_{nl} are orthogonal it follows that

$$\langle H_1 \rangle_{\text{Ex}} = \langle H_2 \rangle_{\text{Ex}} = 0 \ .$$

Thus $\langle H \rangle_{\text{Ex}}$ becomes

$$\begin{aligned}\langle H \rangle_{\text{Ex}} = \frac{1}{2}\Bigg[&\iint \phi_0^*(\mathbf{r}_1)\phi_{nl}^*(\mathbf{r}_2)\frac{e^2}{4\pi\varepsilon_0|\mathbf{r}_1 - \mathbf{r}_2|}\phi_0(\mathbf{r}_2)\phi_{nl}(\mathbf{r}_1)\,\mathrm{d}^3\mathbf{r}_1\,\mathrm{d}^3\mathbf{r}_2 \\ + &\iint \phi_0^*(\mathbf{r}_2)\phi_{nl}^*(\mathbf{r}_1)\frac{e^2}{4\pi\varepsilon_0|\mathbf{r}_1 - \mathbf{r}_2|}\phi_0(\mathbf{r}_1)\phi_{nl}(\mathbf{r}_2)\,\mathrm{d}^3\mathbf{r}_1\,\mathrm{d}^3\mathbf{r}_2\Bigg] \ .\end{aligned} \tag{9.65}$$

The two terms in Eq. (9.65) are equal to each other, and we may write

$$\langle H \rangle_{\text{Ex}} = \iint \phi_0^*(\mathbf{r}_2)\phi_{nl}^*(\mathbf{r}_1)\frac{e^2}{4\pi\varepsilon_0|\mathbf{r}_1 - \mathbf{r}_2|}\phi_0(\mathbf{r}_1)\phi_{nl}(\mathbf{r}_2)\,\mathrm{d}^3\mathbf{r}_1\,\mathrm{d}^3\mathbf{r}_2 \ . \tag{9.66}$$

We refer to Eq. (9.66) as the 'exchange energy of the coulomb interaction between the electrons'. The integral in Eq. (9.66) is dominated by the behaviour when $\mathbf{r}_1 \approx \mathbf{r}_2 \approx \mathbf{r}$ when the integrand becomes approximately

$$|\phi_0(\mathbf{r})|^2|\phi_{nl}(\mathbf{r})|^2\frac{e^2}{4\pi\varepsilon_0 r_{12}} \tag{9.67}$$

where r_{12} is the separation between the electrons. This quantity is always positive, and hence the exchange energy is always positive and space antisymmetric states will always be lower in energy than the corresponding space symmetric states. The approximation Eq. (9.67) also suggests that the value of the integral is

$$\langle H \rangle_{\text{Ex}} = \frac{e^2}{4\pi\varepsilon_0 \langle r_{12} \rangle} \tag{9.68}$$

where $\langle r_{12} \rangle$ is some suitable average of the distance between the electrons. As atomic sizes are of the order of a_0, $\langle H \rangle_{\text{Ex}}$ is expected to be of the order of $e^2/4\pi\varepsilon_0 a_0$, i.e. electron volts, and evaluation of the integral confirms this. It is, therefore, much larger than the effect of spin orbit-coupling.

We would expect the approximation we have used for the cases with $l = 0$ to give answers that are substantially too large, as the range of the wavefunction for the excited electron is only about half what it ought to be. This is indeed the case. For higher l-values we should expect our approximation to be better. We illustrate in Table 9.2 the situation in the case when the excited electron has $n = 2$.

Table 9.2. Level splittings in helium when $n = 2$

l-value	Observed splitting	Calculated splitting using wavefunctions quoted in the text
0	0.79 eV	2.38 eV
1	0.27 eV	0.21 eV

To specify an excited level in helium therefore, it is not enough to specify the quantum numbers of the individual electron states, we must also specify the symmetry of the wavefunction. States with antisymmetric space wavefunctions always lie lower than the corresponding state with a symmetric space wavefunction.

9.15 THE CONNECTION BETWEEN SYMMETRY AND TOTAL SPIN

For our wavefunctions to be totally antisymmetric a symmetric space wavefunction must be combined with an antisymmetric spin wavefunction and vice versa, as can be seen by studying Eq. (9.58). Thus the classification of the states in terms of their space symmetry can equally well be made in terms of their spin symmetry. Thus we can say that for a given pair of values of n and l for the excited electron, those states that correspond to a symmetric spin wavefunction will lie lower than those that correspond to an antisymmetric spin wavefunction.

The symmetric spin wavefunctions are

$$\alpha(1)\alpha(2) \tag{9.69a}$$

$$\frac{1}{\sqrt{2}}\{\alpha(1)\beta(2) + \alpha(2)\beta(1)\} \tag{9.69b}$$

$$\beta(1)\beta(2) \tag{9.69c}$$

where $1/\sqrt{2}$ in Eq. (9.69b) has been included to preserve normalization. Similarly the antisymmetric spin wavefunction is

$$\frac{1}{\sqrt{2}}\{\alpha(1)\beta(2) - \alpha(2)\beta(1)\} \; . \tag{9.70}$$

According to the discussion of section 9.9 we can add the two electron spins together to produce total spins of $S = 1$ or $S = 0$. The states with $S = 1$ will have possible values of the z-component M_S equal to 1, 0 and -1, in units of $\hbar$, while the $S = 0$ state has $M_S = 0$ only of course. The states of Eqs. (9.69a and 9.69c) have M_S equal to $+1$ and -1 respectively and must belong to the state with total spin equal to 1. Now for a given physical situation with a definite value of total spin, the values of the z-component we expect to observe depend on the direction we choose for our z-axis. The exchange symmetry of the situation cannot be affected by the choice of the z-axis, and hence we expect all the z-components of a given spin to have the same exchange symmetry. It follows that Eq. (9.69b) must be the $M_S = 0$ component of the state with S equal to unity. By elimination Eq. (9.70) must represent the state with $S = 0$.

Because there are three states with $S = 1$, corresponding to the three possible values of M_S with spin wavefunctions given by Eq. (9.69), these states are called *triplet* states, and similarly a state with $S = 0$ is called a singlet state.

We may now modify the wording at the end of the previous section to read, *the triplet states* (*with* $S = 1$) *always lie lower than the corresponding singlet state* ($S = 0$).

The difference in energy between the two sets of states is just twice the exchange energy of the coulomb interaction between the electrons, Eq. (9.66). The separation between singlet and triplet states when one electron is in the 1s state, and the other in the 2s state, is 0.79 eV.

9.16 THE DESCRIPTION OF THE EXCITED STATES OF HELIUM

We have been considering those states where only one of the electrons is excited, and so the orbital angular momentum of the state of the atom as a whole will be the same as the orbital angular momentum of the excited electron, which has a definite value. Thus the states we have been considering have definite values for their total spin and their total orbital angular momentum. If we now consider states in which both electrons are excited we see that the arguments of sections 9.14 and 9.15 are still valid with ϕ_0 and ϕ_{nl} replaced by $\phi_{n_1l_1}$ and $\phi_{n_2l_2}$ and these states will also have definite values for the total spin. It turns out that they also have definite values for their total orbital angular momentum. It is conventional to denote the total orbital angular momentum by the symbol L. States formed from a specific pair of single electron states n_1l_1, n_2l_2 but differing in their values of L are *not* degenerate. For a specific pair of values l_1 and l_2 we find, by an extension of the arguments of section 9.9, that

$$|l_1 - l_2| \leqslant L \leqslant l_1 + l_2 \ . \tag{9.71}$$

Just as the specific values 0, 1, 2, 3, etc. for the quantum number l are denoted by the symbols s, p, d, f etc. so the same values for the quantum number L are denoted by the symbols S, P, D, F etc. The total spin is denoted by putting the values of $(2S + 1)$ as a left hand superscript to L. The value of $(2S + 1)$ is called the *multiplicity*. The way in which these conventions are used is best shown by a few examples. The ground state of helium is denoted by

$$(1\mathrm{s})^2\ {}^1\mathrm{S}$$

which means there are two electrons in the 1s state, their total orbital angular momentum is zero and the multiplicity is one, i.e. their total spin is zero. The next two states of helium are denoted by

$$(1\mathrm{s})(2\mathrm{s})\ {}^3\mathrm{S} \quad \text{and} \quad (1\mathrm{s})(2\mathrm{s})\ {}^1\mathrm{S}$$

which means there is one electron in the 1s state, one in the 2s state, the total orbital angular momentum is zero, and in the first case, with multiplicity 3, the total spin is equal to one, and in the second case, with multiplicity 1, the total spin is equal to zero. The next two states are denoted by

$$(1\mathrm{s})(2\mathrm{p})\ {}^3\mathrm{P} \quad \text{and} \quad (1\mathrm{s})(2\mathrm{p})\ {}^1\mathrm{P}$$

and it is hoped the meanings of their descriptions will be clear.

The sequence of states in helium with one electron excited, is shown in Fig. 9.13, with triplet and singlet states shown separated.

It should, perhaps, be mentioned that the separation of the total wavefunction into a spin and a space portion, each with a definite exchange symmetry is only always possible for two identical particles. When we have more than two particles this is not possible except in certain special cases. One of these is the case of a completely symmetric spin state, e.g. of the form $\alpha(1)\alpha(2)\alpha(3)\ldots\alpha(n)$ which has the maximum possible value of $M_S = n\hbar/2$ and hence the maximum possible total spin. The space wavefunction is then completely antisymmetric and the wavefunction vanishes whenever any pair

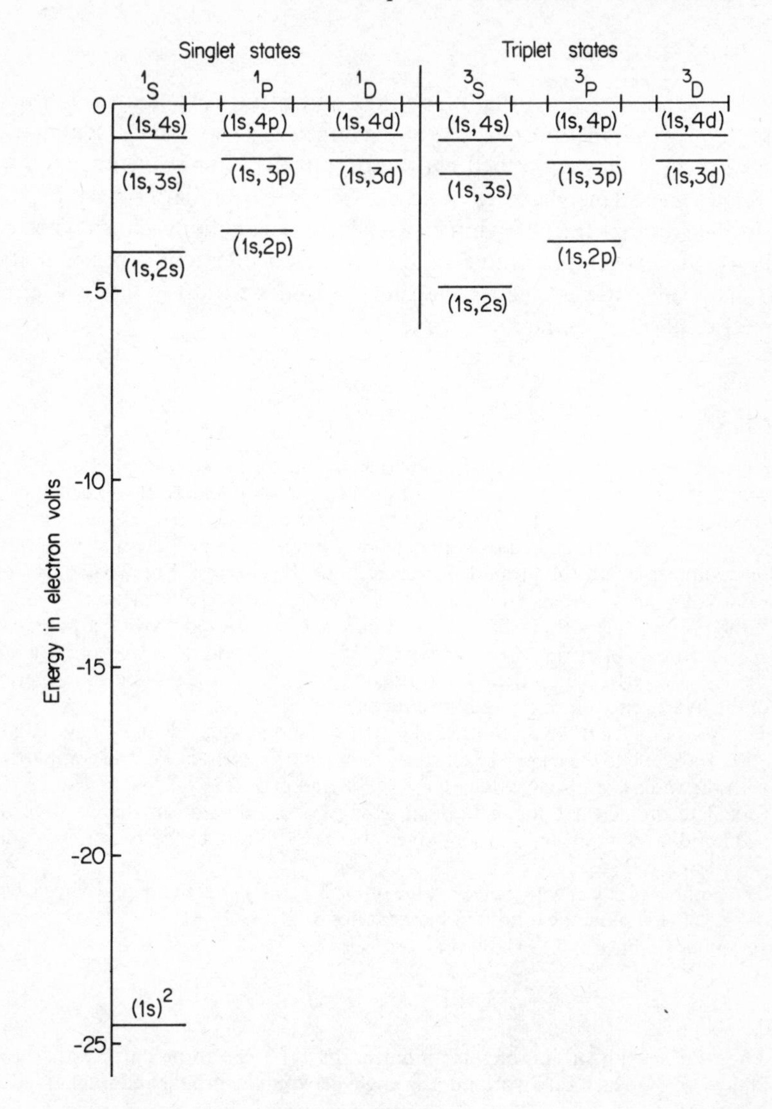

Fig. 9.13. The energy levels of the helium atom due to the excitation of a single electron. The singlet and triplet states are shown separately. Note that each triplet state is slightly lower in energy than the corresponding singlet state.

of particles come together. It thus has the lowest expectation value of the Coulomb interaction, and lies lowest in energy. We thus arrive at *Hund's rule* 'For any configuration the state with the highest spin always lies lowest in energy'.

9.17 SUMMARY

We may summarize the situation on the excited states of helium as follows. As the total wavefunction must be antisymmetric we may have symmetric space wave functions combined with antisymmetric spin wavefunctions or vice versa. The former have total spin $S = 0$ while the latter have $S = 1$. Due to the fact that the antisymmetric space wavefunctions vanish when $\mathbf{r}_1$ equals $\mathbf{r}_2$ the expectation value of the Coulomb interaction between the electrons in this case is smaller than for the states with $S = 0$ which have symmetric space wavefunctions.

PROBLEMS 9

9.1 An atom has three electrons, one in each of the states $\phi_1 = xf(r)$, $\phi_2 = yf(r)$, $\phi_3 = zf(r)$. Show that the electron charge distribution is spherically symmetric.

9.2 An electron moves in the potential $V = e/(4\pi\varepsilon_0 r)$ for $0 \leqslant r \leqslant r_0$, and $V = 0$ for $r > r_0$. By treating the deviation from a pure Coulomb potential as a perturbation, estimate the ground state energy of the electron. For what range of values of r_0 (in units of a_0) do you think you would get a reliable answer?

9.3 If the proton were a thin spherical shell of radius r_0 it would provide a potential energy for an electron $V = -e^2/(4\pi\varepsilon_0 r_0)$ for $r < r_0$ and $V = -e^2/(4\pi\varepsilon_0 r)$ for $r > r_0$. Use perturbation theory to estimate the change in the ground state energy of the hydrogen atom this would produce.

9.4 The transition from the 3p to the 3s level in sodium produces light of wavelength 5890 Å. Calculate the energy spacing between the two levels, in eV, and compare it with the value from Eq. (9.20) with $\delta_0 = 1.38$ and $\delta_1 = 0.88$.

9.5 Calculate the effective magnetic field acting on the electron in the 3p state of sodium due to its orbital motion, given that the splitting between the $3p_{1/2}$ and $3p_{3/2}$ levels is 2.15×10^{-3} eV.

9.6 The number of states with a given value of j is $2j+1$. If j goes in integral steps from $|l-s|$ to $(l+s)$ show the total number of states is $(2l+1)(2s+1)$.

9.7 In analogy with Eq. 9.33 show that $\hat{\mathbf{j}} \cdot \hat{\mathbf{s}} \psi_{jls}(\mathbf{r})$

$$= \tfrac{1}{2}[j(j+1) + s(s+1) - l(l+1)]\hbar^2 \psi_{jls}(\mathbf{r}) \; .$$

9.8 An electron with orbital angular momentum $l = 2$ can form states with $j = \frac{3}{2}$ and $j = \frac{5}{2}$. For each case work out the angle between the orbital and total angular momenta.

9.9 The second ionization potential of lithium is 75.3 volts. Calculate the total binding energy of the electrons in the Li^+ ion.

9.10 The general form of Eq. (9.56) for the electrostatic interaction between two electrons in the 1s state of an atom with nuclear charge Ze is

$$\langle U \rangle = \frac{5Z}{8a_0} \frac{e^2}{4\pi\varepsilon_0} \; .$$

Use this result to calculate the ground state energy of the Li^+ ion and compare it with the result of Problem 9.9.

9.11 A helium atom has both its electrons in the 2p shell. Using an appropriate notation to denote the individual magnetic substates, write down the possible symmetric space wavefunctions and the possible antisymmetric space wavefunctions. List in each case the M_L values that occur and hence deduce the values of the total orbital angular momentum L. Associate each group with the appropriate value of S.

9.12 In Section 9.12 we described the electrons in the helium ground state by 1s wavefunctions. Clearly these will fall off too quickly at large r. We can improve the situation by taking wavefunctions of the form $A_1$1s + $A_2$2s + $A_3$3s etc. and adjusting the constants A_1, A_2, A_3 etc. However we cannot get the right answer this way. Suggest a reason why this is so.

CHAPTER 10

Spectral lines and selection rules

10.1 INTRODUCTION

The experimental information on the energy levels of hydrogen and other atoms, which we have compared with the predictions of quantum mechanics in Chapters 7 and 9, has been largely acquired not through direct observation using experiments of the type described in section 3.9, but through the observation of the spectrum of light emitted by atomic vapours. The reason for this is largely that the information obtained this way is far more precise. For example it is a fairly straightforward matter using a spectroscope to observe that the yellow light emitted by sodium vapour, when excited by an electrical discharge, consists of two spectral lines at 5890 Å and 5896 Å and hence deduce that the first two excited states of sodium are situated 2.11 eV above the ground state and separated by an energy of 2.15×10^{-3} eV. To make the same observation directly would require a beam of electrons and an analysing system each with an energy resolution better than 2×10^{-3} eV. This is very difficult.

The most accurate observations are made when we observe the light emitted when the vapour pressure is low. All our quantum mechanical treatments so far have been for isolated systems, and the individual atoms will be closest to this condition when the vapour pressure is low.

When an atom makes a transition from one energy level to another, with the emission of light, we learn more than the mere fact that there are two levels in the atom separated by the energy of the transition. This is because certain *selection rules* exist which place restrictions on the type of level to which a given level may decay. We presented the evidence for the existence of such selection rules in section 3.9 when we noted that in the controlled excitation of an atom not all the possible spectral lines given by

$$h\nu_{rs} = E_r - E_s$$

are actually observed. It is clearly important for us to know what these selection rules are and how they arise.

10.2 THE SELECTION RULES FOR HYDROGEN AND THE ALKALI METALS

The energy levels of the alkali metals and hydrogen can be specified by the quantum numbers n, l and j. When a transition takes place these quantum numbers change from an initial set n_i, l_i, j_i to a final set n_f, l_f, j_f. We denote the change in the quantum numbers l and j (the selection rules do not depend on n), by

$$\Delta_j = j_f - j_i$$
$$\Delta l = l_f - l_i \ .$$

The observed selection rules can be expressed in terms of these changes as follows

$$\left.\begin{array}{l} \Delta j = 0, \pm 1 \\ j_i = 0 \rightarrow j_f = 0 \text{ forbidden} \end{array}\right\} \quad (10.1)$$

and

$$\Delta l = \pm 1 \ . \quad (10.2)$$

We now have to provide a plausible reason for these rules.

10.3 THE PROPERTIES OF ATOMIC STATES AND THE ANALYSIS OF ELECTROMAGNETIC RADIATION

When an atom makes a transition from a state n_i, l_i, j_i with energy E_i to another state n_f, l_f, j_f with energy E_f we have the replacement of the initial state by the final state plus a radiation field carrying an energy $h\nu$ where

$$h\nu = E_i - E_f. \quad (10.3)$$

As this transition is spontaneous, and not mediated by any outside influence, all the conservation laws must apply. Eq. (10.3) is just an expression of the conservation of energy; and the other conservation law we are particularly concerned with is the conservation of angular momentum. As transitions can and do take place between states of different angular momentum it is clear that the radiation field must carry away the difference between the angular momenta of the two states. We are therefore led to look for an analysis of electromagnetic radiation which brings out the angular momentum properties explicitly. The performance of such an analysis is very complicated, but the results are simple. We find we can analyse an arbitrary radiation field into a series of terms which carry respectively, $1\hbar$, $2\hbar$, $3\hbar$ etc. units of angular momentum per photon. It should be noted that there is no term which carries away *no* angular momentum, this is a consequence of the transversality of the electromagnetic field. In saying that the field carries k units of angular momentum, say, we mean exactly the same as if we were referring to the angular momentum of a particle, namely that the square magnitude of the angular momentum carried by the field is given by $k(k + 1)\hbar^2$, and the component along the z-direction is quantized in integral steps between $-k\hbar$ and $+k\hbar$.

We find that the field which carries away $1\hbar$ of angular momentum radiates energy much more rapidly than the others. We can add plausibility to this statement by considering how a photon carries away angular momentum. Let us first consider a related problem. Suppose we have an atom in a highly excited state of angular momentum J_i, and it decays by emitting an electron leaving the ion in a state of angular momentum J_f (such a process can take place, it is known as autoionization). For the purposes of this discussion we will ignore the attractive force between the ion and the electron, so we can assume that after emission the free electron travels in a straight line with a constant velocity $\mathbf{v}$ determined by the energy of the transition. As the spin of the electron is $\frac{1}{2}\hbar$, if the emitted electron carries away no orbital angular momentum we must have $|J_i - J_f| = \frac{1}{2}\hbar$. If $|J_i - J_f|$ is greater than $\frac{1}{2}\hbar$ the electron must carry away some orbital angular momentum. If the electron were a classical particle this would imply that the path of the electron does not pass through the centre of the atom, as illustrated in Fig. 10.1 but is

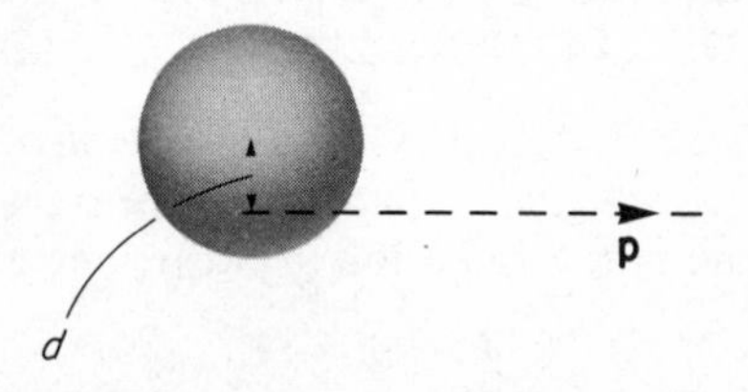

Fig. 10.1. If an electron of momentum $\mathbf{p}$ is emitted by an atom and carries away some orbital angular momentum, classically this implies its final path is offset from the centre of the atom by an amount d where $p \times d$ is the angular momentum carried away.

offset by a distance d so that the angular momentum carried away is given by $p \times d$ where p is the momentum of the electron. Needless to say because of the uncertainty principle we cannot say that an electron is emitted at a distance d from the centre of the atom. Nevertheless if the electron is required to carry l units of angular momentum and we find $l\hbar/p$ is much greater than the radius of the atom, not surprisingly we find the transition is decidedly improbable.

Now let us apply the same ideas to the emission of a photon. In view of the fact that a photon can take away $1\hbar$ of angular momentum or more, but not zero angular momentum, it appears plausible that a photon has an intrinsic angular momentum of $1\hbar$. This is in fact the case. Hence if we have a transition between the two levels J_i and J_f and

$$|J_i - J_f| \leqslant 1\hbar \tag{10.4}$$

the photon can be emitted from the middle of the atom. Suppose $|J_i - J_f|$ is equal to $2\hbar$ and the photon is required to carry away $1\hbar$ of orbital angular momentum. Classically, to do this the photon would have to be emitted at a distance d from the centre of the atom such that

$$pd = \hbar \tag{10.5}$$

where p is the momentum of the photon.

The momentum of a photon is given by

$$p = \hbar/\bar{\lambda} \tag{10.6}$$

where $\bar{\lambda}$ equals $\lambda/2\pi$. Substituting Eq. (10.6) in Eq. (10.5) we get

$$\frac{\hbar}{\bar{\lambda}}d = \hbar \ .$$

Hence

$$d = \bar{\lambda} \ . \tag{10.7}$$

For light in the middle of the visible spectrum $\bar{\lambda}$ is about 10^{-5} cm, that is three orders of magnitude greater than a typical atomic radius. One finds, in fact, that the probability of an atom emitting a photon with $1\hbar$ of orbital angular momentum is smaller than for emitting a photon with no orbital angular momentum by a factor of approximately $(r/\bar{\lambda})^2$ where r is the radius of the atom.

The importance of this analysis lies not merely in the fact that we have found a suitable set of fields for our problem, but that *one of the fields radiates energy much more rapidly than any of the others, and hence is the only one that matters.*

This field is known as the electric dipole field because it has exactly the same radiation pattern as an electric dipole (see Fig. 10.2). It carries $1\hbar$ of

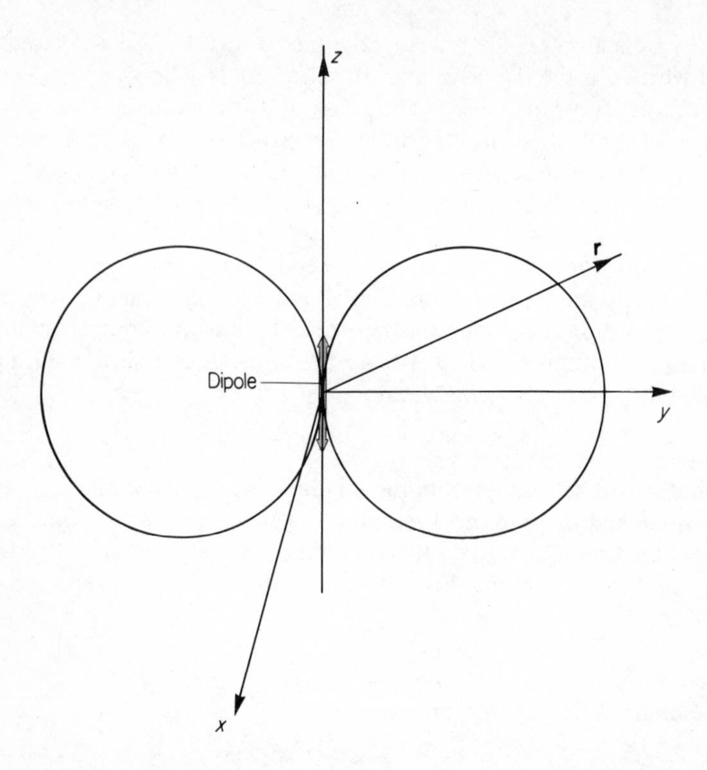

Fig. 10.2. The polar diagram for the radiation emitted by an electrical dipole oscillating along the z-direction. The polar diagram is symmetrical about the z-axis and the section in any plane passing through the z-axis consists of two circles which touch at the origin. At a large distance r the electric and magnetic fields produced by the oscillating dipole are proportional to the intercept on the radius vector made by the polar diagram. There is no radiation in the z-direction.

angular momentum. Thus we can say that *an excited atom will decay by emitting electric dipole radiation unless there is no lower state which can be reached by such an emission process.*

If electric dipole radiation is not possible, then under normal laboratory conditions no radiation is usually observed. Calculated life times for electric dipole radiation are of the order of 10^{-8} s and hence if electric dipole radiation is not possible we would expect lifetimes of the order of 10^{-2} s or greater. In this time many collisions will take place in an atomic vapour and the excited atom will lose its energy during one of its collisions.

Thus we can say that the observed selection rules will be the selection rules for the emission of electric dipole radiation.

10.4 THE SELECTION RULES FOR ELECTRIC DIPOLE RADIATION (1)

We have stated that electric dipole radiation carries one unit of angular momentum in the same way as a particle can have one unit of angular momentum. In the radiation process the initial state with angular momentum j_i in units of $\hbar$, breaks up into a final state with angular momentum j_f and an electric dipole radiation field with unit angular momentum. If we are to conserve angular momentum in the process we must have the vector sum of the final state angular momentum plus the angular momentum of the radiation field equal to the initial state angular momentum, all interpreted quantum mechanically. Hence in order to arrive at the selection rules we have to consider the addition of unit angular momentum to an angular momentum j_f. This is clearly an extension of the case of addition of angular momentum we considered in section 9.9. By exactly similar arguments to the ones we used following Eq. (9.25) we can show that if we have two independent angular momenta j_1 and j_2 they can combine together to produce a resultant angular momentum j whose value lies between $|j_1 - j_2|$ and $j_1 + j_2$ i.e.

$$|j_1 - j_2| \leqslant j \leqslant j_1 + j_2 \tag{10.8}$$

and where j is integral or half integral according to whether $j_1 + j_2$ is integral or half integral.

If we take j_1 as equal to j_f, j_2 as equal to one, and j as equal to j_i we get

$$|j_f - 1| \leqslant j_i \leqslant j_f + 1 \tag{10.9}$$

or

$$\left.\begin{aligned} \Delta j = j_f - j_i = 0, \pm 1 \\ j_i = 0 \rightarrow j_f = 0 \text{ forbidden } . \end{aligned}\right\} \tag{10.1}$$

The last result follows from Eq. (10.9) by putting $j_i = 0$ when we must have $j_f = 1$. We represent these rules vectorially in Fig. 10.3.

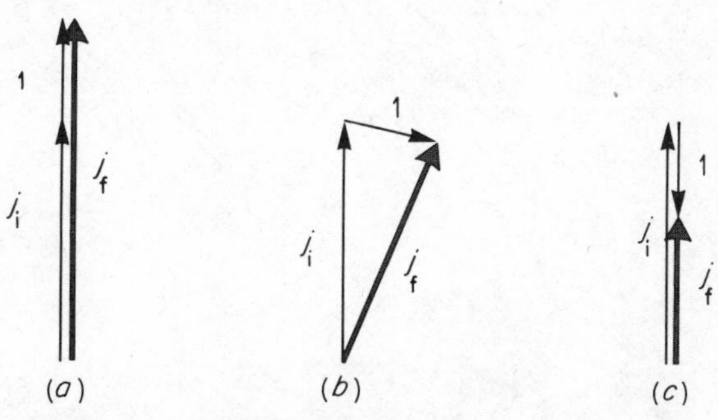

Fig. 10.3. The vectorial representation of the angular momentum selection rules for dipole radiation: (a) $j_f - j_i = 1$, (b) $j_f - j_i = 0$, (c) $j_f - j_i = -1$.

We have thus accounted for the first of our selection rules; we now have to account for the second. The radiation is due to the *motion* of charge and has nothing to do with the spin of the electron. Thus we could have carried through the argument for a spinless particle when j would have been equal to l. This would lead to

$$\Delta l = 0, \pm 1$$

$$l = 0 \rightarrow l = 0 \text{ forbidden,}$$

instead of $\Delta l = \pm 1$. Clearly there is some extra restriction.

10.5 THE ELECTRIC DIPOLE TRANSITION PROBABILITY AND THE CONCEPT OF PARITY

To understand why the electric dipole selection rule is more restrictive for orbital angular momentum than for total angular momentum in atoms with one valence electron we have to consider more details of what happens. We start by writing down the expression for the transition probability per unit time between an initial state ψ_i and a final state ψ_f for the emission of electric dipole radiation. It can be shown that this is given by

$$T = B(v)\left|\int \psi_f^*(\mathbf{r})(-e\mathbf{r})\psi_i(\mathbf{r})\, d^3\mathbf{r}\right|^2 \tag{10.10}$$

where the constant $B(v)$ depends on the energy of the transition. We note that if ψ_f and ψ_i are the same state then the integral in Eq. (10.10) is the expectation value of the static dipole moment for the state ψ_i.

Suppose now we have the system in the state ψ_i and then look at it some time later. Due to the fact it can radiate the system could be found either in ψ_i or in ψ_f. This suggests that the wavefunction at the moment of inspection should be a superposition of these two wavefunctions. As the two wavefunctions refer to states of different energies we must include their time factors when superposing them. Suppose we consider a total wavefunction of the form

$$\psi = a\psi_i\, e^{-iE_it/\hbar} + b\psi_f\, e^{-iE_ft/\hbar} .$$

If we now calculate the dipole moment for this wavefunction we get

$$\int \psi^*(-e\mathbf{r})\psi\, d^3\mathbf{r} = a^*a\int \psi_i^*(-e\mathbf{r})\psi_i\, d^3\mathbf{r} + b^*b\int \psi_f^*(-e\mathbf{r})\psi_f\, d^3\mathbf{r}$$
$$+\, a^*b\, e^{i(E_i-E_f)t/\hbar}\int \psi_i^*(-e\mathbf{r})\psi_f\, d^3\mathbf{r}$$
$$+\, ab^*\, e^{-i(E_i-E_f)t/\hbar}\int \psi_f^*(-e\mathbf{r})\psi_i\, d^3\mathbf{r} .$$

If we write the third term as

$$R\left|\int \psi_i^*(-e\mathbf{r})\psi_f \, d^3\mathbf{r}\right| e^{i\delta} e^{i\omega t}$$

where $R = |ab|$

$$\delta = \arg(b) - \arg(a) + \arg\left(\int \psi_i^*(-e\mathbf{r})\psi_f \, d^3\mathbf{r}\right)$$

$$\omega = (E_i - E_f)/\hbar$$

the fourth term becomes

$$R\left|\int \psi_i^*(-e\mathbf{r})\psi_f \, d^3\mathbf{r}\right| e^{-i\delta} e^{-i\omega t}$$

and the sum of the last two terms is

$$2R\left|\int \psi_i^*(-e\mathbf{r})\psi_f \, d^3\mathbf{r}\right| \cos(\omega t + \delta) = 2R\left|\int \psi_f^*(-e\mathbf{r})\psi_i \, d^3\mathbf{r}\right| \cos(\omega t + \delta) .$$

Thus the expectation value of the dipole moment contains two static terms, which we shall later show to be zero, and a term oscillating at a frequency ω such that $\hbar\omega = E_i - E_f$, with an amplitude proportional to the integral occurring in Eq. (10.10). Now, classically, an oscillating dipole emits radiation at the same frequency, with an intensity proportional to the square of the amplitude. Hence we have provided some small justification for Eq. (10.10) as the electric dipole transition probability.

Taking Eq. (10.10) as correct, let us study the transition probability between two s-states, two p-states, and an s and a p-state. In each case we wish to discover whether the integral

$$\int \psi_f^*(\mathbf{r})(-e\mathbf{r})\psi_i(\mathbf{r}) \, d^3\mathbf{r} \tag{10.11}$$

is finite or zero.

The wavefunctions for s-states are spherically symmetric. If ψ_i and ψ_f are both s-states this means that

$$\left.\begin{aligned} \psi_i(-\mathbf{r}) &= \psi_i(\mathbf{r}) \\ \psi_f^*(-\mathbf{r}) &= \psi_f^*(\mathbf{r}) . \end{aligned}\right\} \tag{10.12}$$

The middle term in Eq. (10.11), $(-e\mathbf{r})$, changes sign when we go from the point $\mathbf{r}$ to the point $\mathbf{r}' = -\mathbf{r}$. This taken together with Eq. (10.12) means the whole integrand of Eq. (10.11) changes sign when we go from the point $\mathbf{r}$ to the point $-\mathbf{r}$, and contributions to the integral from these two points exactly cancel, so the integral vanishes if ψ_i and ψ_f are both s-states.

In section 7.6 we showed we could write p-states in the form

$$\left.\begin{aligned}\Psi_0(\mathbf{r}) &= z\psi(r)\\ \Psi_x(\mathbf{r}) &= x\psi(r)\\ \Psi_y(\mathbf{r}) &= y\psi(r)\end{aligned}\right\} . \tag{7.33}$$

where $\psi(r)$ is spherically symmetric and $\mathbf{r}$ is the point (x, y, z). As at the point $\mathbf{r}' = -\mathbf{r}$, i.e. $z' = -z$ etc., we have

$$\left.\begin{aligned}\Psi_0(-\mathbf{r}) &= -z\psi(r) = -\Psi_0(\mathbf{r})\\ \Psi_x(-\mathbf{r}) &= -x\psi(r) = -\Psi_x(\mathbf{r})\\ \Psi_y(-\mathbf{r}) &= -y\psi(r) = -\Psi_y(\mathbf{r}) .\end{aligned}\right\} \tag{10.13}$$

If both ψ_i and ψ_f behave like Eq. (10.13) then the integrand in Eq. (10.11) at $\mathbf{r}' = -\mathbf{r}$ is equal, but of opposite sign, to the integrand at $\mathbf{r}$; the integral will therefore vanish. Transitions between two p-states are not possible.

If however only one of the wavefunctions changes sign when we go from $\mathbf{r}$ to $-\mathbf{r}$, the other one remaining the same, as occurs if one is an s-wavefunction and the other a p-wavefunction, then the integrand at $-\mathbf{r}$ is equal to the integrand at $\mathbf{r}$, and the integral does not necessarily vanish.

We have thus seen two examples of the rule that $\Delta l = 0$ transitions are not allowed, and in both cases the reason is the same; it is due to the behaviour of the wavefunctions involved when we go from the point $\mathbf{r}$ to the point $\mathbf{r}' = -\mathbf{r}$. The behaviour of a function $f(\mathbf{r})$ when we go from the point $\mathbf{r}$ to the point $\mathbf{r}' = -\mathbf{r}$ is referred to as its *parity*.

10.6 PARITY

Let $f(\mathbf{r})$ be an arbitrary function of the position vector $\mathbf{r}$. If

$$f(-\mathbf{r}) = f(\mathbf{r}) \tag{10.14a}$$

$f(\mathbf{r})$ is said to possess *even* or *positive* parity. If

$$f(-\mathbf{r}) = -f(\mathbf{r}) \tag{10.14b}$$

$f(\mathbf{r})$ is said to possess *odd* or *negative* parity. If neither of the results of Eq. (10.14) occurs $f(\mathbf{r})$ is said to possess mixed parity. The reason for this last description is that quite generally we can write

$$\begin{aligned}f(\mathbf{r}) &= \tfrac{1}{2}\{(f(\mathbf{r}) + f(-\mathbf{r})) + (f(\mathbf{r}) - f(-\mathbf{r}))\}\\ &= \phi_+(\mathbf{r}) + \phi_-(\mathbf{r})\end{aligned}$$

where $\phi_+(\mathbf{r})$ has even parity and $\phi_-(\mathbf{r})$ has odd parity.

We now show that a non-degenerate state has a definite parity. Consider first the one-dimensional Schrödinger equation

$$-\frac{\hbar^2}{2m}\frac{d^2f(x)}{dx^2} + V(x)f(x) = Ef(x) \tag{10.15}$$

where

$$V(-x) = V(x) \ . \tag{10.16}$$

Let us re-label our axis so that everywhere x becomes $-x$. Eq. (10.15) becomes

$$-\frac{\hbar^2}{2m}\frac{d^2f(-x)}{d(-x)^2} + V(-x)f(-x) = Ef(-x) \ . \tag{10.17}$$

Now

$$\begin{aligned}\frac{df(-x)}{d(-x)} &= \frac{df(-x)}{dx}\frac{dx}{d(-x)} \\ &= -\frac{df(-x)}{dx} \ .\end{aligned}$$

Hence

$$\begin{aligned}\frac{d^2f(-x)}{d(-x)^2} &= -\frac{d}{d(-x)}\frac{df(-x)}{dx} \\ &= -\frac{d^2f(-x)}{dx^2}\frac{dx}{d(-x)} \\ &= \frac{d^2f(-x)}{dx^2} \ .\end{aligned}$$

Substituting this and Eq. (10.16) in Eq. (10.17) we get

$$-\frac{\hbar^2}{2m}\frac{d^2f(-x)}{dx^2} + V(x)f(-x) = Ef(-x)$$

showing that $f(-x)$ is also a solution belonging to the same energy eigenvalue E. If the level is non-degenerate, $f(-x)$ and $f(x)$ can at most differ by a constant factor

$$f(-x) = \lambda f(x) \ . \tag{10.18}$$

Consider any pair of points $x = x_0$ and $x = -x_0$. For $x = x_0$, Eq. (10.18) gives

$$f(-x_0) = \lambda f(x_0) \ ,$$

and for $x = -x_0$ it gives

$$f(x_0) = \lambda f(-x_0) \ .$$

Combining these results we have

$$f(x_0) = \lambda^2 f(x_0)$$

for any x_0. Hence $\lambda^2 = 1$ and $\lambda = \pm 1$, that is

$$f(-x) = \pm f(x)$$

and $f(x)$ has definite parity.

This argument can readily be extended to three dimensions to show that, if $V(-\mathbf{r}) = V(\mathbf{r})$ then $f(-\mathbf{r})$ is a solution of the Schrödinger equation belonging to the energy eigenvalue E if $f(\mathbf{r})$ is a solution beloning to this eigenvalue. The reader is reminded that in polar coordinates r is always positive, and if $\mathbf{r}$ is the point (r, θ, ϕ) then $-\mathbf{r}$ is the point $(r, \pi - \theta, \pi + \phi)$. An identical argument to the one above now shows that $f(\mathbf{r})$ has definite parity if the level is non-degenerate.

It should be emphasized that this result is a direct consequence of the symmetry of the potential.

10.7 EXAMPLES OF WAVEFUNCTIONS WITH AND WITHOUT A DEFINITE PARITY

The simplest wavefunction we can take is that of a plane wave travelling in the x-direction

$$\psi = e^{ikx} \ .$$

This wavefunction does not have a definite parity. But

$$\psi = e^{-ikx}$$

is also a solution of the Schrödinger equation in free space with the same energy and we can form from these two solutions a pair of solutions with definite parities

$$\begin{aligned} \phi_+ &= \tfrac{1}{2}(e^{ikx} + e^{-ikx}) \\ &= \cos kx \end{aligned}$$

and

$$\begin{aligned} \phi_- &= \tfrac{1}{2}(e^{ikx} - e^{-ikx}) \\ &= i \sin kx \ . \end{aligned}$$

As a second example consider the s- and p-wave states of the hydrogen atom. We have already noted that s-wavefunctions have positive parity as they are spherically symmetric and p-wave functions have negative parity as they change sign on going from **r** to $-\mathbf{r}$. In hydrogen, neglecting spin orbit effects, the 2s and 2p wavefunctions are degenerate (see Eq. (7.41)), and hence a linear combination of them would also be a solution of the Schrödinger equation belonging to the eigenvalue

$$E = \frac{1}{4}\left(\frac{e^2}{4\pi\varepsilon_0 2a_0}\right) .$$

Such a solution would not have definite parity.

One of the most important set of functions with a definite parity are the angular momentum eigenfunctions (the spherical harmonics $Y_{lm}(\theta, \phi)$) examples of which are quoted in section 7.7. These can be shown to have odd (negative) parity if l is odd, and even (positive) parity if l is even. Denoting the parity by Π, we write this as

$$\Pi = (-1)^l . \tag{10.19}$$

10.8 THE SELECTION RULES FOR ELECTRIC DIPOLE RADIATION (2)

With the results of the last section we are now in a position to explain the electric dipole selection rules for one-electron atoms and for the levels of helium that we have discussed.

To summarize the one-electron case we have

(1) $\Delta j = 0, \pm 1$; $0 \to 0$ forbidden. This is because electric dipole radiation carries away one unit of angular momentum.

(2) The electric dipole operator $-e\mathbf{r}$ does not operate on the spin of the electron, and hence m_s must be unchanged and the argument in (1) could be carried through for spinless particles. This leads us to $\Delta l = 0$, ± 1; $0 \to 0$ forbidden.

(3) Because the operator $-e\mathbf{r}$ changes sign under the parity operation (replacing **r** by $-\mathbf{r}$) the initial and final states must have opposite parities. As the parity of a state with a definite value of l is $(-1)^l$ we have Δl must be odd. Coupled with (2) this gives us $\Delta l = \pm 1$.

The above selection rules may be applied with very little change to the levels of helium we have described in Chapter 9, in which only one electron is excited, the other being in the ground state. The Δl rule applies to the excited electron and since $l_1 = 0$ we have the total orbital angular momentum $L = l_2$ and hence

$$\Delta L = \pm 1 . \tag{10.20}$$

Further, in this case the wavefunctions can be factored into a space and a spin portion. As the dipole operator does not act on the spin, and states with

$S = 1$ and $S = 0$ are orthogonal to each other, we get the additional rule

$$\Delta S = 0 \ . \tag{10.21}$$

The result of this is that in helium we only get transitions within the singlet system and within the triplet system, but not between the two systems.

In chapter 9 we did not discuss the total angular momentum J of the levels. The orbital angular momentum L and total spin S can be combined to form a total angular momentum J such that

$$|L - S| \leqslant J \leqslant L + S \tag{10.22}$$

in anology with Eq. (9.27) and the subsequent arguments.

Once again, because electric dipole radiation carries away one unit of angular momentum, we have the selection rule

$$\left.\begin{array}{l} \Delta J = 0, \pm 1 \\ 0 \to 0 \text{ forbidden} \ . \end{array}\right\} \tag{10.23}$$

In Figs. 10.4 and 10.5 we show the main transitions in sodium and helium consequent upon these selection rules. The notation in Fig. 10.5 is an extension of the notation explained at the end of section 9.9. Thus $^1D_2$ stands for a singlet D-level (i.e. $S = 0$, $L = 2$) with resultant angular momentum $J = 2$, and $^3P_{0,1,2}$ stands for a triplet P-state (i.e. $S = 1$, $L = 1$) which allows three values for the resultant angular momentum: $J = 0, 1$ or 2, as indicated by the right hand suffixes.

PROBLEMS 10

10.1 A sequence of levels in an alkali atom specified by the quantum numbers n and l, where n is variable but l fixed, all decay to a level with quantum numbers n_0 and l_0. Under high resolution the emitted spectral lines are found to be doublets, all with the same spacing in frequency. What are the values of l and l_0?

10.2 What would have been the values of l and l_0 in question 1 if the doublet spacing in frequency had decreased with increasing n?

10.3 The 2s and 2p wavefunctions of hydrogen are degenerate and a linear combination of them is therefore a solution of the Schrödinger equation of mixed parity. Given their normalized forms are

$$\psi(2s) = \frac{1}{\sqrt{(8\pi)}}\left(\frac{1}{a_0^3}\right)^{1/2}\left(1 - \frac{r}{2a_0}\right) e^{-r/2a_0}$$

$$\psi(2p) = \frac{1}{\sqrt{(8\pi)}}\left(\frac{1}{a_0^3}\right)^{1/2} \frac{r}{2a_0} e^{-r/2a_0} \cos\theta$$

show that for the function $a\psi(2s) + b\psi(2p)$, where a and b are real, the probability of the electron being above the x, y plane is $\frac{1}{2} - \frac{3}{4}ab$, and that of being below the x, y plane is $\frac{1}{2} + \frac{3}{4}ab$. ($a^2 + b^2 = 1$.)

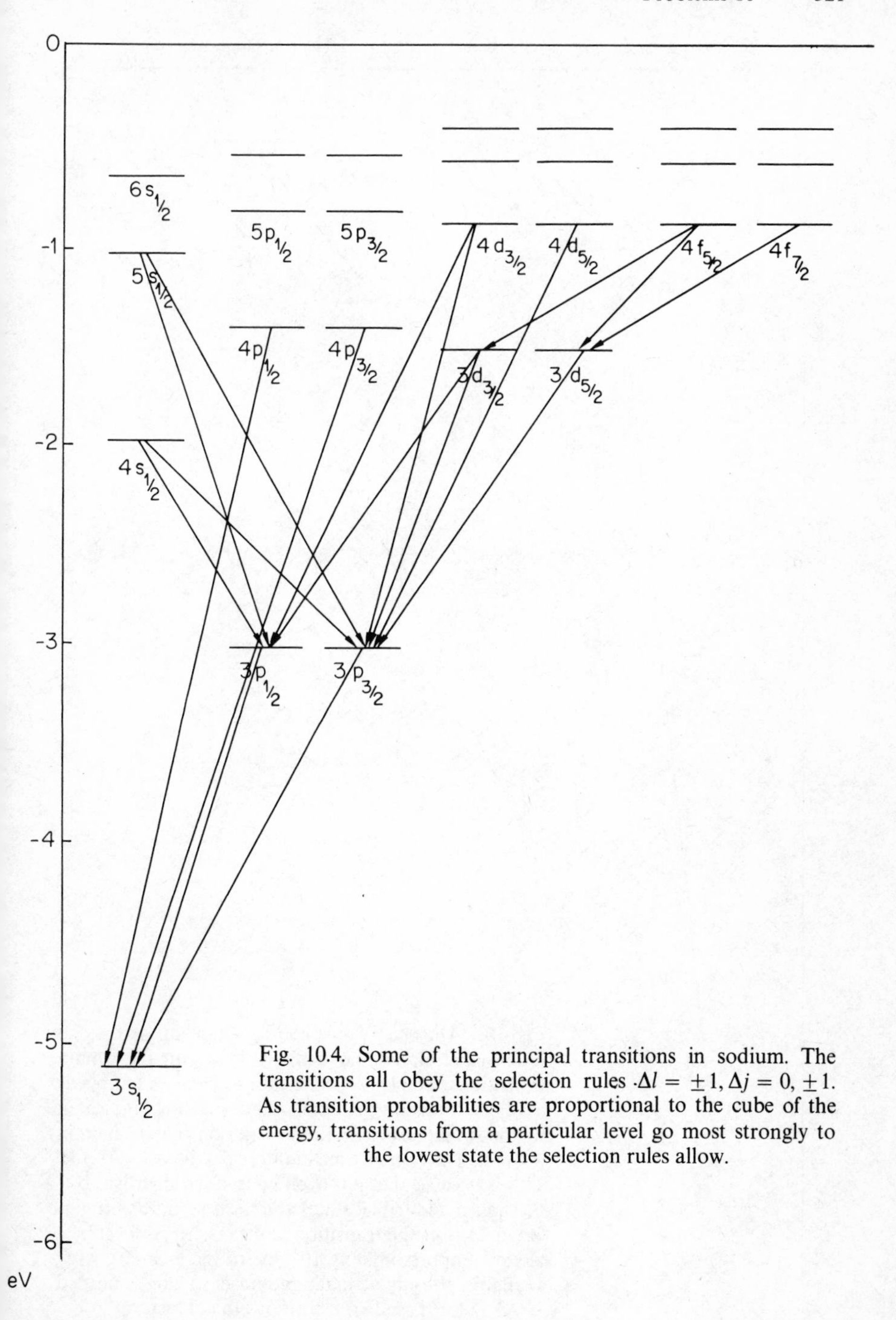

Fig. 10.4. Some of the principal transitions in sodium. The transitions all obey the selection rules $\Delta l = \pm 1, \Delta j = 0, \pm 1$. As transition probabilities are proportional to the cube of the energy, transitions from a particular level go most strongly to the lowest state the selection rules allow.

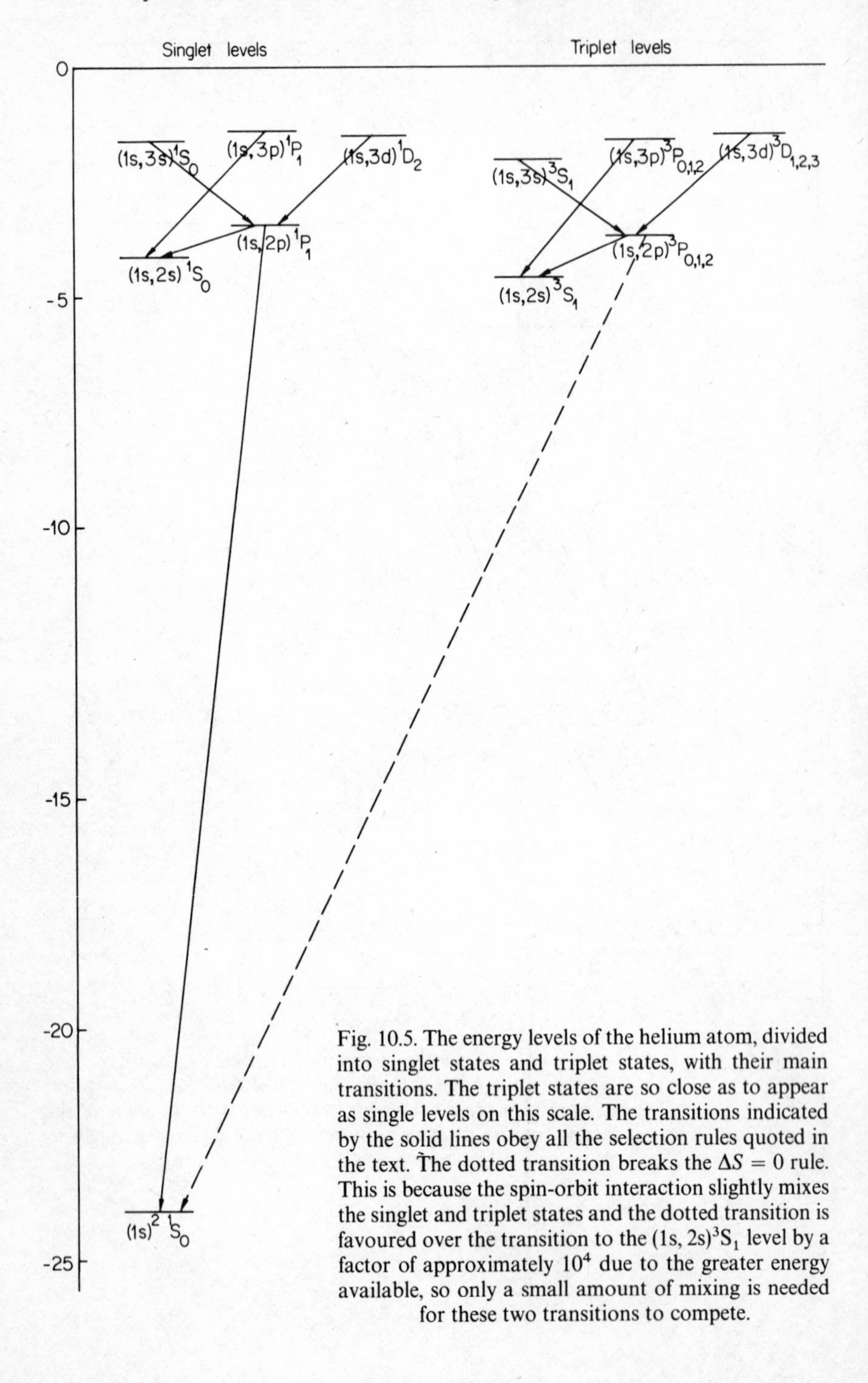

Fig. 10.5. The energy levels of the helium atom, divided into singlet states and triplet states, with their main transitions. The triplet states are so close as to appear as single levels on this scale. The transitions indicated by the solid lines obey all the selection rules quoted in the text. The dotted transition breaks the $\Delta S = 0$ rule. This is because the spin-orbit interaction slightly mixes the singlet and triplet states and the dotted transition is favoured over the transition to the (1s, 2s)$^3S_1$ level by a factor of approximately 10^4 due to the greater energy available, so only a small amount of mixing is needed for these two transitions to compete.

CHAPTER 

Characteristic X-ray spectra

11.1 INTRODUCTION

In Chapters 9 and 10 we have successfully accounted for the broad features of the energy levels of one and two electron atoms up to their ionization potentials and also for the spectral lines they emit when excited by, say, an electrical discharge. However these spectral lines occur mainly in or near the visible region of the spectrum, and it is known that atoms can also emit radiation in the X-ray region, with wavelengths of the order of an angstrom, and which is characteristic of the individual atomic species. These X-rays are known as characteristic X-rays, and were first discovered in 1896.

If a target made from a single element is bombarded by a beam of electrons of sufficient energy the X-ray spectrum produced is as shown in Fig. 11.1. The spectrum consists of two parts; a general background starting at a wavelength λ_0 and extending to longer wavelengths, plus some superimposed lines. The general background is produced by the deceleration of the incident electrons in the target material, and is known as bremsstrahlung. The shortest wavelength produced, λ_0, is given by the relation

$$\frac{hc}{\lambda_0} = eV \tag{11.1}$$

where eV is the energy of the incident electrons.

The superimposed peaks are called characteristic X-rays, as they are characteristic of the element which constitutes the target.

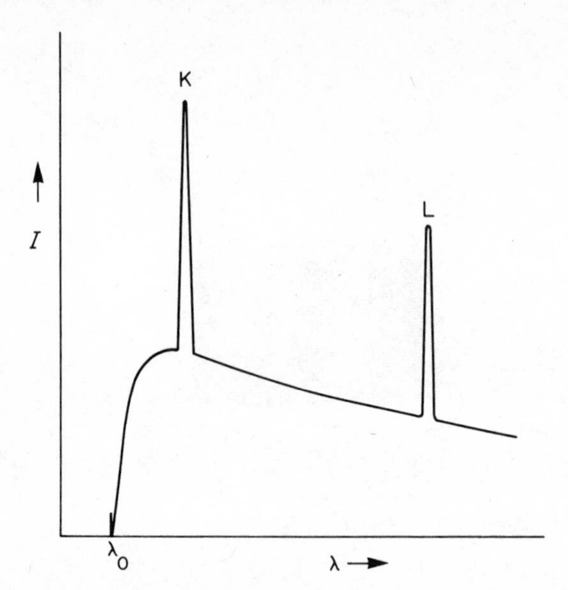

Fig. 11.1. The X-ray spectrum produced by the electron bombardment of an element.

The early experiments were done with low resolution. In these circumstances the characteristic X-rays appeared as well separated single lines which were labelled K, L, M, etc., in order of increasing wavelength. Later experiments have shown that there is considerable structure within each of these lines, but while the whole group of lines in the K X-ray peak covers typically 5% to 10% of the central wavelength, the L X-ray peak is at a wavelength some seven to ten times greater, so the distinction is quite clear.

The characteristic X-rays for a given element, say copper, have the same wavelengths whatever the energy of the electrons used for bombardment. There is a minimum bombarding energy below which a particular line does not appear, which is somewhat in excess of that obtained by equating the electron energy to the energy of the photon, for reasons which will appear later.

Further evidence that the characteristic X-rays are to be associated with the target material comes from the fact that they are produced if the target material is irradiated with X-rays. If the incident X-rays are monochromatic we find the scattered X-rays have three components:

(i) X-rays of the same wavelength as the incident X-rays;

(ii) X-rays at a slightly longer wavelength which are the result of Compton scattering (see section 3.8);

(iii) X-rays of precisely the same wavelength as in the superimposed peaks when the element is bombarded with electrons, i.e. the characteristic X-rays.

Once again we find the characteristic X-rays are only produced if the energy of the photon is above a certain minimum, and this minimum energy is *exactly* the same as the minimum electron energy required to produce the characteristic X-rays.

11.2 MOSELEY'S EXPERIMENTS

The source of the characteristic X-rays was clearly indicated by the systematic work of Moseley in 1913 and 1914 on the relation between the characteristic X-ray wavelengths and the atomic number of the element.

Let us concentrate on the K-line. Better resolution had shown this to consist of several components which were labelled K_α, K_β, K_γ, etc., in ascending order of frequency. The strongest of these is the K_α line.

Moseley measured the wavelengths (and hence the frequencies) of the K_α line for a large number of elements. Some of Moseley's results are shown in Fig. 11.2 in which the square root of the frequency of the K_α line is plotted

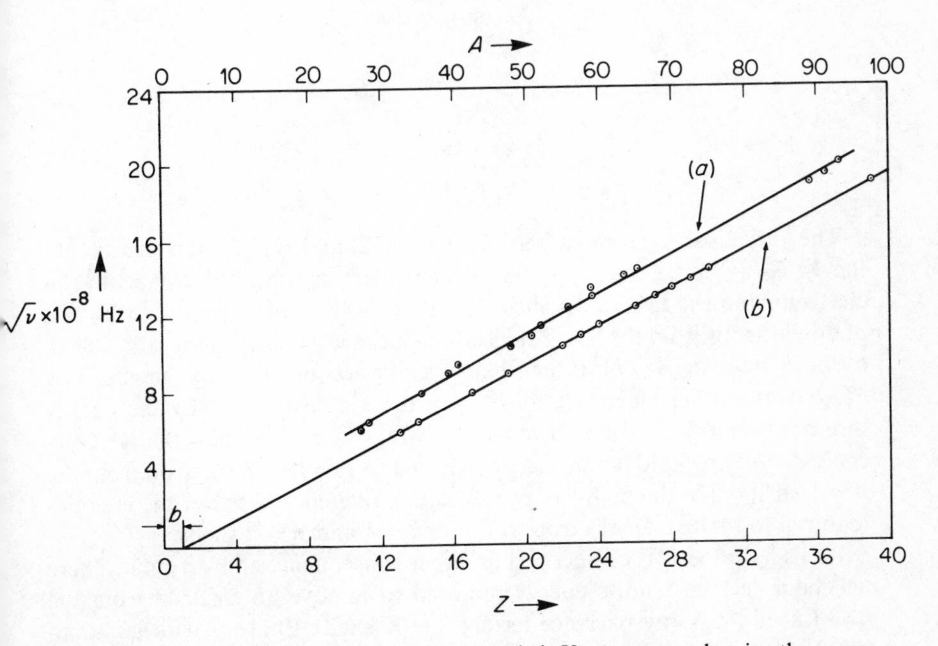

Fig. 11.2. A plot of Moseley's results on characteristic X-ray spectra showing the square root of the frequency times 10^{-8} plotted (*a*) against atomic weight *A*, and (*b*) against atomic number *Z*.

against both atomic weight and atomic number. It will be seen that a much smoother result is obtained when the plot is made against atomic number. In this case the curve can be represented by

$$\nu = 0.248 \times 10^{16}(Z - b)^2 \text{ Hz} \tag{11.2}$$

where b is approximately equal to one.

At the time of Moseley's work Bohr had just published his model of the hydrogen atom in which an electron moved round the positively charged nucleus in a circle. Bohr only allowed those orbits for which the angular momentum is $n\hbar$ where n is an integer. This model gave the same energy levels for hydrogen and other one electron atoms as we have obtained using the Schrödinger equation in Chapter 7, namely

$$E_n = -\frac{1}{n^2}\frac{Z^2e^2}{4\pi\varepsilon_0 \cdot 2a_0} . \tag{7.32}$$

Moseley compared his results with what would be obtained using Eq. (7.32) and assuming the K_α line was caused by a transition between levels with $n = 2$ and $n = 1$. Such a line would have a frequency given by

$$h\nu_{K\alpha} = \frac{3}{4}\frac{Z^2e^2}{4\pi\varepsilon_0 \cdot 2a_0} .$$

Hence

$$\nu_{K_\alpha} = \frac{3}{4}\frac{Z^2e^2}{4\pi\varepsilon_0 \cdot 2a_0h} \tag{11.3}$$

$$= 0.246 \times 10^{16}\, Z^2 \text{ Hz} .$$

The very close agreement between Eq. (11.2) and Eq. (11.3) suggests that the K X-rays arise through the incident electron or photon ejecting an electron from the innermost shell, and that the K_α line is produced by one of the electrons from the $n = 2$ shell falling into the vacancy in the $n = 1$ shell, the occurrence of $(Z - b)^2$ rather than Z^2 being accounted for by the screening effect of the other electrons. Such a picture also explains why the energy threshold for the production of the K X-rays is greater than the K X-ray energy; the threshold is the energy required to remove an electron from the $n = 1$ shell while the K X-ray energy is the difference between the energies required to remove an electron from the $n = 1$ and $n = 2$ shells.

If this hypothesis is correct certain other consequences flow from it. There will be a certain definite energy required to remove an electron from the $n = 1$ shell (in X-ray parlance termed the K shell). Photons with less than this energy will be unable to eject K electrons and thus unable to produce K X-rays. If, however, the photons have enough energy to eject an electron from the K shell not only do we expect to see K X-rays, but we also have a new mechanism by which photons can be absorbed, so we can expect a rise in the absorption coefficient at this point.

If monochromatic X-rays are passed through a foil of uniform thickness x the intensity, I, of the X-rays of the same wavelength which emerge on the far side is related to the incident intensity I_0 by

$$I = I_0 e^{-\mu x}$$

where μ is a constant for any particular material and wavelength, and is known as the absorption coefficient. The quantity most commonly quoted in the literature is the mass absorption coefficient defined by μ/ρ where ρ is the density of the material. In Fig. 11.3 we plot the mass absorption coefficient of lead in the region of X-ray wavelengths where lead K X-rays are

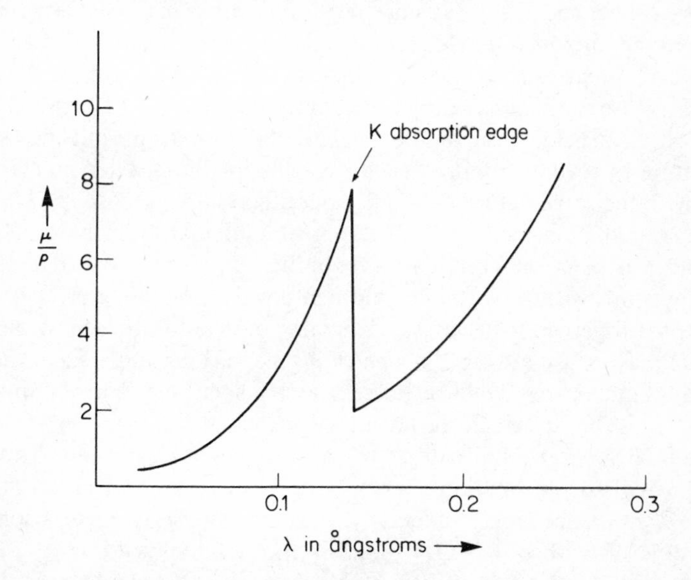

Fig. 11.3. The mass absorption coefficient μ/ρ in $cm^2\,g^{-1}$ in lead as a function of X-ray wavelength. Note the abrupt change at $\lambda = 0.1405$ Å corresponding to the binding energy of a K electron in lead.

produced. This shows a very sharp discontinuity which occurs at just the point where lead K X-rays are first produced. A similar discontinuity is observed at the point where the next group of X-rays, the L X-rays, are produced, and the energy difference between photons which just produce K X-rays and photons which just produce L X-rays is found to be exactly equal to the energy of the K X-rays. These sudden changes in absorption are referred to as 'absorption edges'.

Further, if the K_α line is produced by an electron falling from the $n = 2$ shell into the $n = 1$ shell it is reasonable to suppose that the K_β and K_γ lines which occur at slightly higher frequencies are due to electrons from higher shells falling into the vacancy in the K shell. As the production process merely involves the *removal* of one of the electrons the relative intensities of the various components forming the K X-ray line should not depend on the removal mechanism. This is indeed found to be the case.

11.3 X-RAY ENERGY LEVELS AND SELECTION RULES

Let us assume that the above picture for the production of characteristic X-rays is correct, and discuss in qualitative terms the scheme of energy levels we expect and the X-ray lines resulting from transitions between these levels, taking into account the appropriate selection rules.

The levels are caused by the *absence* of an electron from a shell. We specify the shells close to the nucleus by the quantum numbers n, l and j. Due to the strong electric field close to the nucleus the spin-orbit splitting is large, amounting to several hundred electron volts for the heavier nuclei. Hence states with the same values of n and l, but different values for $j(=l \pm \frac{1}{2})$, are well separated. Now a complete shell has zero angular momentum. Hence, if we remove an electron with quantum numbers (n, l, j, m_j) from the full (n, l, j) shell, by conservation of angular momentum the remaining electrons must be coupled together so as to have angular momentum quantum numbers $(l, j, -m_j)$. Thus aside from the sign of m_j the various states have the same quantum numbers as would be possessed by a single electron occupying the same level as that in which the vacancy occurs.

As the X-ray states are caused by vacancies, the excitation of an X-ray level is equal to the binding energy of an electron in that shell. Thus the level with $n = 1$ lies highest in energy, as shown in Fig. 11.4. Next comes the group of levels with $n = 2$. Of this group the electrons with $l = 0, j = \frac{1}{2}$ are most strongly bound and therefore the $n = 2, l = 0, j = \frac{1}{2}$ level is the next highest X-ray level. As in the atom the spin-orbit coupling puts the level with $j = l - \frac{1}{2}$ lower than that with $j = l + \frac{1}{2}$ so in the X-ray levels the level with $j = l - \frac{1}{2}$ is *higher* than that with $j = l + \frac{1}{2}$. In other words the whole spectrum of levels is just like that of an alkali turned upside down. Due however, to the strong central field of the highly charged nucleus the perturbations caused by the presence of the other electrons are relatively much smaller than in an alkali metal, and the levels occur in well separated groups, each group corresponding to a particular value of n. Also because the electrons are closer to the nucleus they are moving much more rapidly, and in a much stronger electric field than the valence electrons in an alkali. The effective magnetic field acting on a charged particle, in its own reference frame, when it moves through an electric field, is proportional to the product of its speed and the intensity of the electric field. It is this effective magnetic field acting on the magnetic moment of the electron which is responsible for the spin-orbit effect, and so we expect large spin-orbit splittings in the X-ray levels, as we have already pointed out.

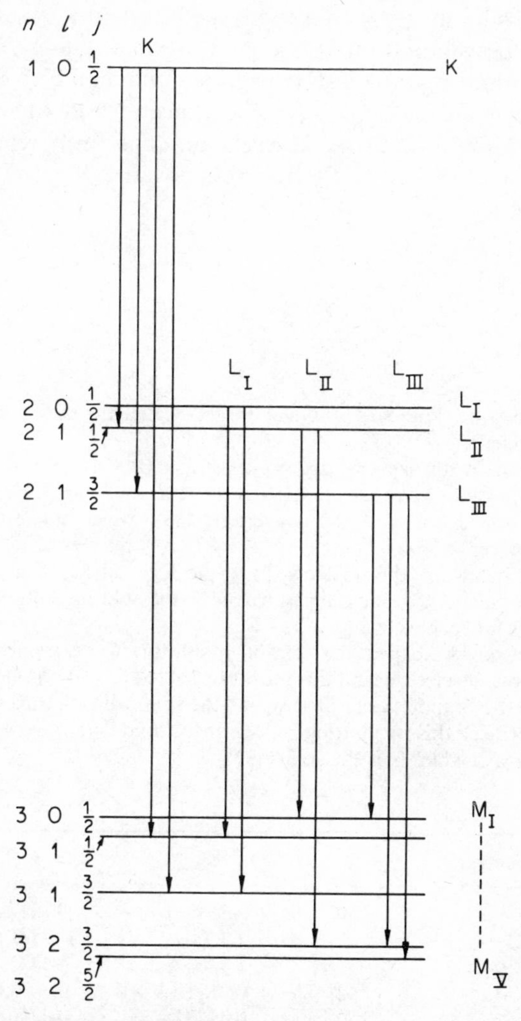

Fig. 11.4. A schematic version of atomic X-ray levels together with the allowed transitions. The vertical scale is approximately logarithmic in energy. Note that the nomenclature K, L_I, L_{II}, etc., is used not only to denote the occurrence of a vacancy in a particular shell, but also to denote the group of X-rays produced when that particular vacancy is filled by an electron from any of the higher shells.

A typical X-ray level diagram is shown in Fig. 11.4 together with the allowed transitions. In X-ray work the $1s_{1/2}$ level is called the K level, and the $2s_{1/2}$, $2p_{1/2}$ and $2p_{3/2}$ levels are called the L_I, L_{II} and L_{III} levels respectively where the symbols $2p_{3/2}$ etc. mean $n = 2$, $l = 1$, $j = \frac{3}{2}$ etc. as in section 9.9. The selection rules for these transitions are clearly the same as for a single electron, as a transition from a state n_1, l_1, j_1 to a state n_2, l_2, j_2 is caused by an electron moving from a state with quantum numbers n_2, l_2, j_2 to one with quantum numbers n_1, l_1, j_1. The transitions in which the vacancy moves from the K level to the L levels are collectively referred to as the K_α X-rays. The transition to the L_{III} level is called K_{α_1} and the transition to the L_{II} level K_{α_2}.

PROBLEMS 11

11.1 90 keV electrons bombard a metal target. Calculate the shortest wavelength X-rays produced.

11.2 For silver the K absorption edge occurs at $\lambda = 0.485$ Å and the L_I, L_{II} and L_{III} absorption edges occur at $\lambda = 3.24$ Å, 3.51 Å and 3.69 Å respectively. What information does this give you concerning the wavelengths of the characteristic X-rays of silver?

11.3 Tabulated below are the wavelengths of the K_{α_1} and K_{α_2} X-ray transitions in a series of elements. Use the data to calculate the splitting between the $2p_{1/2}$ and $2p_{3/2}$ levels for each element.

Theoretically we expect the spin-orbit splitting to be proportional to $\langle E/r \rangle$. If we assume the electric field E is proportional to $(Z - \delta)/r^2$, where δ is the shielding due to the 1s and 2s electrons, we find the spin-orbit splitting is proportional to $(Z - \delta)^4$. Check this by plotting $\eta^{1/4}$ against Z and determine δ. Is the result you get for δ reasonable? (η is the splitting.)

Element	Z	K_{α_1}	K_{α_2}
Ca	20	3.3549 Å	3.3517 Å
Fe	26	1.9360 Å	1.9321 Å
Fe	30	1.4360 Å	1.4322 Å
Mo	42	0.71210 Å	0.70783 Å
Sn	50	0.49404 Å	0.48961 Å
Nd	60	0.33596 Å	0.33128 Å

11.4 The $(Z - 1)$th ionization energy of an atom with nuclear charge Ze is given quite accurately by the expression $E = (Z - 0.3)^2 \times 13.6$ eV. The energy required to remove a K electron from several neutral atoms is given below. Compare these results with the above formula and explain the difference.

Element	Z	K binding energy
Ca	20	4,038 eV
Zn	30	9,659 eV
Mo	42	19,999 eV
Sn	50	29,200 eV

11.5 Over a wide range of elements the intensity ratio $K_{\alpha_1}/K_{\alpha_2}$ is very close to 2. Explain this result.

11.6 The K_α line in copper occurs at 1.539 Å and in zinc at 1.434 Å. These two lines span the K absorption edge of nickel, which has a mass absorption coefficient μ/ρ of 325 cm^2 g^{-1} at 1.434 Å and 48.0 cm^2 g^{-1} at 1.539 Å. What thickness of nickel, in mg per cm^2, is required to make a filter which would change the intensity ratios of the K X-rays from brass by a factor of ten?
(X-rays traversing a layer of a substance of thickness x centimetres are reduced in intensity by a factor of $e^{-\mu x}$. The mass absorption coefficient is μ/ρ where ρ is the density in grams per cc.)

11.7 A lead atom (atomic weight 208) emits a K_{α_1} line of wavelength 0.16980 Å. Calculate the recoil energy of the lead atom due to the emission of a photon of this energy.

APPENDIX 

Fourier Series and Fourier Transforms

Suppose we have a function $f(x)$ that is periodic in x, that is

$$f(x + l) \equiv f(x) \tag{A.1}$$

for all x. We would like to be able to expand such a function in a series of sines and cosines, as these are the simplest periodic functions we know.

If such an expansion is possible, clearly only those sine and cosine terms will occur which themselves satisfy Eq. (A.1), hence it must be of the form

$$f(x) = \tfrac{1}{2}b_0 + \sum_{n=1}^{\infty} a_n \sin\frac{2\pi nx}{l} + \sum_{n=1}^{\infty} b_n \cos\frac{2\pi nx}{l} . \tag{A.2}$$

The first term, $\frac{1}{2}b_0$, is included to represent the average value of $f(x)$, as the average of any sine or cosine function over a whole number of periods is zero, the factor of a half is included for later convenience. The various terms in Eq. (A.2) are orthogonal to each other in the sense

$$\left.\begin{aligned}
&\int_{-l/2}^{l/2} \sin\frac{2\pi nx}{l} \sin\frac{2\pi mx}{l}\, \mathrm{d}x = 0, m \neq n \\
&\int_{-l/2}^{l/2} \cos\frac{2\pi nx}{l} \cos\frac{2\pi mx}{l}\, \mathrm{d}x = 0, m \neq n \\
&\int_{-l/2}^{l/2} \sin\frac{2\pi nx}{l} \cos\frac{2\pi mx}{l}\, \mathrm{d}x = 0 .
\end{aligned}\right\} \tag{A.3}$$

We use this property to find the coefficients in Eq. (A.2). If we multiply both sides by $\sin 2\pi mx/l$ and integrate from $-l/2$ to $+l/2$ we obtain

$$\int_{-l/2}^{+l/2} f(x)\sin\frac{2\pi mx}{l}\,\mathrm{d}x = a_m\int_{-l/2}^{+l/2}\sin^2\frac{2\pi mx}{l}\,\mathrm{d}x$$

$$= \frac{a_m}{2}l\ .$$

Hence

$$a_m = \frac{2}{l}\int_{-l/2}^{+l/2} f(x)\sin\frac{2\pi mx}{l}\,\mathrm{d}x\ . \tag{A.4a}$$

Similarly

$$b_m = \frac{2}{l}\int_{-l/2}^{+l/2} f(x)\cos\frac{2\pi mx}{l}\,\mathrm{d}x \tag{A.4b}$$

and

$$b_0 = \frac{2}{l}\int_{-l/2}^{+l/2} f(x)\,\mathrm{d}x\ . \tag{A.4c}$$

The expansion Eq. (A.2) can be shown to be possible for a wide class of functions. It is sufficient if $f(x)$ is continuous except for a finite number of finite discontinuities. Thus, in practice, any periodic function one will come across in physics can be expanded in the form of Eq. (A.2); it is then said to be expanded in a Fourier series.

We can obtain an alternative expansion to Eq. (A.2) by expressing the sines and cosines in complex exponential form and combining terms with the same value of n. Thus,

$$a_n\sin\frac{2\pi nx}{l} + b_n\cos\frac{2\pi nx}{l}$$

$$= \frac{a_n}{2\mathrm{i}}(\mathrm{e}^{\mathrm{i}2\pi nx/l} - \mathrm{e}^{-\mathrm{i}2\pi nx/l}) + \frac{b_n}{2}(\mathrm{e}^{\mathrm{i}2\pi nx/l} + \mathrm{e}^{-\mathrm{i}2\pi nx/l})$$

$$= \frac{b_n - \mathrm{i}a_n}{2}\mathrm{e}^{\mathrm{i}2\pi nx/l} + \frac{b_n + \mathrm{i}a_n}{2}\mathrm{e}^{-\mathrm{i}2\pi nx/l}$$

$$= c_n\,\mathrm{e}^{\mathrm{i}2\pi nx/l} + c_{-n}\mathrm{e}^{-\mathrm{i}2\pi nx/l} \tag{A.5}$$

where

$$\left.\begin{aligned} c_n &= \frac{b_n - ia_n}{2} \\ c_{-n} &= c_n^* \\ &= \frac{b_n + ia_n}{2} \;. \end{aligned}\right\} \tag{A.6}$$

Thus, Eq. (A.2) becomes

$$f(x) = \sum_{n=-\infty}^{+\infty} c_n \, e^{i2\pi nx/l} \tag{A.7}$$

where the coefficients c_n are given by

$$c_n = \frac{1}{l} \int_{-l/2}^{+l/2} f(x) \, e^{-i2\pi nx/l} \, dx \;. \tag{A.8}$$

Note that two terms $e^{i2\pi nx/l}$ and $e^{i2\pi mx/l}$ are orthogonal in the sense

$$\left.\begin{aligned} \int_{-l/2}^{+l/2} e^{-i2\pi mx/l} \, e^{i2\pi nx/l} \, dx &= 0, \quad n \neq m \\ &= l, \quad n = m \end{aligned}\right\} \tag{A.9}$$

that is, one has to take the product of one with the complex conjugate of the other.

Eqs. (A.2) and (A.7), with coefficients given by Eqs. (A.4) and (A.8) respectively, therefore allow us to give an expansion of a periodic function.

Can we extend these considerations in such a way as to allow us to make an expansion of a non-periodic function? We now sketch, without any pretence at rigour, how this may be done.

We may regard a non-periodic function as a periodic function in which the length of the period has extended to infinity. If, in Eq. (A.7), we denote $2\pi n/l$ by the symbol $k(n)$, Eq. (A.7) becomes

$$f(x) = \sum_{n=-\infty}^{+\infty} c_n \, e^{ik(n)x}$$

where $k(n) = 2\pi n/l$ ranges in steps of $2\pi/l$ from $-\infty$ to $+\infty$ as n goes from $-\infty$ to $+\infty$. As l tends to infinity, the values of $k(n)$ become continuous and the sum becomes an integral

$$f(x) = \int_{-\infty}^{+\infty} c(k) \, e^{ikx} \, dk \;. \tag{A.10}$$

$c(k)\,dk$ is the sum of the amplitudes between k and $k + dk$. In the same range of $k(n) = 2\pi n/l$ prior to letting l go to infinity, there would have been $(l/2\pi)\,\mathrm{d}k$ terms, the coefficients being given by Eq. (A.8). Thus the expression for the amplitude $c(k)$ is

$$c(k) = \frac{1}{2\pi}\int_{-\infty}^{+\infty} f(x)\,\mathrm{e}^{-ikx}\,\mathrm{d}x\ . \tag{A.11}$$

The lack of symmetry between Eqs. (A.10) and (A.11) is usually removed by defining $g(k) = (2\pi)^{1/2}c(k)$ so that Eq. (A.10) becomes

$$f(x) = \frac{1}{(2\pi)^{1/2}}\int_{-\infty}^{+\infty} g(k)\,\mathrm{e}^{ikx}\,\mathrm{d}k \tag{A.12}$$

with the amplitude $g(k)$ given by

$$g(k) = \frac{1}{(2\pi)^{1/2}}\int_{-\infty}^{+\infty} f(x)\,\mathrm{e}^{-ikx}\,\mathrm{d}x\ . \tag{A.13}$$

$g(k)$ is the Fourier transform of $f(x)$.

A weakness in the above argument concerns the extension of Eq. (A.9). As l tends to infinity the integral of Eq. (A.9) becomes

$$\int_{-\infty}^{+\infty} \mathrm{e}^{\mathrm{i}(k'-k)x}\,\mathrm{d}x \tag{A.14}$$

which is undefined. Let us see what conditions Eq. (A.14) must satisfy in order that Eqs. (A.12) and (A.13) make sense.

From the symmetry of Eqs. (A.12) and (A.13) it is obvious that we may either regard $f(x)$ as an arbitrary given function of x and $g(k)$ as its Fourier transform or vice versa providing we note the difference in the sign of the exponential in the two cases.

Inserting Eq. (A.12) into Eq. (A.13) we get

$$g(k) = \frac{1}{2\pi}\int_{-\infty}^{+\infty}\mathrm{d}x\int_{-\infty}^{+\infty}\mathrm{d}k'g(k')\,\mathrm{e}^{\mathrm{i}(k'-k)x} \tag{A.15}$$

where the integration over k' is performed first, followed by integration over x. We now invert the order of integration and obtain

$$g(k) = \int_{-\infty}^{+\infty}\mathrm{d}k'g(k')\left\{\frac{1}{2\pi}\int_{-\infty}^{+\infty}\mathrm{e}^{\mathrm{i}(k'-k)x}\,\mathrm{d}x\right\}\ . \tag{A.16}$$

The integral within the braces in Eq. (A.16) is a function of $(k' - k)$ only and is indeed our required integral, which we denote as $\delta(k' - k)$. Thus (A.16) becomes

$$g(k) = \int_{-\infty}^{+\infty} g(k')\,\delta(k'-k)\,\mathrm{d}k'\ . \tag{A.16a}$$

As $g(k)$ is an *arbitrary* function of k Eq. (A.16a) can only be true if $\delta(k' - k)$ somehow picks out the point $k = k'$. This means it must be zero if k' is not equal to k, and infinite at the point $k' = k$:

$$\left.\begin{aligned} \delta(k' - k) &= 0, \quad k' \neq k \\ &= \infty, \quad k' = k \end{aligned}\right\} \tag{A.17}$$

The nature of the infinity at $k' = k$ is demonstrated by using Eq. (A.17) to limit the range of integration to a small region between k_1 and k_2, containing the point k, over which the variation of $g(k)$ may be neglected. Whence Eq. (A.16a) becomes

$$g(k) = g(k) \int_{k_1}^{k_2} \delta(k' - k)\, \mathrm{d}k'$$

i.e.

$$\left.\begin{aligned} \int_{k_1}^{k_2} \delta(k' - k)\, \mathrm{d}k' &= 1 \quad \text{for} \quad k_1 < k < k_2 \\ &= 0 \text{ otherwise} . \end{aligned}\right\} \tag{A.18}$$

Thus for consistency Eq. (A.14) must read

$$\int_{-\infty}^{+\infty} \mathrm{e}^{\mathrm{i}(k' - k)x}\, \mathrm{d}x = 2\pi\, \delta(k' - k) \tag{A.14a}$$

with $\delta(k' - k)$ defined by Eqs. (A.17) and (A.18). $\delta(k' - k)$ is known as the Dirac delta function.

The importance of Fourier transforms in quantum mechanics, apart from their general utility, lies in the fact that $\phi(k) = \mathrm{e}^{\mathrm{i}kx}$ is an eigenfunction of the momentum operator $-\mathrm{i}\hbar\partial/\partial x$ belonging to the eigenvalue $\hbar k$. Thus Eq. (A.12) represents the expansion of a wavefunction $f(x)$ in terms of momentum eigenfunctions, the amplitude of any component being given by Eq. (A.13).

APPENDIX

Numerical values of physical constants

In Table B.1 we tabulate some useful relations and approximate numerical values which have been found valuable in atomic physics.

In Table B.2 we have tabulated accurate numerical values of the most commonly occurring physical quantities in atomic physics. The first column gives the name of the quantity, the second the more common symbol and the third the value and units. The number in parentheses at the end of the value, but prior to the powers of ten, gives the standard deviation uncertainties in the last digits of the quoted values. The quoted values are from 'Determination of e/h, Using Macroscopic Quantum Phase Coherence in Superconductors. Implications for Quantum Electrodynamics and the Fundamental Physical Constants' by B. N. Taylor, W. H. Parker and D. N. Langenberg, *Rev. Mod. Phys.*, 1969, Vol. 41, No. 3, pp. 375–493.

Table B.1

Some useful relations and numerical values
1 electron volt = 1.602×10^{-19} J
Hydrogen ionization potential = 13.6 V
kT at room temperature (290 K) = 1/40 eV
Rest mass energy of an electron, $m_e c^2$ = 511 keV
Rest mass energy of a proton $M_p c^2$ = 938 MeV
Energy of 1 a.m.u. = 931 MeV
A photon of wavelength 1.24×10^4 Å has $\hbar\omega$ = 1 eV.

Table B.2. Accurate numerical values of physical constants

Quantity	Symbol	Value and Units
Avogadro's number	N_0	$6.022169\,(40) \times 10^{23}\ \mathrm{mol}^{-1}$
Boltzmann's constant	k	$1.380622\,(59) \times 10^{-23}\ \mathrm{J\,K}^{-1}$
Gas constant	R	$8.31434\,(35)\ \mathrm{J\,mol}^{-1}\,\mathrm{K}^{-1}$
Electron rest mass	m_e	$9.109558\,(54) \times 10^{-31}\ \mathrm{kg}$
Proton rest mass	M_P	$1.672614\,(11) \times 10^{-27}\ \mathrm{kg}$
Ratio of proton mass to electron mass	M_P/m_e	$1836.109\,(11)$
Faraday	F	$9.648670\,(54) \times 10^{7}\ \mathrm{C\,mol}^{-1}$
Electron charge	e	$1.6021917\,(70) \times 10^{-19}\ \mathrm{C}$
Electron charge to mass ratio	e/m_e	$1.7588028\,(54) \times 10^{11}\ \mathrm{C\,kg}^{-1}$
Velocity of light	c	$2.9979250\,(10) \times 10^{8}\ \mathrm{m\,s}^{-1}$
Planck's constant	h	$6.626196\,(50) \times 10^{-34}\ \mathrm{J\,s}$
	$\hbar$	$1.0545919\,(80) \times 10^{-34}\ \mathrm{J\,s}$
Bohr radius	a_0	$5.2917715\,(81) \times 10^{-11}\ \mathrm{m}$
Bohr magneton	μ_B	$9.274096\,(65) \times 10^{-24}\ \mathrm{J\,T}^{-1}$
Electron magnetic moment in Bohr magnetons	μ_e/μ_B	$1.0011596389\,(31)$
Fine structure constant	α	$7.297351\,(11) \times 10^{-3}$
	$1/\alpha$	$137.03602\,(21)$
Stefan–Boltzmann constant	σ	$5.66961\,(96) \times 10^{-8}\ \mathrm{W\,m}^{-2}\,\mathrm{K}^{-4}$

APPENDIX 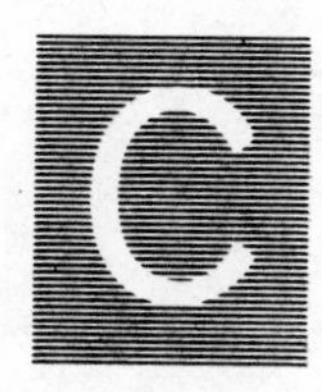

Solutions to problems

CHAPTER 1

1.1 Use $\frac{1}{2}kT = \frac{1}{2}C\overline{\theta^2}$ where C is the torque constant of the fibre, and $C\theta = nAiB$ where n is the turns in the coil, A the area of the coil and i and B are the current and magnetic field. We get R.M.S. deviation equals 2.2×10^{-3} mm.

1.2 2.815×10^{-10} m.

1.3 3.38×10^{-6} m, 3.38×10^{-5} m, 4.75×10^{-3} m, 3.36×10^{-1} m. The last two results require the use of a relativistic formula.

1.4 1.67×10^{-3} T.

1.5 Radius equals 1.44×10^{-6} m, $q = 4.85 \times 10^{-19}$ C, i.e. 3 electron charges.

1.6 1.05×10^{-3} ms^{-1}. (There are 3 degrees of freedom.)

1.7 5.93×10^5 ms^{-1}, 5.93×10^6 ms^{-1}, 2.82×10^8 ms^{-1}.

CHAPTER 2

2.1 $\frac{1}{3}e^2x_0^2\omega^4/c \times 10^{-7}$ J s^{-1} where x_0 is the amplitude and ω the angular frequency. 2.76×10^{-12} J s^{-1}.

2.2 Fraction emitting equals 5×10^{-8}.

2.3 2.18 cm.

2.4 6.02 V s^{-1}.

2.5 11.1 kV, 7.15×10^{-2} T.

2.6 2.1×10^7.

2.7 Let the impact parameter be b and the initial velocity u. Let the closest distance the α-particle gets to the nucleus in *this* collision be r, and let its velocity at this point be v. At this point v will be perpendicular to r. From conservation of energy we have

$$\tfrac{1}{2}mu^2 = \tfrac{1}{2}mv^2 + \frac{zZe^2}{4\pi\varepsilon_0 r},$$

and from conservation of angular momentum

$$ub = vr.$$

Hence, $\frac{1}{2}mv^2 = \frac{1}{2}mu^2b^2/r^2$ and therefore

$$\frac{1}{2}mu^2\left(1 - \frac{b^2}{r^2}\right) = \frac{zZe^2}{4\pi\varepsilon_0 r}$$

$$= \frac{zZe^2}{4\pi\varepsilon_0 a} \cdot \frac{a}{r}$$

$$= \frac{1}{2}mu^2 \cdot \frac{a}{r}$$

where a is the closest distance of approach in a head-on collision. Hence

$$(1 - b^2/r^2) = a/r \ .$$

Now $b = a/2 \cot \phi/2 = a/2$, for $\phi = 90°$.
Therefore, we have

$$1 - a^2/4r^2 = a/r$$

which gives

$$r = a/2 \pm \sqrt{a^2/2} \ .$$

As r must be positive we take the positive sign.
From the data $a = 4.76 \times 10^{-13}$ cm.
Therefore $r = 5.65 \times 10^{-13}$ cm.

2.9 The scattering from the second foil must be 1.1 times that from the first foil. As the scattering probability is proportional to $1/E^2$ the mean energy in the second foil is $(1/1.1)^{1/2}$ times that in the first foil. The energy loss in one foil is the difference between these two energies, i.e. 0.05 E. Hence energy loss is 250 keV.

2.10 A velocity diagram will demonstrate the first point. We obtain a laboratory differential cross-section from a centre of mass cross-section as follows

$$\left.\frac{dP(\theta)}{d\Omega}\right)_{\text{lab}} = \left.\frac{dP(\phi)}{d\Omega'}\right)_{\text{C.M.}} \frac{d\Omega'}{d\Omega} \ .$$

As $\phi = 2\theta$, we have

$$\frac{d\Omega'}{d\Omega} = \frac{\sin\phi \, d\phi}{\sin\theta \, d\theta} = 4\cos\theta \ .$$

As $E_{\text{lab}} = 2E_{\text{C.M.}}$ the formula follows.

2.11

$$\frac{dE}{dt} = -\frac{2}{3}\frac{\mu_0}{4\pi}\frac{e^2a^2}{c}$$

$$\frac{dE}{dt} = \frac{dE}{dr} \cdot \frac{dr}{dt} \ .$$

Therefore

$$\frac{dr}{dt} = \frac{dE}{dt} \Big/ \frac{dE}{dr} \ .$$

For a particle moving in a circle under an inverse square law of force it may readily be shown that the kinetic energy is minus half the potential energy. Hence the total energy, E, of the electron in the present case is given by

$$E = -\frac{e^2}{8\pi\varepsilon_0 r}, \qquad \frac{\mathrm{d}E}{\mathrm{d}r} = \frac{e^2}{8\pi\varepsilon_0 r^2} .$$

Inserting the given expression for the acceleration a we get

$$\frac{\mathrm{d}r}{\mathrm{d}t} = -\frac{4}{3}\frac{\mu_0}{4\pi}\frac{e^2}{c}\frac{e^2}{4\pi\varepsilon_0}\frac{1}{m^2r^2}$$

whence $t = \frac{\pi}{\mu_0}\frac{4\pi\varepsilon_0 c}{e^4}m^2r^3$

$= 1.2 \times 10^{-10}\,\mathrm{s}$.

CHAPTER 3

3.1 $9.66\ \mu$, $0.484\ \mu$.

3.2 1.0 eV.

3.3 1.5 eV.

3.4 19.773 keV, 227 eV.

As the electron energy is so low we use $p^2 = 2mE$ to determine the electron momentum as $8.1 \times 10^{-24}\,\mathrm{kg\,ms^{-1}}$. The direction of electron recoil is determined by its momentum components, which are in turn determined by the change in momentum components for the photon. Remembering $p = E/c$ for photons, we have

Change in forward momentum is $(20 - 19.773\cos\theta)/c$.
Change in transverse momentum is $19.773\sin\theta/c$ where $\theta = 45°$.
Therefore

$$\tan\phi = \frac{19.773\sin\theta}{20 - 19.773\cos\theta} = 2.32 .$$

Therefore $\phi = 66.7°$.

3.5 Let the incident photon have energy $h\nu_1$, the reflected photon $h\nu_2$. Let E_1, p_1, E_2, p_2 be the initial and final *total* energies and momenta of the electron. Conservation of energy gives us

$$h\nu_1 + E_1 = h\nu_2 + E_2. \tag{i}$$

Assuming photon and electron both reverse their directions of motion, conservation of momentum gives us

$$\frac{h\nu_1}{c} - p_1 = p_2 - \frac{h\nu_2}{c} . \tag{ii}$$

Hence we have

$$E_2 = (h\nu_1 - h\nu_2) + E_1 \tag{iii}$$

and

$$p_2 c = (h\nu_1 + h\nu_2) - p_1 c . \tag{iv}$$

Square Eqs (iii) and (iv), subtract, and cancel $E_2^2 - p_2^2c^2$ with $E_1^2 - p_1^2c^2$ both being equal to $m_0^2c^4$.

$$0 = (h\nu_1 - h\nu_2)^2 - (h\nu_1 + h\nu_2)^2 + 2E_1(h\nu_1 - h\nu_2) + 2p_1c(h\nu_1 + h\nu_2)$$

hence

$$4h\nu_1 h\nu_2 = 2h\nu_1(E_1 + p_1c) - 2h\nu_2(E_1 - p_1c) \ .$$

Divide through by $2h\nu_1 h\nu_2$ to obtain

$$\frac{1}{h\nu_2}(E_1 + p_1c) = 2 + \frac{(E_1 - p_1c)}{h\nu_1} \ .$$

For the stationary electron case $p_1 = 0$ and $E_1 = m_0c^2 = 511$ keV.
Hence

$$\frac{1}{h\nu_2} = \frac{2}{511} + \frac{1}{1{,}000}$$
$$= 4.91 \times 10^{-3}\ \text{keV}^{-1} \ ,$$

and

$$h\nu_2 = 204\ \text{keV} \ .$$

When $E_1 = (511 + 100)$ keV $= 611$ keV, we obtain p_1c from

$$p_1c = (E_1^2 - m_0^2c^4)^{1/2}$$
$$= (611^2 - 511^2)^{1/2}$$
$$= (1{,}122 \times 100)^{1/2}$$
$$= 334\ \text{keV} \ .$$

Thus

$$\frac{945}{h\nu_2} = 2 + \frac{277}{1{,}000}$$
$$= 2.277 \ .$$

Hence

$$h\nu_2 = 415\ \text{keV} \ .$$

3.6 Take area of pupil as 4 mm^2 and the wavelength of light as 6,000 Å. This gives 1,000 photons per second.

3.7 $hc/\lambda = 2.11$ eV for $\lambda = 5893$ Å.

3.8 2.105 eV, 2.103 eV.

3.9 The photon energies are 8.85 eV, 4.89 eV and 3.96 eV. Hence we assume levels are excited at 8.85 eV and 4.89 eV, the 3,132 Å transition being between these two. (The lower level cannot be at 3.96 eV as the 2,536 Å line could not then be stronger than the 3,132 Å transition.) Hence we expect electron energies of 10 eV, 5.11 eV and 1.15 eV.

3.10 The cross section $\sigma = \pi \times 10^{-20}$ m^2. The number of atoms per m^3, N, may be determined from $N\lambda\sigma = 1$. Hence, as 1 mole occupies 22.4 litres, we get pressure equals 9.0×10^{-2} torr.

CHAPTER 4

4.1 26°1′28″ and 61°21′0″. From $2d \sin\theta = n\lambda$ we get $d\theta = \tan\theta\, d\lambda/\lambda$. For the last figure to be accurate $d\lambda/\lambda \approx 2 \times 10^{-5}$. Hence $d\theta$ is of the order of 2″ and 7″ of arc for the two angles above.

4.2 0.8766 Å, 12°24′50″ and 59°18′50″.

4.3 From $2d \sin\theta = n\lambda$ we have $2d \cos\theta\, d\theta = \lambda\, dn$. For $dn = 1$, $\lambda = 2d \cos\theta\, d\theta$ which is smallest for the largest θ which is 60° in this case. Hence $\lambda = d \times d\theta = 3.5 \times 10^{-4}$ Å. Hence

$$n = 2d \sin\theta/\lambda = \tan\theta/d\theta = 5.7 \times 10^3 .$$

4.4 355 MeV, 355 MeV, 65 MeV.

4.5 0.39 Å, 0.038 Å, 8.8×10^{-3} Å, 1.2×10^{-3} Å.

4.6 $\omega = ku$ and $\lambda = 2\pi/k$, hence $\dfrac{d\omega}{dk} = u + k\dfrac{du}{dk}$

$$= u + k\frac{du}{d\lambda}\frac{d\lambda}{dk}$$

$$= u - \lambda\frac{du}{d\lambda} .$$

4.7 400 eV.

4.8 Angular spread equals $\pm\lambda/a$ from diffraction. If we take $\Delta p \Delta x \approx h$ and $\Delta x = a$, then $\Delta p \approx h/a$. Hence from the uncertainty principle, we obtain the angular spread $\pm\dfrac{\Delta p}{2p} \approx \pm\dfrac{\lambda}{2a}$.

4.9 Spread due to geometry of slits $= \pm 2a/l$ where a is slit width and l their separation. The spread due to diffraction is $\pm\lambda/a$. The total spread equals $\pm(2a/l + \lambda/a)$ which has a minimum at $a = (\lambda l/2)^{1/2}$. For the values given $a_{min} = 3.5 \times 10^{-5}$ m.

4.10 6.6×10^{-8} eV, 6.6 eV, 6.6 MeV.

4.11 0.4 mm.

CHAPTER 5

5.1 1.74 Å, 1.83 Å.

5.2 12.8°.

5.3 Bragg's law is $2d \sin\theta' = n\lambda'$ where θ' and λ' are measured inside the specimen $\lambda = \mu\lambda'$ and $\cos\theta = \mu \cos\theta'$. Hence

$$\frac{n\lambda}{\mu} = 2d\left(1 - \frac{\cos^2\theta}{\mu^2}\right)^{1/2}$$

or

$$n\lambda = 2d(\mu^2 - \cos^2\theta)^{1/2} .$$

Put $\mu^2 = 1 + 2\delta$ and expand the square root to get

$$n\lambda = 2d \sin\theta(1 + \delta/\sin^2\theta) .$$

5.4 Assume the particle is incident from the left.

For $x < 0$ put $\psi = e^{ik_1x} + B\,e^{-ik_1x}$ where $k_1 = (2mE/\hbar^2)^{1/2}$.
For $x > 0$ put $\psi = C\,e^{ik_2x}$ where $k_2 = (2m(E - V_0)/\hbar^2)^{1/2}$.
Matching amplitudes and derivatives at $x = 0$ gives us

$$B = \frac{1 - k_2/k_1}{1 + k_2/k_1}. \qquad \text{Hence } BB^* \text{ equals the expression given .}$$

5.5 ψ must vanish at $\pm a/2$. Hence $\psi = \sin 2p\pi x/a$ or $\psi = \cos(2q - 1)\pi x/a$, in both cases p and q are integers going from 1 to infinity. The energies are $2p^2\pi^2\hbar^2/ma^2$ and $(2q - 1)^2\pi^2\hbar^2/2ma^2$ respectively. By putting $2p$ equal to the even n values and $(2q - 1)$ equal to the odd n values

$$E = \frac{n^2\pi^2\hbar^2}{2ma^2} = \frac{n^2h^2}{8ma^2} .$$

5.6 The normalizing factor is $\sqrt{(2/a)}$ for each wavefunction.

5.7
$$\langle x^2 \rangle = \frac{2}{a}\int_{-a/2}^{+a/2} x^2 \sin^2 \frac{2\pi px}{a}\,dx \tag{i}$$

and

$$\langle x^2 \rangle = \frac{2}{a}\int_{-a/2}^{+a/2} x^2 \cos^2 \frac{(2q-1)\pi x}{a}\,dx . \tag{ii}$$

Eq. (i) may be written

$$\langle x^2 \rangle = \frac{2}{a}\int_{-a/2}^{+a/2} \frac{x^2}{2}\left[1 - \cos\frac{4\pi px}{a}\right]dx$$
$$= \frac{a^2}{12} - \int_{-a/2}^{+a/2} \frac{x^2}{a}\cos\frac{4\pi px}{a}\,dx .$$

Two integrations by parts lead to

$$\langle x^2 \rangle = \frac{a^2}{12} - \frac{a^2}{8\pi^2p^2} .$$

For (ii) we get similarly

$$\langle x^2 \rangle = \frac{a^2}{12} - \frac{2a^2}{(4q-2)^2\pi^2} .$$

Putting $2p$ equal to the even values of n and $2q - 1$ equal to the odd values of n we get

$$\langle x^2 \rangle = \frac{a^2}{12} - \frac{a^2}{2\pi^2n^2} .$$

A classical particle would move to and fro and be equally likely to be found anywhere between $\pm a/2$. Hence the probability of finding it between x and $x + dx$ will be $1/a\,dx$. The average value of x^2 is therefore

$$\int_{-a/2}^{+a/2} \frac{1}{a}x^2\,dx = \frac{a^2}{12} .$$

CHAPTER 6

6.1 From the conservation of energy we have

$$\tfrac{1}{2}mv_1^2 - \alpha/r_1 = E \tag{i}$$

$$\tfrac{1}{2}mv_2^2 - \alpha/r_2 = E \tag{ii}$$

where v_1 and v_2 are the velocities at r_1 and r_2. From conservation of angular momentum we have

$$mv_1r_1 = mv_2r_2 = J\ . \tag{iii}$$

r_1^2 times Eq. (i) minus r_2^2 times Eq. (ii) gives us

$$\tfrac{1}{2}m(v_1^2r_1^2 - v_2^2r_2^2) - \alpha r_1 + \alpha r_2 = E(r_1^2 - r_2^2)\ .$$

By Eq. (iii) the first term is zero, hence

$$E = -\alpha(r_1 - r_2)/(r_1^2 - r_2^2)$$
$$= -\alpha/(r_1 + r_2)\ .$$

From r_1^2 times Eq. (i)

$$\tfrac{1}{2}mv_1^2r_1^2 - \alpha r_1 = Er_1^2$$

i.e.

$$\frac{J^2}{2m} - \alpha r_1 = -\frac{\alpha r_1^2}{r_1 + r_2}\ .$$

Hence

$$J^2 = 2m\alpha\left(\frac{r_1r_2}{r_1 + r_2}\right)\ .$$

6.2 By Eq. (6.4)

$$|\mu| = \frac{e}{2m}|l|\ ,$$

and

$$\frac{mv^2}{r} = \frac{e^2}{4\pi\varepsilon_0r^2}\ .$$

Hence

$$m^2v^2r^2 = l^2 = e^2mr/4\pi\varepsilon_0\ .$$

Hence

$$|\mu| = \frac{e^2}{2}\left(\frac{r}{4\pi\varepsilon_0m}\right)^{1/2}\ .$$

6.3 9.27×10^{-23} newtons, 1.71×10^{-24} newtons.

6.5 $[\hat{l}_x, \hat{l}_y] \equiv (\hat{l}_x\hat{l}_y - \hat{l}_y\hat{l}_x) = (y\hat{p}_z - z\hat{p}_y)(z\hat{p}_x - x\hat{p}_z) - (z\hat{p}_x - x\hat{p}_z)(y\hat{p}_z - z\hat{p}_y)$

$$= y\hat{p}_zz\hat{p}_x - z\hat{p}_yz\hat{p}_x - y\hat{p}_zx\hat{p}_z + z\hat{p}_yx\hat{p}_z - z\hat{p}_xy\hat{p}_z + z\hat{p}_xz\hat{p}_y + x\hat{p}_zy\hat{p}_z$$
$$-x\hat{p}_zz\hat{p}_y\ .$$

Because $[z, \hat{p}_x] = 0$, etc., it is only necessary to preserve the order where we have a coordinate and momentum referring to the same component. Thus in the first term any order will do provided $\hat{p}_z$ always comes before z, and we may write the second term in any order we please. As a result the second and third terms cancel with the sixth and seventh terms respectively. Hence

$$[\hat{l}_x, \hat{l}_y] = y\hat{p}_x(\hat{p}_z z - z\hat{p}_z) + x\hat{p}_y(z\hat{p}_z - \hat{p}_z z) .$$

From the results of the previous problem the two terms in brackets are $-i\hbar$ and $+i\hbar$ respectively, hence

$$[\hat{l}_x, \hat{l}_y] = i\hbar(x\hat{p}_y - y\hat{p}_x) = i\hbar\hat{l}_z .$$

6.6 $\sqrt{6}\hbar$, 35.26°, 65.91°, 90°, 114.09°, 144.74°.

6.7 A particle moving towards an opening in a container will escape through that opening in a time $\mathrm{d}t$ providing it is initially within a distance $c\,\mathrm{d}t$ of the opening, where c is its velocity. In an ideal gas the probability of a particle having a velocity c is proportional to $c^2\,\mathrm{e}^{-mc^2/2kT}$. Hence the number of particles emerging with a velocity c is proportional to $c^3\,\mathrm{e}^{-mc^2/2kT}$. The most probable velocity of emergence v is therefore $(3kT/m)^{1/2}$. For silver at 1,300 K this is 548 m s^{-1}.

From Problem 3 the force F on the atom in the magnet is 9.27×10^{-23} N. Within the magnet an atom moves on a parabolic path and emerges having acquired a transverse momentum $p_\perp = Fl/v$ where l is the length of the magnet. The deflection on the screen is $(p_\perp/p_\parallel) \times L$ where L is the distance from the centre of the magnet to the screen and $p_\parallel$ the momentum along the magnet axis. The result is 0.53 mm.

6.9 $4.29 \times 10^{30}\,\hbar$, $1.36 \times 10^{3}\,\hbar$.

CHAPTER 7

7.1 $\frac{3}{2}a_0, 3a_0$.

7.2 $\langle V \rangle = -e^2/4\pi\varepsilon_0 a_0$. $\langle T \rangle = e^2/8\pi\varepsilon_0 a_0$.

7.3 $l_y = -i\hbar\left(z\frac{\partial}{\partial x} - x\frac{\partial}{\partial z}\right)$, $\hat{l}_y f(r) = 0$.

Hence

$$\begin{aligned}\hat{l}_y(x + iz)f(r) &= f(r)\hat{l}_y(x + iz) \\ &= f(r)\{-i\hbar(z - ix)\} \\ &= -\hbar(x + iz)f(r) .\end{aligned}$$

7.4 $\langle r \rangle = \frac{1}{2}n(2n+1)a_0$, $\langle r^2 \rangle = \frac{1}{2}n^2(n+1)(2n+1)a_0^2$.

Spread equals

$$1/(2n+1)^{1/2} .$$

7.5 The potential at any point is proportional to Z. The range parameter is proportional to $1/Z$. Hence E is proportional to Z^2.

7.7 $$\int xf(r)\phi(r)\,\mathrm{d}^3\mathbf{r} = \int_{-\infty}^{+\infty}\int_{-\infty}^{+\infty}\int_{-\infty}^{+\infty} xf[(x^2+y^2+z^2)^{1/2}]\phi[(x^2+y^2+z^2)^{1/2}]\,\mathrm{d}x\,\mathrm{d}y\,\mathrm{d}z .$$

The integrand is an odd function of x and hence the integral over x vanishes.

CHAPTER 8

8.1 $l = 0, 1, 2, 3$ and 4.
2, 6, 10, 14 and 18. Total = 50.
8.2 2×10^{-12} m, 8×10^{-12} m.
8.4 He. It has the highest ionization potential.
8.6 4, 20.

CHAPTER 9

9.1 The charge distribution is proportional to $\{|\phi_1|^2 + |\phi_2|^3 + |\phi_3|^2\}\,\mathrm{d}^3\mathbf{r}$ which equals $[f(r)]^2(x^2 + y^2 + z^2)\,\mathrm{d}^3\mathbf{r} = [rf(r)]^2\,\mathrm{d}^3\,\mathbf{r}$.
9.2 The perturbation is $U = 0$, $0 \leqslant r \leqslant r_0$, $U = e^2/4\pi\varepsilon_0 r$, $r_0 \leqslant r \leqslant \infty$.
The change in the ground state energy is, therefore,

$$\delta E = \frac{4}{a_0^3}\int_{r_0}^{\infty} r^2\,\mathrm{e}^{-2r/a_0}\frac{e^2}{4\pi\varepsilon_0 r}\,\mathrm{d}r$$

$$= \frac{4\,\mathrm{e}^2}{4\pi\varepsilon_0 a_0^3}\int_{r_0}^{\infty} r\,\mathrm{e}^{-2r/a_0}\,\mathrm{d}r\ .$$

Partial integration gives us

$$\delta E = \frac{e^2}{4\pi\varepsilon_0 a_0}\,\mathrm{e}^{-2r_0/a_0}\left(1 + \frac{2r_0}{a_0}\right).$$

For this to be less than, say, 10% of the ground state energy $2r_0/a_0$ must be greater than five.
9.3 The perturbation is

$$U = \frac{e^2}{4\pi\varepsilon_0}\left(\frac{1}{r} - \frac{1}{r_0}\right), \quad r < r_0$$

and $U = 0$ otherwise. For this we have to evaluate

$$\frac{4}{a_0^3}\int_0^{r_0}\left(r - \frac{r^2}{r_0}\right)\mathrm{e}^{-2r/a_0}\,\mathrm{d}r$$

which can be done using the result

$$\int_0^{x_0} x^n\,\mathrm{e}^{-x}\,\mathrm{d}x = n!\left[1 - \mathrm{e}^{-x_0}\left(1 + x_0 + \frac{x_0^2}{2!} + \cdots + \frac{x_0^n}{n!}\right)\right]$$

Collecting terms one obtains

$$\delta E = \frac{e^2}{4\pi\varepsilon_0}\left[\left(\frac{1}{a_0} - \frac{1}{r_0}\right) + \left(\frac{1}{a_0} + \frac{1}{r_0}\right)\mathrm{e}^{-2r_0/a_0}\right].$$

9.4 From the wavelength the energy spacing is 2.11 eV. From Eq. (9.20) and the given values we get 2.16 eV.

9.5 The energy shift is $\langle \hat{\mu} \cdot \hat{\mathbf{B}} \rangle$. For an electron $\hat{\mu} = -(e/m)\hat{\mathbf{s}}$. Let $\hat{\mathbf{B}} = K\hat{\mathbf{l}}/\hbar$. Then

$$\langle \hat{\mu} \cdot \hat{\mathbf{B}} \rangle = -\frac{eK}{m\hbar}\langle \hat{\mathbf{s}} \cdot \hat{\mathbf{l}} \rangle .$$

$\langle \hat{\mathbf{s}} \cdot \hat{\mathbf{l}} \rangle$ equals $\frac{1}{2}l\hbar^2$ for $j = l + \frac{1}{2}$ and equals $-\frac{1}{2}(l+1)\hbar^2$ for $l = j - \frac{1}{2}$. Hence the energy splitting is equal to

$$\frac{eK}{m\hbar}\frac{(2l+1)}{2}\hbar^2 = 2.15 \times 10^{-3}\,\text{eV} .$$

Hence $K = 70$. The magnitude of $\mathbf{l}$ is $\sqrt{(l(l+1))}$ which equals $\sqrt{2}$ in the present case..Thus $B = 99$ T.

9.8 $$\cos\theta = \frac{\langle \hat{\mathbf{j}} \cdot \hat{\mathbf{l}} \rangle}{[j(j+1)l(l+1)]^{1/2}} .$$

For $j = \frac{3}{2}$, $\theta = 18.2°$, for $j = \frac{5}{2}$, $\theta = 14.9°$.

9.9 197.7 eV.

9.10 193.8 eV.

9.11 We specify the states by quoting the m_l values of the two electrons, the m_l value for the first electron always being placed first. Thus an unsymmetrized state with $M_L = 1$ formed by the first electron having $m_l = 1$ and the second having $m_l = 0$ is denoted by (1, 0). The symmetric and antisymmetric states with $M_L = 1$ would be (1, 0) + (0, 1) and (1, 0) − (0, 1) respectively. With this notation the possible symmetric states for the various values of M_L are:

$M_L = 2$	(1, 1)
$M_L = 1$	(1, 0) + (0, 1)
$M_L = 0$	(1, −1) + (−1, 1) and (0, 0)
$M_L = -1$	(−1, 0) + (0, −1)
$M_L = -2$	(−1, −1)

There are thus six symmetric states, two with $M_L = 0$ and one for each of the other values of M_L. There must be a state with $L = 2$ to account for the states with $M_L = \pm 2$. This state will also account for the states with $M_L = \pm 1$ and one of the states with $M_L = 0$. There is one remaining state with $M_L = 0$, and hence it must have $L = 0$. These states will be associated with $S = 0$. (It should be mentioned that the state with $L = 0$ and the $L = 2$ state with $M_L = 0$ are both linear combinations of the two states quoted above with $M_L = 0$, neither of which has L as a good quantum number.)

The possible antisymmetric states are

$M_L = 1$	(1, 0) − (0, 1)
$M_L = 0$	(1, −1) − (−1, 1)
$M_L = -1$	(−1, 0) − (0, −1).

These clearly are the three magnetic substates of a state with $L = 1$, and are to be associated with $S = 1$.

9.12 Due to the mutual repulsion between the two electrons, they tend to be on opposite sides of the nucleus. One cannot achieve such a spatial correlation by using s-wavefunctions only as they are spherically symmetric.

CHAPTER 10

10.1 As spacing is fixed the final level must be a doublet and the initial levels singlets. Hence $l = 0, l_0 = 1$.

10.2 $l = 1, l_0 = 0$.

10.3 The probability of the electron being above the (x, y) plane is

$$\int_0^\infty \int_0^{\pi/2} \int_0^{2\pi} (a\psi(2s) + b\psi(2p))^2 r^2 \sin\theta \, dr \, d\theta \, d\phi$$

$$= \int_0^\infty \int_0^{\pi/2} \int_0^{2\pi} (a^2|\psi(2s)|^2 + b^2|\psi(2p)|^2 + 2ab\psi(2s)\psi(2p)) r^2 \sin\theta \, dr \, d\theta \, d\phi .$$

As the wavefunctions are normalized the first two terms are $\frac{1}{2}a^2$ and $\frac{1}{2}b^2$ respectively and therefore contribute a total of $\frac{1}{2}$. The final term is

$$\int_0^\infty \int_0^{\pi/2} \int_0^{2\pi} \frac{2ab}{8\pi a_0^3} \frac{1}{2a_0} \left(r^3 - \frac{r^4}{2a_0} \right) e^{-r/a_0} \cos\theta \sin\theta \, dr \, d\theta \, d\phi$$

$$= \frac{ab}{8a_0^4} \int_0^\infty \left(r^3 - \frac{r^4}{2a_0} \right) e^{-r/a_0} \, dr$$

$$= \frac{ab}{8a_0^4} \left(3!\, a_0^4 - \frac{4!\, a_0^4}{2} \right)$$

$$= -\tfrac{3}{4} ab .$$

Hence the probability of electron being above the x–y plane is $\frac{1}{2} - \frac{3}{4}ab$.

CHAPTER 11

11.1 0.138 Å.

11.2 For K_{α_1}, $\lambda = 0.5583$ Å. For K_{α_2}, $\lambda = 0.5627$ Å.

11.3 The splittings in eV are: Ca 3.54, Fe 12.98, Zn 23.03, Mo 105.3, Sn 227, Nd 522. You should get a good straight line with $\delta \approx 4$.

11.4 The calculated values are, respectively, 5,300 eV, 12,000 eV, 23,700 eV and 33,600 eV. The difference is due to the presence of the outer electrons reducing the field at large distances.

11.5 There are four $p_{3/2}$ electrons and only two $p_{1/2}$ electrons.

11.6 8.33 mg cm^{-2}.

11.7 Photon momentum $p = h/\lambda$. Lead atom must take on an equal and opposite momentum. Therefore the recoil energy is

$$E = h^2/2m\lambda^2$$

where m is the mass of the lead atom. We find

$$E = 14.3 \times 10^{-3} \text{ eV} .$$

APPENDIX 

Reading list

Below we give a brief list of some books at about the same level as the present text.

M. Born, *Atomic Physics*, 8th edn., Blackie, London, 1969.
This is a classic text. The first edition appeared in 1935. Although some might consider it a little old fashioned the quality of the author's mind shows through and it will repay study. There are many penetrating and delightful comments. (For example his demonstration that $E = mc^2$ without mention of relativity.)

R. M. Eisberg, *Fundamentals of Modern Physics*, Wiley, New York, 1961.
This book gives a comprehensive account of modern physics including about one hundred and fifty pages on nuclear physics. It is strongly historical and gives clear accounts of the origins of relativity and of the development of quantum mechanics from Bohr's original model, through Sommerfeld and de Broglie to Schrödinger. The mathematically weak student should find it helpful as mathematical derivations are given in full.

U. Fano and L. Fano, *Basic Physics of Atoms and Molecules*, Wiley, New York, 1959.
Although not always an easy book it contains all the basic material at a high level of penetration. It pays great attention to the results of superposition and to analysis of a situation in terms of eigenstates and probabilities.

W. Heitler, *Elementary Wave Mechanics*, 2nd edn., Oxford University Press, Oxford, 1956.
This elegant little book manages to give a clear introduction to the subject plus several examples in a mere 90 pages. Every physics student should have a copy.

F. K. Richtmyer, E. H. Kennard and J. N. Cooper, *Introduction to Modern Physics*, 6th edn., McGraw-Hill, New York, 1969.
This is another classic text, which is particularly convincing and detailed on the experimental side. The first edition by F. K. Richtmyer alone appeared in 1932.

E. H. Wichmann, *Quantum Physics* (Berkeley Physics Course, vol. 4), McGraw-Hill, New York, 1971.

This book is largely devoted to the ideas of quantum physics, and contains comparatively little mathematics. Its approach is very modern and emphasizes strongly the significance of invariance and symmetry properties and order of magnitude calculations of physical quantities from the fundamental constants of nature.

Index

The more important references are indicated by italic page numbers. Where a topic is treated on several successive pages only the first page number is given.

Some Useful Relations and Numerical Values

1 electron volt = 1.602×10^{-19} J

Hydrogen ionization potential = 13.6 V

kT at room temperature (290 K) = 1/40 eV

Rest mass energy of an electron, $m_e c^2$ = 511 keV

Rest mass energy of a proton $M_p c^2$ = 938 MeV

Energy of 1 a.m.u. = 931 MeV

A photon of wavelength 1.24×10^4 Å has $\hbar\omega = 1$ eV

63 AA JADI – ASAHARA'S DAUGHTER – TO BE HIS SUCCESSOR.

95 ANTHRAX

94 BOTULINUS ~~AUTH~~ TOXIN

279 HYDROGEN CYANID GAS

60 BRAIN-WAVE HATS

102 WORKED THRU THE INTERNET

3 --- RELIGIOUS SECT

99
62 SANSCRIT HOLY NAME

NOTES

BOUGHT JULY 28, 96!
REVIEW 8-3-98 - STUCK [illegible]